Hands-On Nature

Information and Activities for Exploring the Environment with Children

Edited by Jenepher Lingelbach and Lisa Purcell
Illustrated by Susan Sawyer

Published by Vermont Institute of Natural Science
Woodstock, Vermont

In association with University Press of New England
Hanover and London ·

Revised and Expanded Edition

Published by Vermont Institute of Natural Science, Woodstock, VT 05091
In association with University Press of New England, Hanover, NH 03755
©2000 by Vermont Institute of Natural Science
First edition, edited by Jenepher Lingelbach, published in 1986 by Vermont Institute of Natural Science

Printed in the United States of America
5 4 3 2 1

ISBN: 1–58465–078–8
Library of Congress Card Number: 00–104923

Hands-On Nature Revision Team: Lisa Purcell, Christina Runcie, Susan Sawyer, Deb Parrella, Bonnie Ross, Jenepher Lingelbach, Elizabeth Cooper, Amy Powers, Karen Murphy, Charmaine Kinton, Melissa Kendall, Jennifer Guarino, Nicole Conte, Carol Curran

Designed by Mirabile Design, Montpelier, Vermont

Illustrated by Susan Sawyer, with additional illustrations by Edward Epstein (pages 37–38, 51, 104, 122, 129, 176, 178, 184, 191, 225, 234, and 303) and Barbara Carter (pages 193, 205–206, 211, 215, 217, 236, 239, 244, 247, 252, 273, 278, 280, 288, 296)

Photographs by Cecil Hoisington, Jill Kurash, Judy Cohen, Steve Faccio, Catherine Metropoulos, Dianne Pedley, Dennis Coello, Peter Wrenn

Snowflake designs (page 212) are derived from W. A. Bentley and W. J. Humphreys, *Snow Crystals* (New York: Dover, 1962). Wind Force Scale (page 292) is adapted from Massachusetts Audubon Society's version of the Beaufort Scale (Massachusetts Audubon Society, Lincoln, MA).

Illustrations on pages 215 and 217 adapted from photographs by Paul Rezendes, from *Tracking and the Art of Seeing,* © 1999 by Paul Rezendes. All rights reserved. HarperCollins Publishers, Inc.

Although the information in this document has been funded in part by the U.S. Environmental Protection Agency under assistance agreement NE991561-01-0 to Vermont Institute of Natural Science, it may not necessarily reflect the views of the Agency and no official endorsement should be inferred.

Printed on recycled paper

ACKNOWLEDGMENTS

The Board of Trustees of the Vermont Institute of Natural Science would like to thank all who have made possible this second edition of *Hands-On Nature*.

The generous contributions of many individuals, businesses, and organizations supported this effort. Our heartfelt thanks to: Byrne Foundation, Burton G. Bettingen Corporation, Emily Fisher, Vermont Country Store, Terry Ehrich, Hemmings Motor News, International Paper Company Foundation, Dewing Foundation, Mascoma Foundation, and the U. S. Environmental Protection Agency.

The Board also wishes to acknowledge with sincere appreciation the work of the *Hands-On Nature* revision team and other VINS staff members who participated in this project. Special thanks to the many professionals from other organizations who reviewed the various units of the book for scientific accuracy. These included:

Trish Hanson, Ph.D., VT Department of Forests, Parks and Recreation;

Mark Breen, meteorologist, Fairbanks Museum and Planetarium;

Kimberly Royar, wildlife biologist, VT Department of Fish and Wildlife;

Russell C. Hansen, retired physics teacher, Hotchkiss School, and *Birds in Flight* photographer;

Larry Sherk, Ph.D., Vermont College and Landmark College;

Meade Cadot, Ph.D., director, Harris Center for Conservation Education;

Ronald Rood, author of more than 30 nature books including *A Land Alive* and *How Do You Spank a Porcupine?* and commentator on Vermont Public Radio;

Nate Fice, Rutland County forester, VT Department of Forests, Parks and Recreation;

Everett Marshall, non-game and natural heritage program, VT Agency of Natural Resources;

Frank Watson, professor emeritus, University of Vermont – College of Education.

The VINS Board appreciates your willingness to contribute your time and expertise to this project.

Finally, to the ELF staff and volunteers, past and present, THANK YOU! You have taken the goal of this book, "connecting children with the natural world," and made it a reality.

Signed,

Deborah Granquist
President, VINS Board of Directors

THANKS FROM THE EDITOR

*There was a child went forth every day
And the first object he look'd upon, that object he
      became,
And that object became part of him for the day or a
      certain part of the day,
Or for many years or stretching cycles of years.*
      – Walt Whitman from "Autumn Rivulets"
      in **Leaves of Grass**

The first edition of **Hands-On Nature,** edited by Jenepher Lingelbach and illustrated by Edward Epstein, was published in 1986 as an outgrowth of the Vermont Institute of Natural Science Environmental Learning for the Future (ELF) program. Many people contributed to and encouraged the production of that first edition — ELF educators; VINS trustees, members and staff; scientific advisors; financial contributors; and of course, the thousands of Vermont volunteers who taught the workshops to children in their communities. In that first edition, Jen gave her sincere thanks to those involved, and I heartily second all she wrote to that creative and committed group. I can certainly hear their voices in the pages of this book.

In much the same way, this second edition has been a mighty team effort. I sincerely appreciate the help of all who contributed time, energy, and thought to updating and revising the information and activities in this book. It has been an effort of love, one made for children and for the natural world.

Bonnie Ross and Jenepher Lingelbach, both former directors of the ELF program, sowed the seeds for this second edition and brought together a remarkable team of writers and educators. The lead authors, Christina Runcie, Susan Sawyer and Deb Parrella, researched and revised each unit, and a stronger core I can't imagine — intensely inquisitive, exacting, and always eager. Equally as important to the success of this project were fellow ELF trainers Karen Murphy, Elizabeth Cooper, Amy Powers, and Charmaine Kinton who helped field test the new activities — their stories kept the work grounded, focused, and fun. Susan Sawyer illustrated each chapter, with additional drawings by Barbara Carter and Edward Epstein, and each page reflects the joy and attention with which these artists view nature. VINS staff members Jennifer

Guarino, Melissa Kendall, Nicole Conte, Chip Darmstadt, Mike Clough, Steve Faccio, Ned Swanberg, Heather Behrens, and Peter Watt were involved in a variety of ways, from reviewing drafts to finding resources; my thanks to each of them for their ready assistance and good humor. Thanks, too, to the teachers, librarians, and **Hands-On Nature** users whose helpful suggestions we've incorporated into these chapters. And, to the ELF volunteers and children who jumped in and got their feet wet trying out each new activity with enthusiasm, patience, and heart — this book couldn't have been written without you.

Once the words were down on paper, other folks offered their guidance and assistance through the production process. Sue Kashanski, red pencil in hand, thoughtfully and graciously considered every word in every chapter. Linda Mirabile brought the pages to life, artfully blending the written text with Susan Sawyer's illustrations. And Patsy Fortney gave the entire document a final, careful, thorough read. My thanks, too, to Trish Hanson, Mark Breen, Kim Royar, Russell Hansen, Larry Sherk, Meade Cadot, Ronald Rood, Nate Fice, Everett Marshall, and Frank Watson who shared their expertise and reviewed chapters. And finally, as VINS Board President Deborah Granquist notes in her acknowledgments, a project of this scope would be impossible to complete without significant financial assistance; the generosity of those contributing to this effort was both affirming and heartwarming.

Didn't I say it was a mighty team effort?! My sincere thanks to all who contributed their talents, skills, and financial support to this project. And to my family: group hug.

Lisa Purcell

Lisa Purcell

TABLE OF CONTENTS

Introduction.. 1

Adaptations.. 11

Amazing Insects: Adapted and Adaptable... 12

Grasses: Slender Stalks with Seeds that Nourish the World.............................. 19

Hunter-Hunted: A Complex Relationship.. 24

Teeth and Skulls: Dentition Determines the Diet... 31

Beaks, Feet and Feathers: Fantastic Flying Machines.. 38

Owls: Silent Predators of the Night.. 45

Thorns and Threats: Plants' and Animals' Strategic Defenses......................... 51

Frogs and Polliwogs: Miraculous Transformation... 58

Habitats 65

Life in a Field: Exposed to the Elements... 66

Forest Floor: Home of the Hidden Workers.. 73

Rotting Logs: Temporary Homes on the Forest Floor... 80

Animals in Winter: Many Ways to Cope with Winter... 87

Snug in the Snow: Snow Is a Welcome Blanket for Many.................................. 95

White-tailed Deer: Elusive Beauty of Woods and Field 100

Streams: The Challenge of a Moving, Watery World... 107

Ponds: Life in Still Waters.. 114

Cycles.. 121

Insect Lives: Surviving the Seasons in Stages.. 122

Meet a Tree: The Sum of Many Parts... 129

Seed Dispersal: Ingenious Ways to Get Around... 135

Fly Away or Stay?: Where Do the Birds Go and Why?....................................... 141

Galls: Small Homes for Tiny Creatures.. 150

Winter Twigs: Signs of Four Seasons... 157

Bird Songs: Musical Messages... 164

Inside a Flower: Making a New Generation... 171

Dandelions: Survivors in a Challenging World.. 176

Designs of Nature.. 183

Spiders and Webs: Webs and Their Weavers.. 184

Variations on a Leaf: The Great Producers... 191

Cones: Cradles for the Conifers.. 198

Snow and More: Crystals in the Clouds... 205

Tracks and Traces: Clues that Tell a Tale.. 215

Winter Weeds: Rugged Remnants of Summer Flowers....................................... 222

Camouflage: Designed to Conceal... 228

Honeybees: Hives and Honey.. 234

Earth and Sky 243

    Finding Your Way: Clues that Give Us Direction................................ 244

    Erosion: Shaping the Landscape.. 251

    Pebbles and Rocks: Archives of Earth's History................................ 258

    Breath of Life: Earth's Invisible Blanket.. 266

    Sound Symphony: Good Vibrations.. 273

    Water, Water Everywhere: Journey through the Water Cycle............. 278

    Wind and Clouds: Reading the Sky.. 286

    Sun Power: Energy for Life.. 295

Appendices:

    A: Environmental Learning for the Future (ELF) – *in your community* 302

    B: Equipment – Where to Get It or How to Make It............................ 303

Glossary... 305

General Bibliography 310

Suggested Reading for Children.. 312

Index.. 321

INTRODUCTION

Hands-On Nature invites children and those who teach them to discover a world of simple wonders. One needn't travel much farther than one's own doorstep to watch a spider gracefully spinning its silken web or to hear the songs of birds seemingly celebrating the rise of a new day. Children are naturally curious about the world around them. When invited outdoors to learn, children become explorers and questioners, poets and artists. Adults can help children gain an understanding of nature by giving them the time and skills needed to look more closely and make their own discoveries. Exploring the natural world together can be a wonderful learning experience for both children and adults.

The purpose of this book is twofold: to provide solid background information and creative activities for new and experienced educators and to enable parents, community members, and other nonformal educators to successfully teach about nature. *Hands-On Nature* is not an educational curriculum. It is a collection of activities that allow children to develop important skills they can use throughout their lives, skills such as asking meaningful questions, making careful observations, finding ways to test their ideas, and sharing their thoughts and observations with others.

Ultimately, we hope to encourage children's curiosity and concern about the natural world and to provide experiences from which they can gain an understanding of the way it functions. The activities and information offered in this book are the first step in what we hope will be a lifetime of learning about and caring for all of nature. What children learn and how much they care will affect us all.

How This Book Came into Being

Hands-On Nature was first published in 1986 as an outgrowth of a successful environmental education/ natural science program designed and taught by the staff of the Vermont Institute of Natural Science (VINS). As part of VINS' Environmental Learning for the Future (ELF) program, the chapters in *Hands-On Nature* have been presented as workshops to teachers, community volunteers, and children in elementary schools throughout Vermont over the past four decades.

The first edition of *Hands-On Nature* was quite successful, with tens of thousands of books distributed. Since that first printing, ELF volunteers and other *Hands-On Nature* users have offered suggestions for improving the activities or adjusting them to various settings and developmental abilities. This second edition of *Hands-On Nature* contains updated background information, plus many new activities and explorations, all of which have been used successfully with thousands of children in classrooms, summer camps, home-school groups, and after-school programs.

Guiding Philosophy

Children, like adults, learn best when they're actively engaged in the learning process and when they are having a good time. The activities in these chapters encourage inquiry and allow for discovery. Children are invited to look more closely, ask questions, and explore freely. We've included activities of many kinds – drawing, writing, creating models, sorting and classifying, role-playing, and more – with the hope that every child will feel the joy of success.

In developing each chapter, there were a number of thoughts that guided us through the labyrinth of possible activities and differing approaches:

~ Focus each topic so the information and activities clearly relate to each other, the activities have a logical flow, and the learning builds on previous knowledge.

~ Allow children to make their own discoveries while the adult leaders serve as co-explorers and co-learners.

~ Give children firsthand experience by including an outdoor field excursion, preferably nearby. The best way to learn about nature is firsthand, by seeing it, hearing it, touching it, and smelling it.

~ Help children to look for differences without emphasizing identification. Encourage them to make observations that could help them identify the object themselves, if they choose to do so.

~ Model respect for nature and each other. This includes releasing all creatures and leaving natural settings undisturbed. It also means encouraging the children to listen to and learn from one another.

A Variety of Audiences

Anyone who works with children or who wants to learn more about the natural world can use this book. The background information in each chapter is written for the adult reader and provides a solid overview of each topic. The activities are easily adapted for various age groups, for use in different parts of the country and in a variety of educational settings. Although originally designed for children at the K-6 level, many of the activities have been adapted for older and younger children as well.

Step back and take a look at the needs, backgrounds, and interests of the children you are working with as you plan your activities. Some children learn best when working individually; others need to share their discoveries in group discussions. By presenting different types of activities, you can address the individual talents and needs in each group so that every child will gain knowledge, according to his or her own abilities.

HOW TO USE THIS BOOK

Goals and Objectives

The goals of *Hands-On Nature* are to spark children's curiosity about the natural world, to give them opportunities to explore and make their own discoveries, and to encourage children to assume some responsibility in caring for our environment.

The specific objectives of the learning activities are to give children skills that help them learn and discover and to provide them with basic natural science concepts upon which they can build an understanding of the natural world and the way it functions.

Overall Design

The format of this book is designed to be simple, clear, and consistent. The information should be easy to find, easy to use, and easy to adapt to a variety of uses. Our hope is that it will encourage creative teaching and spirited learning.

The chapters are grouped into five separate themes: Adaptations, Habitats, Cycles, Designs of Nature, and Earth and Sky. The information and activities in each chapter are focused around the overarching theme. For example, within the Habitats theme we look at white-tailed deer and how their habitat needs change through the seasons. As part of the Designs of Nature theme, we examine the form and function of spider webs in one chapter and the different kinds of snow crystals in another.

Chapter Layout

Within each chapter, there are two main parts: an informational essay and an activity section. The informational essay is written with the activities in mind and covers the basic facts and concepts the children can learn through the activities. For some readers, the essay will be enough background information. For those who wish to read further, we have listed works consulted and suggested reference materials at the end of each essay. (Suggested children's books on each topic are listed in the bibliography.) Words set in boldface in the background essays are defined in the glossary at the end of the book. Following the background essay are the chapter's activities, arranged in sequence to offer a logical progression of information from an introductory to a concluding activity. Puppet show scripts and sometimes an additional fact or study sheet follow the activities.

The Many Options

The chapters are designed to be thematically connected, often in seasonal sequence. The activities themselves are arranged so that the information and concepts presented build upon each other. However, there are many ways the book can be used.

~ Each chapter can stand by itself, or individual activities may be used as a supplement to other natural science programs.

~ The order of the activities within a workshop may be changed, and activities can be added or deleted.

~ Many of the activities are appropriate for use as learning stations, and leaders can rotate smaller groups of children through them rather than lead all the children simultaneously through each activity.

~ By themselves, the background essays in each chapter make interesting and informative natural history reading.

Some Features of the Activity Sections

The activities are the heart of *Hands-On Nature* and were designed to encourage discovery and learning. There are a number of features included in each activity section to help leaders and teachers use the book to its fullest. For clarity's sake, they are described here in order of appearance.

Focus – a brief statement explaining the central concept of the workshop.

Opening question – a broad question to prompt the children to start thinking about the topic, to give them a chance to share what they already know, and to introduce the first activity. This question could be presented again after the children have completed the activities to assess their understanding of the information.

Objective – an explanation of what the children will be doing and learning in each activity.

Activity instructions – a step-by-step explanation of how to set up and teach each activity.

Materials – a list of needed equipment or supplies for each activity.

Stations – where appropriate, we have listed activities that work as activity stations through which small groups of children rotate. The activities selected are generally short explorations that take about the same length of time so that one group is not waiting for another to complete its task. An adult at each station can help the group get started and then lead a review discussion, as time permits.

Extensions – suggestions to extend the scope of the learning; usually these include at least one long-term project or experiment such as researching local predator species or growing grass in a variety of conditions.

Teaching aids – these are the puppet show scripts or other supplemental information needed to carry out the listed activities.

What's Not Included

Length of time – the time it takes to complete an activity varies greatly depending on the leader, the children, and the time limitations.

Suggested grade levels – although we occasionally include options for different grade levels, we have found learners of all ages to be responsive to a wide range of hands-on experiences depending on how the activity is presented.

The Puppet Shows

From kindergartners to high school teachers, audiences really remember what they've learned in a puppet show. Whether adults put the shows on for the children, or the children put them on for each other, puppet shows are effective teaching tools. You will find the puppet shows in *Hands-On Nature* loaded with information about each topic.

Be sure to read the script ahead of time to become familiar with both the information and the puppet characters. When presenting the puppet shows, be willing to be goofy. The right combination of funny voices, movement, and timing helps to make the show a success. Decide on a special voice for each character – and then remember to use it throughout the show. You wouldn't want Benjy Bear to suddenly sound like Wendy Worm! It helps to use highlighting pens of different colors to mark the various characters.

For the actual puppets, we use simple paper cutouts glued to sticks or dowels. You'll find many of the animals that appear as puppet show characters illustrated throughout the book. If a character is a plant, such as Mary Milkweed, we often use the actual plant. The stage can be a table turned on its side, two chairs or a desk covered with a cloth, or a cardboard screen. The puppet show script can be taped up for puppeteers to read as they perform.

A few good review questions after the performance may help children digest what they've heard. For example, bring out the puppet and ask, "What made Charlotte and the other spiders different from insects?"

Equipment and Materials

Most of the supplies called for in *Hands-On Nature* activities are simple to find or easy to procure, such as crayons, paper, pipe-cleaners, leaves, cups, and cones, to name a few. Appendix B lists materials that may be more difficult to obtain and suggests ways to make them or places to find them.

~ The activity descriptions for field investigations and scavenger hunts can be photocopied right from the book and either taped to cards or used on clipboards.

~ Lightweight, inexpensive clipboards can be made using corrugated cardboard (cut 9" x 12") and large paperclips.

~ Diagrams can be enlarged from the book and colored to use as visual aids.

~ A blackboard, whiteboard, or big pad of paper can be used to bring a static diagram alive, if you create the drawing as the children watch and advise you.

~ Hand lenses (magnifying glasses) are useful for making detailed observations and are therefore listed in the materials section of many activities. Using a hand lens may be difficult for very young learners. Explain that the lens should be held near the eye and then the object brought up to the lens until the object is in focus.

Suggested Reading for Children

At the end of the book, you will find a list of recommended books for children that complement each chapter. Compiling this bibliography was a great challenge as there are many more children's books than we could possibly obtain and review. The lists given represent the recommendations of the VINS staff, ELF volunteers, teachers, librarians, and other environmental educators who offered their suggestions. We have given brief descriptive comments for those books that we ourselves have seen and recommend. You as the teacher/leader should determine the best use of any book. Many of the books listed for children may be helpful for adult reference. And your local library and bookstore are sure to have other wonderful resources.

Assessing Learning

The skills children practice through participation in *Hands-On Nature* activities include observing and describing objects and events, asking questions, comparing and sorting objects, sharing information, and

many others. These are important science skills, and, in fact, the regular use of *Hands-On Nature* activities has been credited with raising the science scores in some Vermont elementary schools.

Hands-On Nature is filled with activities that could serve as assessment tools to gauge children's understanding of the topic. Children can summarize what they've learned by drawing pictures, as they do in the activity Monster Mouthfuls in the Thorns and Threats unit, or by building models, as in the Teeth and Skulls unit's Complete a Jaw activity. The Icy Acts skit each small group creates as part of the Snow and More unit helps demonstrate the group's understanding of the weather processes involved.

Often group discussions reveal other information children have learned during a unit of study. The opening questions presented in each chapter might be asked again days or weeks after the material was first presented. For example, the opening question in Life in a Field asks "What basic needs must a habitat meet?" As children begin their study of the Habitats theme, they might answer this question with a simple list of requirements. Later in their exploration of the concept, however, children may provide responses that demon-

strate an understanding of the ways in which different creatures alter their habitats and how habitats change over time. Older children could be asked to create concept webs for the word "Habitat" both before beginning and after finishing a unit of study as a way of documenting any additional connections the students have made.

Teachers may have students compile a portfolio of field notes and drawings each child has created through the year that demonstrate his or her ability to observe and describe. Journals can also be used as an alternative assessment tool. Ask children to record their thoughts or observations after each activity or lesson. What new questions do they have about the concept? What were they surprised to learn or observe? This type of writing helps children process information and synthesize their learning.

In the same way, drawing can be used to enhance, direct, and record observations or to express ideas imaginatively and visually. If using drawing to gauge children's understanding of a topic, teachers may need to clearly explain the difference between these different drawing functions. If they are recording the things they see, like drawing a flower or an insect, children should be reminded to draw only the things they actually observe rather than what they think they should be seeing. If the activity is a creative drawing, like illustrating a Super Hunter, encourage the children to be as free and imaginative as possible.

By assessing children's learning, we as educators are also assessing our instruction. We can use this feedback to improve our teaching. What did the children know about this topic to begin with? How did these activities broaden their understanding of the concept? What new connections did the children make? Were all children equally engaged in the activities?

Children's enthusiasm for and interest in the activities give a good indication that learning is happening. Certainly the number of times children refer to the ideas and activities can give an indication of whether the goal of sparking curiosity for and love of the natural world is being met.

You as Environmental Educator

You are the leader, and the children will follow your example. Your attitude toward the environment will register clearly with them as you carefully replace a log rolled over for investigation or pick up trash left by people there before you. You don't even need to discuss your actions; they speak louder than words.

How often do children get to see adults taking the time to listen carefully, look closely, or smell deeply? How you feel about nature should and will come through to the children when you stop suddenly to listen to a favorite bird song or pause to watch an ant laboring under a heavy load. Curiosity and caring are contagious.

Respect for All

Whether indoors or out, a respect for each other helps engender a respect for nature. We all have different tolerances for commotion, but none of us need tolerate meanness or thoughtless infringement on the rights of others. Most important is for you to be clear in your own mind about which behaviors are acceptable and which are not and to explain your expectations to the children. Encourage them to discuss their expectations of each other. You may want to create a written "learning contract" with the children that outlines these. Then when you have to discipline a child, you are reviewing behavior codes, not initiating them. Reasonable behavior and clear expectations translate into a lot more fun for everyone.

Everyone Is Afraid of Something

Most people fear or "hate" some things in nature. Learning about something you fear is often the best way to overcome those feelings. This is a worthwhile goal for our own sake and for the sake of the children we influence.

Should one express fear or try to conceal it from the children? You would have to be a very good actor to hide your fear of snakes when you're startled by one during a field trip. So you might as well be honest. When the situation arises, explain that you are afraid of snakes, or spiders, or mice, and that you are trying to increase your knowledge about them so that you will become less fearful. This admission may lead to a good discussion in which children can admit their fears and be encouraged to realize they need not be trapped forever by them. Many leaders who hated spiders have come away from the Spiders and Webs workshop still wary, but with admiration for and curiosity to know more about these creatures.

Some fears may be quite valid for safety reasons; however, many can be dispelled by accurate information that is explained simply. What a favor you will have done for a child if you can dispel a fear. But remember, children rely on you to keep them safe, physically and emotionally. If they choose not to hold the viewing jar containing the spider, respect their decision. Simply by talking about spiders, you are presenting them with information that may allow them to rethink their fear.

Sense of Humor

Children learn best when they're having a good time; your playfulness and sense of humor will keep them on their toes. Most youngsters relish corny jokes and ridiculous riddles – they will laugh at yours and feel great when you laugh at theirs. Keep some jokes or stories up your sleeve for times when things drag a bit or children are tired.

Don't expect everything to be perfect. Be cheerful, and you can turn a mistake or a minor accident (like losing a shoe in the mud) into a comical situation. Laughter is good for the soul as well as for the brain.

Questions and Stories

There are sample questions included in nearly every activity. Use these as springboards for discussion. Asking open-ended questions is meant to encourage higher-level thinking, not simply knowing the "right" answer. One question should inspire more questions from the children.

Sometimes children ask a question because they are curious to know the answer, but often their "questions" are actually lengthy stories and anecdotes. Even carefully planned discussions with the children may open a Pandora's box of tales from their own or their family's experiences. For many children, sharing their stories helps them build understanding and make connections.

However, when you have a large group of children, asking others to sit through a recounting of someone else's adventures may mean you lose the interest of much of the group.

Although it feels mean to cut short a child's story, children understand time limitations. Clarify the difference between a question and a story. If you explain that there will not be enough time to go outdoors or play a game if they share their stories at this time, children will be willing to move on. Tell the children they will have a chance to tell you their stories after the workshop, and then don't forget to give them the chance. A brief, focused sharing circle may be all the children need in order to feel listened to. If you don't have time, perhaps the children can be encouraged to write their stories later.

Identification

Often children ask "What is it?" before they have taken the time to really study the object, and having someone answer that question can halt the inquiry.

On the other hand, names allow us to share information and observations about specific plants and animals. So, if a child asks and you know the name of something, you can share that information, but don't stop there. Say, "Oh, that's a black-capped chickadee. See how it seems to be wearing a little black hat? Look at how it holds those sunflower seeds between its toes!" For most of us, names are only remembered when a personal connection is made.

Running Wild

Exuberance and pent-up energy, especially for children who have just emerged from a school bus or a classroom, can be channeled. Your plans may call for a sit-down discussion or a controlled scavenger hunt, but if you feel the lid about to pop, stop and organize a relay race or a tag game or a "hop like a _____" tour. Sometimes children need to stretch out, try their speed, express their joy at being let loose. The trick is knowing when to let it happen.

Then, when the kinks are out, you can go over your behavior expectations. Let the children know that you expect their attention and respect and that you will give them yours. Minimize disruptions by separating children who you know can egg each other on. If one child is unable to behave according to the expectations you've agreed upon, that child may need to be removed from the group. The other children have the right to learn and explore without being distracted or interrupted.

Boundaries

Outdoor activities often erupt into a joyous explosion of energy, with children dispersing into the far reaches of an outdoor area. Before you give the children their activity instructions or equipment and send them off, clearly define the boundaries within which they may explore. It may be specific limits like a fence, or it may be more general such as within sight of your red kerchief or within earshot of your whistle. Give them a time limit, too, and be sure to specify where to meet when time is up. If they know the ground rules ahead of time, children can concentrate more fully on finding, looking, and investigating while outdoors.

Finders Keepers?

Children love to pick, catch, and keep what they find. Collecting is one of the hardest natural inclinations to temper. And yet, each tiny hand has an impact, and a class full of young children all picking leaves from the same corner of the playground can quickly denude the plants struggling to grow there. We want to give children a hands-on experience, but we also want to respect and nurture all living things.

You, as leader, will have to decide when it is okay to collect and when it isn't. Discuss the dilemma with the children; together you are making a decision about how you care for the natural world. A general rule of

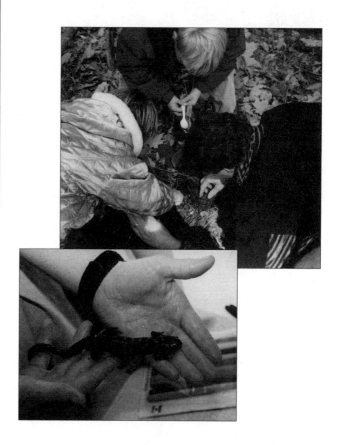

thumb is that it's okay to pick one when there are so many you couldn't possibly count them – picture a field full of dandelions! Another guideline might be, "We won't pick – but it's okay to pick UP." So, for example, children could gather cones, fallen leaves, spent seedheads, but would leave growing things alone.

Once you've decided, make sure the children understand. If they may collect, set firm guidelines on how big and how many: one leaf from a plant, one flower head, or one insect. When living creatures are collected, a release ceremony at the end of the session is very helpful. The leader can give a brief farewell address, and then all the children release their creatures back to the places in which they were found. Our feeling is that no child should be allowed to keep a creature found during a group expedition; possibly the leader may decide to keep one temporarily for group study and observation.

Most of the outdoor explorations in this book say "find" rather than "collect," hoping to encourage children to leave their discoveries where they find them. For those activities that do use natural materials like cones, grasses, and leaves, you may want to gather these ahead of time, especially if your on-site resources are limited.

Adding a Slice of Silence

Children's lives frequently feel as hectic as our own. If you can inject a little serenity into their time with you, you will help them enjoy and understand both the natural world and themselves a little better.

Our noisy intrusion into a field, forest, or wetland brings natural activity to a temporary halt as creatures freeze or scurry into hiding. Encourage the children to be silent and still occasionally, wherever they are, and let the natural flow of life resume around them.

LEARNING OUTDOORS

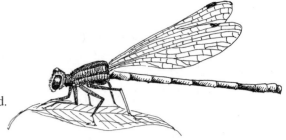

Why Go Out?

Simply stated, outdoors is where a child can become part of the natural world, watching it, wondering about it, and experiencing it firsthand. It's where one can see tracks in the snow, hear a dragonfly's wings, smell honeysuckle or sage or balsam fir. These personal connections with the natural world are ones we want children to experience.

Taking children outdoors to learn about nature can be a challenge. It is little wonder that schoolteachers hesitate to go outside with 20 or 30 children. In this book, the outdoor scavenger hunts and explorations suggested with almost every unit have enough specific tasks for the children to do that being outside is not only manageable but fun. If possible, ask additional adults to come along for outdoor activities. It is much easier for children to focus outside when the groups are smaller. (See appendix A.)

Be prepared before heading outside. If you are going to have children do one of the field investigations or searches included in the *Hands-On Nature* activities, have copies of the activity for the children (or adult leaders) to follow. Be sure to explain the outdoor activities while you are inside; it can be difficult to give detailed instructions outside when the children just want to explore. And be ready to change your plans altogether if something else – thunderheads forming, a hawk chasing a pigeon, a cluster of ladybugs hidden in the tall grass – captures the children's attention. You never know what you may see when you are outdoors.

Which Activities to Do Outdoors

Most of the units have at least one activity that invites children outdoors to explore. Take the time to enjoy the outdoor sensory experiences – listening to the noises of a pond or smelling the richness of the forest floor. These can set the tone for the other outdoor investigations.

Outdoor scavenger hunts are excellent ways to challenge the children while focusing their searches. Remember, the items suggested in these search and find activities are meant to ignite the children's curiosity. The goal isn't to record the greatest number of items found but instead to note the interesting observations and connections made during your explorations.

There are very few activities that cannot be done outdoors. Certain active games requiring a lot of space are much more fun if played outside. Role-playing and other skits are often more realistic and less inhibited when conducted outside. Show and tell activities work well outdoors if there's a good place to sit, look, and listen. Creative projects often become more imaginative when done outside, but materials need to be accessible

and organized. If it's very windy, either find a lot of stones to weight things down or retreat indoors. The weather, the chemistry of the group, and the available outdoor facilities will all help determine how much to do outside.

Where to Go

Usually one need go no farther than the backyard, schoolyard, or nearby open area. Visiting these areas sends a message to children that nature explorations can and should happen right outside their own back doors.

Some units may involve taking field trips to special places, such as a forest or pond. If such a trip is planned, make a point of scouting the area out ahead of time. That way you will know where the children can find what they're looking for, how long they will need to spend, the best route to the area, and the time it will require to get there and back.

Comfort and Safety

Think ahead about the basic physical needs of children. Do the children need to visit the restroom before you head out for your exploration? If you plan to be outside for a while, you may want to bring along a snack for the children and include a rest stop or two on your walk.

Weather happens, but with good planning, weather can be enjoyed not just endured. Make sure that everyone is dressed appropriately to be outside. Footwear and headwear are especially important. If it is cold or rainy outside, give instructions for any outdoor activities while the group is still inside.

Ahead of time, consider what you would do in case of emergency and who would help take charge. And point out obvious dangers or irritants to the children before anyone encounters them unwittingly; barbwire, poison ivy, stinging nettles, and the like can upset the best of plans.

The Trip Itself

Have a good time. Allow the children to set the pace, but stay on schedule. It's better to stop the experience while the children are having fun rather than prolong the exploration and risk ending up with restless or oversaturated children.

Successful outdoor explorations with children take planning as well as a certain degree of letting go. Be flexible. Often the unplanned happenings and the spur-of-the-moment inspirations turn out to be the best parts of the outdoor excursion.

Adaptations

Why do ducks have webbed feet and rose bushes have thorns? Why does a snowshoe hare freeze rather than flee when a predator approaches? Adaptations in the natural world are the structural or behavioral characteristics of a species that have evolved over time in response to the changing demands of the environment.

The chapters in this theme introduce and illustrate how different adaptations contribute to the success and survival of various plants and animals. The well-camouflaged snowshoe hare remains perfectly still when it senses danger, a behavioral adaptation to avoid detection. A rose bush's thorns are actually modified leaves, structural adaptations that protect the plant from being eaten. To best meet the demands of life in a watery habitat, ducks evolved the webbed feet that help them swim efficiently as they look for food and safety. Now imagine those same feet on a bird of prey! Webbed feet would be laughable on a hawk that spends its time in the air and captures prey with its talons.

Learning about adaptations involves examining the characteristics of organisms and comparing their similarities and differences. Children will learn that an individual organism changes throughout its life cycle, while species change over time through the process of natural selection. The study of adaptations invites a child to ask the question: Why? Looking for answers to this question offers an exciting introduction to the world of science.

Amazing Insects

ADAPTED AND ADAPTABLE

No other group of animals is as large and diverse as the insects. Worldwide, more than 1,000,000 different species of insects have been identified, approximately 100,000 of which can be found in North America. They live on top of mountains, in underground caves, in deserts and rivers, and in fields and forests. Although each species of insect is adapted to meet the demands of its own particular environment, all insects have the same basic body structure.

Insects are **arthropods,** related to spiders, crabs, lobsters, millipedes, and centipedes, and like these cousins they have segmented bodies with jointed legs and an exterior skeleton. What differentiates an insect from the other arthropods is that the insect's body consists of three main parts: head, **thorax,** and abdomen, with two **antennae** on the head and six legs on the thorax. The wings, which most adult insects have, also attach to the thorax. (A spider is not an insect because it has two main body parts, eight legs, no antennae, and no wings.)

The head of an insect contains the eyes, the antennae, and the mouthparts. Many insects have both simple and **compound eyes**. The simple eyes sense only light while the compound eyes, with their numerous lenses, detect images. The keen-sighted dragonfly has more than 25,000 lenses in each compound eye. A honeybee's or butterfly's two compound eyes are sensitive to both ultraviolet light and visible light. The other sensory receptors on an insect's head are the antennae. These finely tuned sense organs are capable of feeling, tasting, smelling, detecting temperature, and receiving chemical stimuli. Some are long and slender, some quite feathery, and others club-like. The mouth of an insect also varies from species to species because the mouthparts are adapted to the food each insect eats. Thus a grasshopper's mouthparts are adapted for biting and chewing, a house fly's for lapping, a butterfly's for sucking, and a mosquito's for piercing and sucking. A few species, such as the mayfly, live as adults only long enough to mate, so they have no mouthparts with which to eat.

The middle section, or thorax, has three pairs of legs and usually two pairs of wings (sometimes one, and occasionally none). For this reason, many muscles are located in the thorax. Insect musculature is highly specialized in many species – grasshoppers have about 900 muscles compared to humans' 800, and an ant can carry 50 times its own weight. Legs are adapted in as many ways as mouthparts. The grasshopper has hind legs specialized for jumping; the house fly has sticky pads on its feet, allowing it to walk up vertical walls; honeybees have specialized hairs on their hind legs that form "baskets" in which to carry pollen; and the butterfly's feet can "taste" sweet liquids that cause its sucking mouthpart to uncoil. Insect wings also vary in design. They may be long, short, narrow, wide, leathery, or quite delicate, depending on the type and amount of flying the insect does.

The segmented abdomen contains the heart, the digestive system, and the reproductive organs. On females of some species, the egg-laying device, called the **ovipositor**, protrudes noticeably from the end of the abdomen. Stingers are also located at the end of the abdomen on the few insects that sting: bees, wasps, and some types of ants. When an insect stings, the stinger penetrates the victim's skin and then pumps a poison (usually formic acid) into it. Insects sting for a couple of very good reasons: to kill or stun **prey**, as many wasps do, or to defend themselves or their colonies from danger. Some insects, like honeybee workers, can sting only once and then die; others can sting repeatedly. It's good to remember that these insects usually won't sting if unprovoked, and you can safely spend hours watching them go about their business.

While studying an insect closely, you might be able to see a series of holes in the thorax and abdomen. These are the **spiracles**, which lead to the tubes that carry air throughout the body. Air is pumped in and out by the swelling and relaxing of the abdomen. The challenge of breathing under water has produced some remarkable adaptations, including the breathing tube of the mosquito larva and the water scorpion; the gills of immature caddisflies, damselflies, and mayflies; and a system for carrying an air bubble used by many water beetles.

The success of insects as a group is due to their having several major assets: flight, adaptability, external skeleton, small size, metamorphosis, and the ability to produce multiple offspring rapidly.

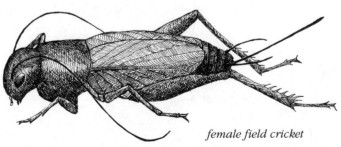

female field cricket

Flight has given insects an advantage over land-bound animals in being able to search actively for their mates, forage widely for food, escape from their enemies, and leave areas that no longer provide for their needs.

Adaptability is another advantage the insects have. With short life cycles and rapid reproduction, insects can change, if necessary, in a relatively short time to meet the requirements of a changing environment. For instance, when smoke from the Industrial Revolution darkened buildings and vegetation in England, one light-colored species of moth became more visible to **predators** in this changed environment, and many of these moths did not survive. However, a few dark individuals born by chance in each generation helped the species adapt to the new landscape by eventually producing generations of the darker moths. More recently, pollution controls have led to a cleaner environment in many areas of England, and the number of light-colored moths has increased due to natural selection.

The external skeleton, called an **exoskeleton**, is vital to insects as a suit of protective armor. Its essential ingredient is **chitin**, which is flexible, lightweight, tough, and very resistant to most chemicals. A growing insect simply sheds its old skeleton or skin, and the new one inside quickly dries and hardens.

The small size of insects acts to their advantage. The speck of food that is a feast for an insect is usually much too small to be noticed by a larger animal. Being small, most insects can readily find shelter, even in tiny cracks, thus escaping detection and predation.

Timing is also important for insect survival. Insect eggs are timed to hatch when the proper food and living conditions are available for the young. Some eggs may lie dormant for years until the conditions are right. For some species, the immature insect can use one food supply while the adult nourishes itself on something completely different.

Insects are fun to study and exciting to watch. There are far too many different species to be able to identify every kind, but a close look can reveal their common characteristics as well as a great variety of remarkable adaptations.

box elder bug

Suggested References:

Borror, D. J., C.A. Triplehorn, and N. F. Johnson. *An Introduction to the Study of Insects*. Orlando: Harcourt-Brace, 1989.

Burton, Maurice. *Insects and Their Relatives*. New York: Facts on File, 1984.

Farb, Peter. *The Insects*. New York: Time-Life Books, 1968.

Milne, Lorus, and Margery Milne. *The Audubon Society Field Guide to North American Insects and Spiders*. New York: Alfred A. Knopf, 1980.

Stokes, Donald. *A Guide to Observing Insect Lives*. Boston: Little, Brown, 1983.

Zim, Herbert, and Clarence Cottam. *Insects*. New York: Golden Press, 1987.

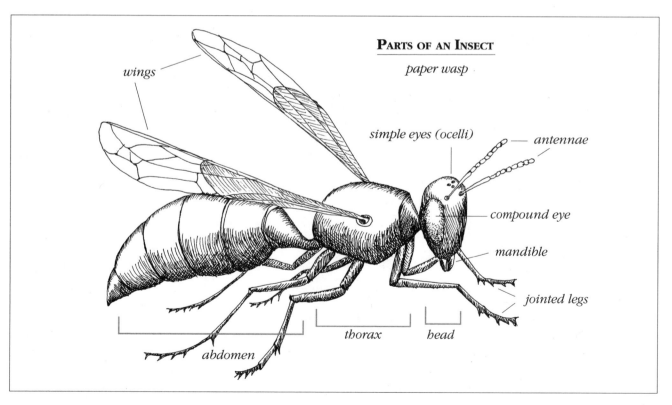

PARTS OF AN INSECT

paper wasp

wings

simple eyes (ocelli)

antennae

compound eye

mandible

jointed legs

thorax   head

abdomen

Amazing Insects

FOCUS: Insects share a basic structural design, while special adaptations, both physical and behavioral, allow each species to meet the demands of its specific habitat.

OPENING QUESTION: *What are some characteristics common to all insects?*

ALIKE YET DIFFERENT

Objective: To become familiar with the basic body structure of insects.

Give each child a picture of an adult insect to examine. Then, in pairs, have the children compare their insects. How are the insects similar? As a whole group, share observations to create a list describing basic insect anatomy: 3 body sections, exoskeleton, eyes, antennae, mouth, wings, 6 legs, spiracles, ovipositor. Discuss what each part does. What differences do children observe among the insects?

Materials:
- magazine pictures of insects or dead insects in view boxes
- insect anatomy chart

PUPPET SHOW

Objective: To discover some of the special adaptations, behavioral and physical, of various insects.

Perform, or have the children perform, the puppet show. Afterward, use the puppet characters to review some insect adaptations. Which are physical (part of their body structure) adaptations and which are behavioral (related to their actions)?

Materials:
- puppets
- props
- script

INSECT SEARCH

Objective: To find a variety of insects outdoors.

In small groups, go to a garden area or a place with tall grass or weeds where the children can look for insect life. Have the children:

Listen quietly for different insect noises.
How many can be heard? Can children find any of the insects making the noises?

Turn over any flat stones or logs and observe the creatures there. Replace carefully!

Look for insects: flying, on flowers, eating leaves, walking on the ground.

Spread out a white cloth in an area with tall plants. Walk through the weeds toward the cloth and observe the insects that hop and crawl onto the cloth.

Carefully watch one insect in its natural environment for at least two minutes. What is the insect doing? How does the insect move?

Gently collect one of the insects discovered while doing this search to inspect more closely.

Afterward, gather the children together and have them share any interesting observations.

Materials:
- hand lenses
- small jars or bug boxes
- pieces of 2' x 3' white cloth

Note: Check to make sure that children who are allergic to bee stings have their emergency kit with them, just in case.

INSPECT AN INSECT

Objective: To look closely at one insect to see its structure and adaptations.

Ask the children to use a hand lens to inspect an insect they've carefully collected in a see-through container.

Place younger children (grades K-2) in small groups with an adult leader. Have each child look at the insect closely, noting how it moves and how it's built. The leader can review the body parts and use the following questions to help the children observe their insects.

Older children (grades 3-6) can work alone or in pairs to answer the following questions.

INSPECT AN INSECT QUESTIONNAIRE

What do you notice first about your insect?

Can you see wings?

Can you see legs? How many pairs can you see? Are all legs the same length? Which, if any, are longer?

Is its shell shiny?

Are there hairs on the insect's body?

Does it have antennae?

Look at its eyes. Do they seem large or small compared to the size of the insect?

Can you see its mouthparts?

Is any part of the insect moving? Which part(s)?

From what you have observed, does the insect have any useful or special adaptations? What are they?

If time allows, draw your insect.

RELEASE CEREMONY

Objective: To return the insects to their proper homes.

Gather in a circle and allow the children to share one observation about the insect they collected. Then, as a whole group, recite the following poem, with the leader reading a line and the children repeating the line as a response.

Fly away, crawl away, run away, hop –
You're free to go, I'm not going to stop
You from living your life. You deserve to be free.
Thank you for sharing this time with me.
Fly away, crawl away, run away, hop –
You're free to go. Goodbye!

Have the children go in their same groups to the places where the insects were collected so they can release them in the appropriate habitats.

Materials:
- small jars or bug boxes
- hand lenses
- Inspect an Insect questionnaires
- pencils
- clipboards
- blank paper

MAKE AN INSECT

Objective: To review the basic body structure of insects and imagine some special adaptations.

Using a 3-part section from an egg carton, pipe cleaners, and plant parts, each child or pair of children fashions an insect that has all the necessary body parts: eyes, mouth, antennae, wings, legs, spiracles, and possibly ovipositor. The children should be sure to include any special adaptations their insects might need to survive, and then each insect should be named. Have the children introduce their insect to the whole group and explain its special adaptations (in small groups to save time).

Materials:
- egg cartons (bottom sections divided into quarters, with three cups in a row; punch holes for younger children to attach eyes, antennae, mouth, six legs, four wings, and ovipositor)
- sharpened pencils as hole-punchers for older students
- pipe cleaners, glue, tape
- large assortment of leaves, grasses, flowers, and seed heads

EXTENSIONS

Flower Neighbors: Locate a clump of conspicuous field flowers. Identify the flowers and observe the insects that frequent them. Take note of nectar and pollen gatherers, leaf eaters, and the predators (such as spiders or ambush bugs) that eat them.

Insect Orders: Working in pairs, children choose an order of insects to research. What similarities do these insects share? Make a poster describing this group of insects that includes a collage of magazine pictures of the various species in the order.

Ant Work: Find an anthill near the schoolyard. Observe the ants. Place cookie crumbs and a sugar cube nearby. What happens when an ant finds it? What do the ants do with the food? If you move the food farther away from the nest, how far will the ants go to get it? Can you find different kinds of ants? If so, do they prefer different foods? Try birdseed and bread as well as sugar and cookie crumbs.

Invite an Entomologist: Invite someone with an insect collection or special knowledge about insects to talk about insect adaptations, insects' usefulness to the environment, and some of the problems they can cause. (A forester, extension agent, bee keeper, college professor, or state entomologist might be available.)

Amazing Insects

Characters: Greg Grasshopper, Maria Monarch, Aunt June — the Insect Fairy, Danny Dragonfly, Wendy Walking Stick, Wally Woolly Bear

Props: branch (to demonstrate walking stick camouflage), insect field guide

[on stage, Greg Grasshopper jumps around, chirping loudly; Maria Monarch appears]

Maria Monarch
What's all the noise about?

Greg Grasshopper
Oh, sorry, Maria Monarch. I'm so excited about the contest today, I guess I got carried away.

Monarch
What contest?

Grasshopper
The Best-Adapted Contest! They're giving an award to the insect with the best adaptation.

Monarch
What are adaptations?

Grasshopper
You know, adaptations are the special things about your body or ways you act that help you to stay alive.

Monarch
Oh, you mean the things that help me escape from predators, or get food, or find a mate – things like that?

Grasshopper
Exactly. I think it's going to be a tough contest, because we insects are all well adapted. Why, we've been around for 350 million years, and we're still doing fine!

Monarch
Who's judging the contest?

Grasshopper
Aunt June, the Insect Fairy, of course. And here she comes now! I'm going to get in line – I don't want to lose my place.
[Grasshopper exits; Insect Fairy enters]

Insect Fairy
Hello, dear.

Monarch
Hi, Auntie! Can I be in the Best-Adapted Contest?

Insect Fairy
Of course, of course. Just get in line with the other insects. When I call on you, tell me about your very best adaptation and how it helps you to live.

Monarch
But, I have **lots** of adaptations. I've got…

Insect Fairy
No, no, dear, don't tell me now. You have to wait your turn.

Monarch
Oh, OK, but it's going to be hard to decide which of my adaptations is best…
[Monarch exits, grumbling]

Insect Fairy
[to the audience] Welcome to the Best-Adapted Contest. I hope you all will help me decide which insect has the best adaptation. First in line is Danny Dragonfly.
[Dragonfly appears]

Danny Dragonfly
Hello, Aunt June. My best adaptation is the way I gather food. You see, when I fly, I fold my hairy legs inward to form a basket so I can scoop up insects while I'm flying. When I'm hungry, I just tip my head into my basket of food!

Insect Fairy
Delicious! A built-in picnic basket is a great adaptation! Thank you, Danny.
[Dragonfly exits] Now it's your turn, Gregory Grasshopper. *[Grasshopper appears]*

Grasshopper
My best adaptation is stridulation.

Insect Fairy
What-you-lation?

Grasshopper
Stridulation! It's the way I sing for a mate. You see, I have these pegs on my legs, and when I rub them on a vein in my wing, it makes a noise like this: *Screek, screek, screek.* It's music to a lady grasshopper's ears.

Insect Fairy
How romantic! A grasshopper serenade. What a wonderful adaptation for finding a mate. Thank you, Gregory. *[Grasshopper exits]* And now let's hear from Wendy Walking Stick.

Wendy Walking Stick
[Walking Stick appears, perched on branch prop]
Hi, Auntie!

Insect Fairy

Wendy? Where are you? I can hear you, but I can't see you. *[Walking Stick separates from the branch]* Oh, there you are!

Walking Stick

Fooled you, didn't I? Well, that's how I avoid predators. You see, when I hold very still, I look just like a twig. Why, I can be either green or brown to match the plant I'm in, and if I move my body back and forth, I look just like a twig swaying in the breeze.

Insect Fairy

Wendy, what wonderful camouflage! Thank you for showing us. *[Walking Stick exits]* Wally Woolly Bear, you're next. *[Woolly Bear enters and brushes against Insect Fairy]* Oooh, stop, you're tickling me! You sure are a fuzzy one, Wally!

Wally Woolly Bear

That's because I'm all covered with these hairs. And I think they are my best adaptation.

Insect Fairy

And why is that, Wally?

Woolly Bear

Because when a bird or something scares me, I just curl up into a ball of fuzz. My coat protects me – and I don't look so good to eat, either!

Insect Fairy

What a useful fur coat! Thank you, Wally. *[Woolly Bear exits]* Well, I wonder who's left. *[Monarch enters]* Why, Maria Monarch. It's your turn, at last. Don't you want to tell me about your best adaptation?

Monarch

Yes – I mean – no! – I mean – yes – I mean, I still can't decide which to tell you about. I'm good at getting food because I have a mouth that works like a straw. I use it to drink sweet nectar from flowers. It even coils up below my head so it's out of the way when I'm not using it.

Insect Fairy

That's very useful, Maria.

Monarch

Yes, but I can also taste things with my feet so I'll know whether to drink or not...

Insect Fairy

Wonderful!

Monarch

And I'm not bothered by predators because my orange and black colors tell other animals that I'm poisonous to eat.

Insect Fairy

Oh! You have warning colors to keep predators away!

Monarch

Yes, plus I have a really good adaptation for staying warm in the winter. I fly south to Mexico.

Insect Fairy

To enjoy that southern sunshine!

Monarch

Yes, Auntie. And all the other monarchs go there too, so it's easy to find mates.

Insect Fairy

Well! You **do** have many successful adaptations!

Monarch

I know. I just can't decide which is the best one.

Insect Fairy

Well, neither can I. In fact, each of you insects has such great adaptations that I think you **all** deserve to win.

Monarch

Yippee, hooray! What do we win, Auntie?

Insect Fairy

Why, the prize is...**survival.**

Monarch

Oh goody! That means I'll be able to have children and grandchildren and great grandchildren and...

Insect Fairy

Yes. **And** you'll all get your pictures in my book of insects! *[hold up field guide to insects]* Now let's all give a big cheer for the insects and their useful and effective adaptations.

Monarch

Hooray, yippee, hooray! Bye, Auntie – I've got to go tell the others!

Grasses

SLENDER STALKS WITH SEEDS THAT NOURISH THE WORLD

Measured in terms of geographic distribution and number of individual plants, the grass family is the most successful flowering plant family in the world. Though the daisy and orchid families have more species, neither grows in such numbers as the grass family. The almost 8,000 species of grasses cover one-third of the planet (and half of the area of the United States), from tropic to tundra and from marsh to desert. This wide distribution offers everyone the opportunity to explore the wonderful variety of adaptations of the grasses growing right in their own neighborhood.

Studying grasses is often intimidating to the inexperienced because at first it's hard to tell true grasses from look-alikes. However, if the round stem has swollen joints, or **nodes,** with each leaf starting its growth at a node, sheathing the stem before angling away to form a blade, then the plant almost certainly belongs to the grass family.

One key to the success of the grasses is the simplicity and adaptability of their basic design, which allows them to resist drought, hold onto soils, and thrive in frequently disturbed areas. The root system may make up as much as 90 percent of a grass plant's weight, an important feature for plants that frequently grow in dry areas. Dense and spreading roots enable grasses to use all available moisture. This thick mat often prevents other plants from growing up through it, so the plant does not have to share water or nutrients. The root mat also helps grasses hold on to loose, sandy, or muddy soils such as those found in marshes and on sand dunes where these plants are apt to grow.

Rhizomes, or underground stems, are another important feature of many grasses. Like roots, rhizomes help grasses hold on to soil and prevent **erosion.** A second important function of rhizomes is propagation. These underground stems send up numerous shoots. If you have ever tried to eliminate quack grass from a garden, you are familiar with the white runners that send up seemingly endless plants. This adaptation is one of the factors that enables grasses to thrive in prairie areas where lightning fires are common. While surface vegetation is burned, the underground stems are unharmed and quickly send up replacements.

The grass family has also adapted to being browsed by a wide variety of animals. Grass leaves are very tough and fibrous, reducing their palatability to many insects and mammals. (Many **herbivores** have evolved mouth parts, teeth, and digestive systems just to cope with the toughness of grasses.) Unlike most other plants, grass plants grow from the base, not from the tip, which helps them to survive grazing, mowing, and burning. Both wild and cultivated grasslands are important to many species, and vitally important in holding and building soil and conserving moisture. Despite the grass plants' special adaptations for surviving browsing, overgrazing by animals can occur, damaging natural grasslands and leaving the soil to blow or wash away.

More adaptations for survival can be seen in the stem of a grass plant. The stems are jointed, with elongated, round, mostly hollow sections interspersed with compact solid nodes. The nodes provide some rigidity for the plant and are the points from which the leaves originate. The long, narrow leaves encircle the stem, one above each node, and then protrude in opposite directions so they do not shade each other. Flexible, slender leaves and stems allow wind and rain to pass without breaking or tearing them. Some stems grow vertically; others may trail along the ground and root anew at each node.

At the tip of the stems are the flowers of the grass plant. Unlike larger, more colorful flowers, grass flowers are small, mostly green, and easy to overlook. While many other plants are pollinated by insects that must be attracted by sight and scent, grasses are pollinated by the wind, so they can spend their energy producing prodigious amounts of **pollen** rather than showy flower parts. The tiny pollen is shed from dangling **anthers,** allowing for easy transportation by the wind. The **pistils** of grass flowers are feathery and sticky, to catch the blowing pollen.

quack grass rhizome, roots and shoots

The suffering of hayfever victims is in part a testament to the pollen productivity of grass.

Grasses also produce seeds in large numbers, and there are many different strategies for seed dispersal in the grass family. Wind carries many grass seeds to new locations. These seeds are often equipped with fine hair-like structures similar to those found on dandelion seeds. Some grasses have seeds that are only lightly attached to the plant. These seeds are designed to be knocked to the ground by rain or passing animals. In fact, animals are instrumental in dispersing many kinds of grass seeds. The seeds may spread by sticking to an animal's body, as is the case with many barbed and pointed seeds, by being swallowed and left in droppings, or by being stored for future use and forgotten. Even ants bury stashes of seeds. People, through the advent of agriculture, transportation corridors, landscaping, and man-made waste places, have also been responsible for much of the dispersion of grass seeds, both intentionally and inadvertently.

People depend heavily on grasses. We use grasses for lawns and playing fields, as building material in bamboo and sod houses, and for thatched roofs. Grasses are used to make a wide variety of useful objects, like baskets, mats, flutes, and brooms. Most important, grass seeds are the basis of most human diets. Over two-thirds of the crops that we cultivate are cereal grasses. Wheat, rice, corn, oats, barley, rye, millet, and sugar cane are all in the grass family. Pasture grasses, hay, and grain comprise most of the diet of the animals we raise for meat and milk.

Human civilization began with the cultivation of the grasses. They are a diverse, hardy, and fascinating family of plants. Though often overlooked, the simple beauty of the different grasses deserves our attention.

Suggested References:

Brown, Lauren. *Grasses: An Identification Guide*. Boston: Houghton Mifflin, 1979.

Clark, Lynn G., and Richard W. Pohl. *Agnes Chase's First Book of Grasses: The Structure of Grasses Explained for Beginners*. 4th edition. Washington, DC: Smithsonian Institution Press, 1996.

Harrington, H. D. *How to Identify Grasses and Grass-Like Plants*. Chicago: Swallow Press, 1977.

Martin, Alexander C. *Weeds*. New York: Golden Press, 1987.

Outwater, Alice. *Water: A Natural History*. New York: Basic Books, 1996.

Pohl, Richard W. *How to Know the Grasses*. Dubuque, IA: William C. Brown, 1978.

PARTS OF A GRASS PLANT

foxtail grass

seed head

round stem —

leaf blade

leaf sheath (surrounds stem) —

node —

fibrous roots

Grasses

FOCUS: Grasses are uniquely adapted to withstand challenges of climate and browsing animals. Members of the grass family produce grains, a vital food source.

OPENING QUESTION: *What characteristics make grasses different from other kinds of plants?*

FOCUS ON FEATURES

Objective: To notice firsthand some of the unique characteristics of grasses.

Gather in one or more circles, each with a leader. Pass a grass plant around each group. Each person should examine it and describe one characteristic. Make sure the following characteristics are noticed:

~ Overall tall skinny shape of the plant
~ Long, narrow shape of the leaves
~ Thick mat of roots
~ Hollow stems
~ Stems have solid joints (nodes)
~ Leaves spread out in alternate directions from the stem
~ Clusters of flowers or seeds at top of stem
~ Each leaf starts at a node and wraps around the stem before spreading open (take off a leaf and unroll it so all can see that the sheath is split)

How might these adaptations benefit the plant? After noting the characteristics of grasses, share the following rhyme that may be helpful in telling grasses from their closest relatives:

Sedges have edges
While rushes are round,
And grasses have joints
From their tops to the ground.

Pass around a few different kinds of grass plants plus some other plants that might be confused with grasses. Possibilities include: cattails (no joints, minuscule seeds), sedges (triangular stems and unsplit sheaths), rushes (round stems but no joints), plantain (no joints), iris (big showy flower). Have the children determine which are grasses. How can they tell?

Materials:
- freshly dug grass plants with the intact roots rinsed off – use corn (with both tassel and ear), *Phragmites*, and/or other large grasses with seed heads
- other grass-like plants such as cattails, iris, sedges, rushes

GRASSES IN PICTURES

Objective: To review the special adaptations of grass plants and the functions these perform.

Show pictures or slides of grasses, highlighting the characteristics that contribute to the success of these plants: flexible stems, narrow leaves, nodes, flowers, roots, rhizomes, sod-forming ability, and abundant seeds. How might these adaptations contribute to the survival of the plant?

Materials:
- slide show or photos and drawings of a variety of grasses, illustrating grass plant adaptations

BUILD A GRASS PLANT *(grades 3-6)*

Objective: To illustrate the characteristic structure of grass plants.

In small groups, have the children use only paper, tape or glue, and scissors to create a three-dimensional grass plant showing the long, narrow leaves, the nodes, the seed heads, and the roots. (Provide several live grass species for groups to refer to.) For additional structural support, the children may tape their grass to a wall or mount to poster board. Label the different features. Afterward, have groups compare their plants and discuss what makes them grasses.

Materials:
- bond paper or newsprint
- tape, glue
- scissors
- poster board (optional)

GRASS MIX AND MATCH (grades K-2)

Materials:
- hand lenses
- sets of 5-6 different grasses with seed heads
- sets of cards with photocopies of the same 5-6 grasses

Objective: To identify different kinds of grasses by their seed heads.

In advance, collect five or six different kinds of grasses with seed heads, one set for each small group. Use a photocopier to create pictures of each type of grass showing the overall shape and pattern of each seed head. Mount these pictures on individual cards, making one set for each small group.

Divide the children into small groups and give each group a set of grasses and grass cards. Each child chooses a grass and then matches it to the correct picture. Later you may want to challenge children to find matching grasses outdoors. How does the ability to produce many seeds benefit the plant? How does it benefit other creatures?

For older children: include an extra picture card of a grass not in the set, increase the number of species, or include a plant that's not a grass and ask them to pick it out.

STALK THE GRASS

Materials:
- Stalk the Grass cards
- paper
- pencils
- clipboards
- hand lenses
- sharp shovels

Objective: To observe variation and adaptations in grasses growing outdoors.

Divide the children into small groups and provide each group with a Stalk the Grass card. Review the tasks with the whole group and explain that each team will choose (or be assigned) one or two tasks to complete, more if time permits. For younger children, have an adult in each group to lead them through a couple of the tasks. Before sending the groups off, be clear about whether they should or should not pick specimens. You may want the children to collect some of the common grasses and other plants for the Grass Weaving project.

When all are done, ask each group to decide on one interesting (or surprising) observation to share with the whole group.

STALK THE GRASS CARD
Choose one or two of these tasks to do with your partner. Be ready to share the results with the rest of the group.

1. Choose one tall grass plant. With eyes closed, run your fingers up the stem from the ground to the tip. How many swollen joints (nodes) are there? How many leaves? Sketch the plant with its nodes and leaves in the right places.

2. Look for grass that is mowed. Have any formed flower heads? See where the newest leaves are. See if you can find nodes. Compare a mowed grass plant with one that has not been mowed.

3. Look for grasses growing in difficult habitats: sidewalks, asphalt, roadsides, pastures. How have they responded to the problems there?

4. Look for grass plants in different stages of flowering: before opening, in bloom (are any shedding pollen?), with seeds, with seeds gone.

5. Look for plants that are grass-like but are not grass. How can you tell? How could you explain the difference to someone new to the subject?

6. Dig up several different types of grass. Shake off the dirt and examine the roots. Do they form a shallow mat or are they in clumps? Are there any rhizomes? Look at the edge of a disturbed area or in a garden for these. How long are the rhizomes? Fill in holes, and try to leave the area looking the way you found it.

GRASS WEAVING

Objective: To use the special features of grasses to create something beautiful or interesting.

Make a simple framework for weaving using forked sticks with yarn wound across the Y, wrapped twice at each pass to hold firm; or create one large stick frame for the whole group. The children can weave the flexible stems of a variety of grasses, with seed heads attached, into the framework. Include other natural finds the children collect, then hang to display.

Materials:
- frames for weaving
- yarn
- grasses and other decorative plants

SNACK

Objective: To understand how humans depend on the grass family for food and other products.

Pass out foods made of grains for the children to eat. Point out that they are eating the seeds of grass plants, which make up a major portion of every person's diet. If possible, also show a selection of unprocessed grains like wheat berries, rice, and corn (often available in bulk from a health food store). Have the children brainstorm a list of other ways we use grasses, for example, in baskets, br

Materials:
- a variety of foods made from grains and grasses, such as popcorn, small rice cakes, breakfast cereals, crackers; (optional) unprocessed grains

the start of a grass weaving

EXTENSIONS

Watch It Grow: Enc   rass and keep a daily
journal of all they o   row? What visitors are noted?
What other plants are present.

Experiments with Grass: Grass plants grow quickly and are perfect for designing controlled experiments. Have the children brainstorm questions about how grass plants grow, and then design tests to explore these questions. Some suggestions: how does mowing affect the plants; can grass grow in the dark; will grass plants grow better as a monoculture or with other species mixed in; are some soils better than others; how does fertilizer affect the growth of grass plants? Results can be determined by measuring the comparative length of the grass shoots, the length of the shoots and roots combined, or by rinsing off the experimental and control plants and weighing each group.

Uses of Grasses: Discuss, or have the children research, the many uses of grass in their homes and around the world. They could draw pictures to illustrate their findings, or bring in samples of different products made from grasses.

Word Games with Grains:
a) Post a list of grains and a list of foods made from them. As a class, connect the foods with the grains.
b) Make up a word search or crossword puzzle using names of grains and foods made from them.

Sprouts: Have the children sprout wheat berries then put them on a salad or sandwich to eat.

A Complex Relationship

All animals share a common problem. They must get enough nourishment to keep their bodies going or else face death. **Herbivores** are animals that depend on plants for their food, **carnivores** catch and eat other animals, and **omnivores** eat both plants and animals. We focus here primarily on the mammals, though insects, fish, **amphibians**, **reptiles**, birds, and all other kinds of animals can be classified by these same ecological roles, that of hunter or hunted.

The problems of getting enough to eat are quite different for herbivores and carnivores. An herbivore's food source is stationary and often fairly abundant. The challenge is how to get enough of it while avoiding attack by a hunter. Many herbivores have to spend most of their waking time eating. Carnivores have to work harder to find and catch their food, and, once caught, it may escape or fight back. However, meat is a more concentrated energy source, so carnivores do not have to eat as often as herbivores do.

Because of their different diets, herbivores and carnivores play different ecological roles. Herbivores are usually the hunted, while carnivores are the hunters. Adaptations, both structural and behavioral, have evolved in both hunters (**predators**) and hunted (**prey**) to help them meet the challenges specific to their role.

Acute senses are critically important for the hunted species. The sense of hearing is usually well developed, and their ears are often large and able to swivel around to catch passing sounds. A deer or rabbit can hear the approach of something long before the intruder can be seen. The eyes of most hunted animals are placed toward the sides of their heads to enable them to get a broader view of their surroundings. **Nocturnal** or **crepuscular** animals that rely in part on vision, like flying squirrels or deer, tend to have large eyes that gather the maximum amount of light. The sense of smell of many hunted animals is also well developed. The sensitive nose of a plant eater helps it tell nutritious leaves from poisonous ones and also alerts it to danger. Of course, different species rely on different senses to varying degrees, and many have one or two senses that are especially keen and others that are less developed.

The ability to make a rapid escape is obviously helpful to hunted animals. A jackrabbit being pursued is able to make quick, unexpected turns and twists, sometimes changing direction in midair. Many herbivores are quick and agile, nervous in exposed places, and acutely aware of the location of the nearest cover. They tend to react with fear and avoid anything unfamiliar. Some herbivores, especially those in flocks or herds, take off whenever any member shows signs of fear. In this way they all benefit from the many pairs of alert eyes and ears.

For several species of prey animals, a speedy escape is not possible or is so costly in terms of energy that it's not the first line of defense. Many have evolved other protective adaptations over time. The opossum plays dead when threatened, hoping to discourage hunters that prefer live prey. The snowshoe hare will often sit still when confronted by danger, hoping to remain undetected. The ungainly porcupine has a protective covering of sharp, penetrating quills for defense. Good prey defenses challenge predators to hone their skills and, generation by generation, develop better senses and hunting strategies. In turn, animals that successfully avoid predation live to pass on their genes.

A predator interacts with the world quite differently than its prey. While prey species tend to avoid the unknown, predators investigate every possible hole and crevice, sound, or smell that might yield food. A hungry predator's movements are often quick, restless, and inquisitive. Carnivores are apt to gorge themselves when food is available and then, if necessary, go for long periods without eating.

Many predators have a well-developed sense of smell. Animal trails provide a rich map of valuable information to the sensitive nose. A predator's sight is also important as even slight movements in the grass may reveal a busy mouse to a hungry fox. A carnivore's eyes are located toward the front of its head, providing the three-dimensional, binocular vision essential for hunting. A leap for moving prey must be timed and placed precisely. Like many prey species, many predators also hear very well. A barn owl with its excellent hearing can locate mice in total darkness. The bat follows and captures flying insects using **echolocation;** the mole in its tunnel feels the vibrations of crawling earthworms. As well as having keen senses, predators are equipped with the tools to catch and eat their prey – powerful shoulders and jaws, wide mouths, sharp teeth, and sharp claws.

Once a predator locates prey, it uses a particular, characteristic strategy to capture it. The bobcat may sit patiently on a ledge above prey travel corridors and wait for an animal to run by. At just the right moment, it pounces for the kill. If it misses, the bobcat will chase the prey for a very short distance before giving up and going back to its perch. Coyotes often work in pairs to chase down prey, relieving each other as needed. Wolves travel and hunt in packs. The fisher, a large weasel relative, crisscrosses huge areas exploring under logs and inside cavities where prey might hide. An otter in a lake can swim faster than the fish it eats; sometimes it will drive fish into a cove to trap them. Otters also turn over rocks in the water to catch crayfish, tear open rotting logs searching for salamanders or insects, or even enter beaver lodges to prey on young beavers. Its natural curiosity has led it to exploit many food sources – this is called opportunistic hunting, that is, making a meal of whatever's available. Sometimes the hunter becomes the hunted, especially the vulnerable young of a species. For example, a bobcat kit, left alone while its mother hunts, can become a meal for an alert great horned owl. In nature's daily struggle for survival, anything is fair game.

Predation is as important to the prey species as it is to the predator. "To understand the value of predation it is important to remember that among all animals the fiercest competition is between members of the same species – between muskrats and muskrats – rather than between members of different species – mink and muskrats. Animals of the same kind need exactly the same kind of food, shelter, nesting areas, mates, and these resources are limited" (Bil Gilbert, *The Weasels*, p. 61). When predators and prey are in balance, their habitat remains healthy and supports a broad array of plants and animals. When carnivores are removed and herbivore numbers grow too large, the herbivores strip their habitat of vegetation, which leads to a collapse of the herbivore population and damage to the whole ecosystem.

In a healthy ecosystem, there are many more herbivores than carnivores, reflecting the energy relationships in a food pyramid. Many prey animals bear numerous offspring, thereby increasing the chances that at least a few will survive. Generally, the smaller the animal, the more numerous the young. Predators, on the other hand, reproduce in much smaller numbers, and the young may take longer to become independent.

The adaptations of both predator and prey are successful only if they enable adequate numbers to survive and reproduce. However, in order for a predator to live, some prey must die. This interaction between predator and prey, the hunter and the hunted, is a vital part of the complex network of relationships in the natural world. Neither the hunter nor the hunted is right or wrong, evil or good. Each plays a critical role in the web of life.

Suggested References:

Alcock, John. *Animal Behavior: An Evolutionary Approach.* Sunderland, MA: Sinaver, 1989.

Gilbert, Bil. *The Weasels: A Sensible Look at a Family of Predators.* New York: Pantheon, 1970.

Rezendes, Paul. *Tracking and the Art of Seeing: How to Read Animal Tracks and Sign.* 2nd edition. New York: Harper Collins, 1999.

Whitfield, Dr. Philip. *The Hunters.* New York: Simon & Schuster, 1978.

red fox

Hunter – Hunted

FOCUS: Animals that hunt other animals for their food, and the animals that are hunted, have special adaptations to meet the challenges of their particular roles.

OPENING QUESTION: *What are some differences between the adaptations of animals that hunt and those that are hunted?*

RABBITS AND FOXES

Objective: To understand the shifting balance and interdependence of groups of predators and their prey.

Sitting in a large semi-circle, the children listen and watch as you read the "Rabbits and Foxes" story aloud. Have an assistant (with own copy of script) follow the stage directions to help you portray the story's characters on felt board with cut-outs. Afterward, use the story's follow-up questions to discuss the changes in the number of rabbits and foxes in relation to the seasons and available food supply.

Materials:
- felt board
- felt cutouts: 15 rabbits, 14 foxes, trees, grass and small shrubs, leaves
- quilt batting for snow
- "Rabbits and Foxes" story

WHAT'S THE DIFFERENCE?

Objective: To identify adaptations that are specific to hunted animals or hunting animals and note those common to both.

Divide into two groups. Using pictures or mounted specimens, have one group study animals that are hunted (prey) and the other group study animals that are the hunters (predators). Each group should make a list of words that describe adaptations of their predators or prey. Compare lists. You might ask the children in one group to read an adaptation and the children in the other group to then read an adaptation that would counter it. Do any animals show both predator and prey adaptations? Why might this be?

Materials:
- mounted specimens or pictures of hunter and hunted animals (may be assembled as two large posters with collaged magazine photos)

GETAWAY SKITS

Objective: To learn a variety of different hunting and escaping strategies.

Divide the children into groups so that each child receives a role to act out in one of the following skits. Give each group a description of their scenario and the appropriate character name signs to wear. Have the children read through and practice their skits. (Props and sound effects are allowed.) Then, one group at a time, the children perform their skits for the other children. After the performance, the audience tells what they saw acted out in the skit.

Materials:
- Scenario cards
- signs with animals' names to tape to players

SCENARIOS:

Bobcat and Hare

Bobcat crouches on ledge and watches Snowshoe Hare nibbling evergreen twigs. Bobcat tenses and prepares to pounce on Hare. Hare senses danger and sits motionless, then runs at the last moment. Bobcat chases for a few feet, then gives up and goes back to ledge, where it washes its face.

Characters: Bobcat, Hare

Polar Life

Seal swims, chases Fish, then comes up to hole in the ice for air. After Seal goes down again, Polar Bear comes along and waits by hole. The next time Seal comes up, Polar Bear takes a big swipe, but just misses Seal, which dives back down into the water.

Characters: Seal, Fish (3-4), Polar Bear

Bird of Prey

Hawk circles in sky, looking down at the ground. Squirrel eats nuts not noticing Hawk until Blue Jays see it and start calling "jay, jay!" Squirrel runs up a tree.

Characters: Hawk, Squirrel, Blue Jays (2)

Northwoods Encounter

A group of Moose, two cows and three calves, is browsing on tree branches. Three or four Wolves gather at a distance, watch Moose, then separate and try to sneak up on the Moose. One Moose smells the Wolves. The two cow Moose face the Wolves keeping their calves behind them. Moose drive Wolves off with hooves.

Characters: Wolves (3-4), Moose (2 mothers, 2-3 young)

Fox and Mice

Mice nibble seeds in tall grass. Fox follows their trail with nose. Mice hear Fox coming, stop nibbling and wait. Fox pounces where it thinks a Mouse is, but just misses.

Characters: Mice (2-3), Fox

SNACK TRACK

Objective: To experience the roles of both predator and prey through a game of hiding and searching.

Divide the children into an even number of teams; half the teams are "Prey," the other half "Predators." Each Prey team has five minutes to hide a snack, leaving a trail for one of the Predator teams to follow. (Give each Prey team a different type of dry pasta to use as trail markers; in undisturbed snow, the trail could be footprints.) When the Prey have completed their task, assign each team of Predators one trail to follow. The Predators should find the snacks and bring them back for the whole group to share. Make sure trail markers get picked up afterward.

Note: If the area is big enough, all groups can initially be Prey and each team can hide a snack, then all become Predators, switch places, and hunt for the hidden snacks; OR, while half the group is busy with its part of this activity, the other half can start the Superhunter activity.

Materials:
- a variety of uncooked pasta (rotini, bowties, macaroni, etc.)
- snack(s) for the whole class

SUPERHUNTER

Objective: To invent a predator that is adapted to catch a specific prey.

Materials:
- prey cards
- large sheets of paper
- crayons, pencils and markers
- tape

With older children in pairs or younger children in small groups with a leader, pass out cards describing a kind of prey and where it can be found. Each team should first discuss the special challenges of finding and catching that prey, and the senses and other adaptations a predator would need to do so. Then the children draw their own imaginary predators hunting their prey.

- ~ Fish in fast-flowing stream
- ~ Spiders hiding in the rocks
- ~ Turtle eggs buried in sand
- ~ Deer in woods on an island
- ~ Snake sunning on a stone
- ~ Ducks swimming in the ocean
- ~ Termites in a rotting log
- ~ Porcupine in a treetop
- ~ Bats hanging on walls of a cave
- ~ Ants in tunnels inside an ant hill
- ~ Beetles on a lily pad in a pond
- ~ Green frog in a grassy swamp
- ~ Snails on underwater plants
- ~ Mice in tunnels under deep snow

When all are done, the children can introduce their animals and describe their adaptations. Display drawings.

EXTENSIONS

Write a Story: Tell the children to pretend to be a hunted or hunting animal. Write a story about how it might feel to be that animal.

Who Lives Here?: Have the children use reference books and community members, including naturalists, trappers, hunters, and game wardens, to learn what top predators live in the area and what they eat. Can they come up with some possible food chain relationships? Create a mural that illustrates these relationships.

Ask the Game Wardens: Have the children write to their Fish and Wildlife Department asking about the hunting and protective laws pertaining to animals they are interested in.

cottontail rabbits

Hunter – Hunted

[To begin, the felt board should have a few hardwood trees and shrubs with green removable leaves, one or two small evergreens, a couple clumps of grass, and one fox and one rabbit, placed far from each other. As the story progresses, you'll need an additional 14 rabbits and 10 foxes, plus some white cotton batting.]

This is a story of a young fox named Freddie and a young rabbit named Roberta in a land called Faraway Woods. Our story starts on a warm and sunny summer day. Freddie and Roberta are both young. They were both born in the spring and are each beginning to learn how to find their own food.

Roberta is having a wonderful summer. There are lots of good things to eat – grass and tender leaves, twigs and buds – and the days have been warm all summer long. *[put up 14 more rabbits]* She has lots of brothers and sisters, aunts, uncles, cousins, grandparents, and friends, and there's plenty of food for all of them. The only thing Roberta worries about is the big danger. In the middle of the field is a fox den *[put up 3 foxes]*, and she's been told many times to be careful of the new fox family that lives there.

Meanwhile, Freddie, the young fox, is learning how to catch grasshoppers, pick berries, lunge after mice, and keep his nose open for any sign of rabbits. He's been told many times that the very best thing for dinner is a nice young rabbit.

Things start changing around Faraway Woods when October comes. The days get shorter, the weather turns chilly, the leaves fall off the trees and shrubs, and the grass dies back. *[remove the leaves and grass]* Food is harder to find. Some rabbits have to go far away from home to look for food. *[remove 3 rabbits]* They often bring back very bad news: it seems that there isn't much food anywhere, and there is another fox family in the area. *[add 3 foxes]*

Meanwhile Freddie is ready to leave his home. He decides not to go too far away from the area since there are so many nice, fat rabbits around that would make good meals all winter long. If only he could catch one!

One day in late fall after spending many hours looking for something to eat, Freddie catches the scent of a rabbit. Crouching down, he slowly approaches a low bush. There, nibbling, is an older rabbit. Freddie feels a little scared since he's never caught anything so big, but his hungry stomach pushes him on. He moves in quickly and springs. Surprisingly, the rabbit is caught unaware, and Freddie manages to kill it. It turns out to be a very old, weak rabbit, but it's food to Freddie. His first successful rabbit hunt! *[remove 1 rabbit]*

Winter finally moves in, *[add batting]* covering Faraway Woods with a layer of white fluffy stuff, something completely new to Freddie and Roberta. Life becomes much more difficult for the rabbits. Food is even harder to find. The farther the rabbits have to go to look for food, the more dangerous it is. *[remove 1 rabbit]* Traveling in the snow is not much fun, plus it is very tiring. *[remove 1 rabbit]* Often they hear that another rabbit has been eaten by a fox. *[remove 2 rabbits]*

By the end of the winter *[remove batting]*, the rabbit colony is not in very good shape. Some rabbits have even starved to death. *[remove 4 rabbits]*

Finally spring returns *[replace grass and leaves on trees]*, but the rabbits that survived, including young Roberta, look pretty thin. Not very many baby rabbits are born this spring. *[add 3 rabbits]*

Freddie and the rest of the foxes have had a great winter dining on rabbits. The foxes are all well fed, and new litters of fox kits are born to their healthy fox mothers. *[add 4 foxes]*

The summer goes well for everybody. All the young foxes grow up eating lots of mice, berries, and grasshoppers. The older foxes notice that there are not as many rabbits this summer as there were last year. Young rabbits, usually so easy to catch, are few and far between. But right now it doesn't matter to the foxes because there is plenty of other food.

Roberta and the rabbits are recovering from their hard winter, with new greens and twigs available and not as many rabbits to feed. They all look a lot better and are getting plenty to eat. The rabbits are extra wary of the foxes since there are so many around these days.

All the animals are surprised at how soon winter comes this year. *[leaves off, batting on]* By Thanksgiving, there is snow on the ground. As winter continues, it becomes obvious that there are too many foxes and not enough food. Freddie and his friends find themselves fighting over hunting territories. Many of the foxes are even forced to leave Faraway Woods. *[remove 8 foxes]* If only there were as many rabbits as last year!

As spring slowly comes *[batting off]*, Roberta notices that almost all the rabbits are alive and well. There may not be too many, but all of them are well fed and looking forward to the new litters of baby rabbits that will soon be born. *[add leaves]*

On the last day of our story, Freddie is lying on a sun-warmed rock looking over the field. He notices how few foxes stayed and survived in Faraway Woods all winter. But as he looks over to the far end of the field, he sees lots of small rabbits hopping around. *[add 8 rabbits]* "Hmm," he says, "look at all those tasty little rabbit meals over there waiting for me. Maybe things aren't going to be so bad after all."

AFTERWARD

Ask the children what might happen next in the story.

What happened to the number of rabbits when there was plenty of food for all of them?

What happened to the food supply when there were lots of rabbits?

What might happen to the rabbits if there were no foxes?

What might happen to the foxes if there were no rabbits?

templates for rabbit and fox felt cutouts

Teeth and Skulls

DENTITION DETERMINES THE DIET

For those who possess them, teeth play a vital part in the game of success and survival. Over time, clear patterns of size, shape, position, number, and distribution of teeth, related closely to each animal's diet, have developed. In fact, dentition (tooth arrangement) and skull shape are so distinctive that they are clear keys to identification of each mammalian species. Mammals have a variety of food preferences, ranging from a bat's specific diet of flying insects to a raccoon's willingness to eat just about anything. Dentition is usually a strong indication of these eating habits.

In relation to their diets, animals can generally be broken into four main groups: **carnivores**, eating meat; **herbivores**, eating plants; **insectivores**, eating insects; and **omnivores**, eating a variety of foods. Specific kinds of teeth are arranged and shaped to fit the needs of members of these groups. **Incisors**, in the front of the mouth, are used for cutting. The dagger-like **canines**, next to the incisors, are for tearing and shredding meat. **Molars**, in the back of the mouth, do the grinding. Many young mammals are toothless at birth, get "milk teeth" first, and then later grow permanent adult teeth. The age of an animal can often be determined by the eruption of the adult teeth and by wear on those teeth.

Carnivores, who are largely meat eaters, have all three kinds of teeth but depend only marginally on their small incisors for nipping and biting. As hunters, they use their dagger-like canines for grabbing, puncturing, and holding onto their **prey**, and sharpened molars for cutting and tearing their food. Powerful sets of muscles control the jaws and provide the force for the scissoring action of these specialized sharp teeth. A bobcat's canines make surgical punctures between the vertebrae of its prey's neck, stopping the prey in its tracks. All members of the cat family are carnivorous, from the laziest house cat to the great lions, as are most members of the weasel and dog families.

Moles, shrews, and bats are familiar insectivores, possessing a mouthful of sharp little teeth, which are used in seizing and crushing hard-shelled insects and other small animals. Although insects of various descriptions make up much of their diet, shrews are well known for attacking and eating small mammals that often surpass the shrews in size.

On the other side of the dietary coin are the plant eaters, or herbivores. These animals often serve as food for the carnivores and are an important link in the **food chain**. They include a wide variety of animals from deer, cows, horses, sheep, and goats to the smaller mice, rats, woodchucks, rabbits, and beavers. They lack canine teeth entirely and possess only incisors with which to bite off their food and molars with which to grind it up. Squirrels and other nut-eating rodents use their incisors as a vice to grip and pressure open or to puncture the hard shells of nuts – or to gnaw through the heaviest, hardest ones like butternuts. Deer lack upper incisors but may use their lower ones to strip bark from trees in winter.

The incisors of rodents as well as those of hares and rabbits are ever-growing and must be used continually to maintain their edges. The layer of enamel on the front of the teeth is harder than the layer of dentine on the back of the teeth; the dentine thus wears away faster, leaving the slightly extended enamel as the sharp cutting edge. If the upper and lower incisors do not meet, due to a broken tooth or some other malformation, the growing teeth may not be ground down by use and may eventually cause the animal's death. Of all the rodents, the most exceptional gnawer must be the beaver, whose work may be seen in the pointed, chewed-off stumps of trees cut down for dam building and food. A beaver's incisors may grow as much as four feet in a year.

While the teeth of carnivores and the teeth of herbivores are similar in arrangement to others in their group, omnivores may be hard to pick out just by looking at their teeth. Skunks, raccoons, pigs, muskrats, bears, and squirrels are very different creatures, yet they

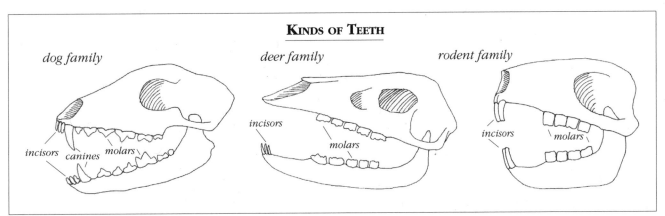

KINDS OF TEETH

dog family

deer family

rodent family

incisors   canines   molars

incisors

molars

incisors   molars

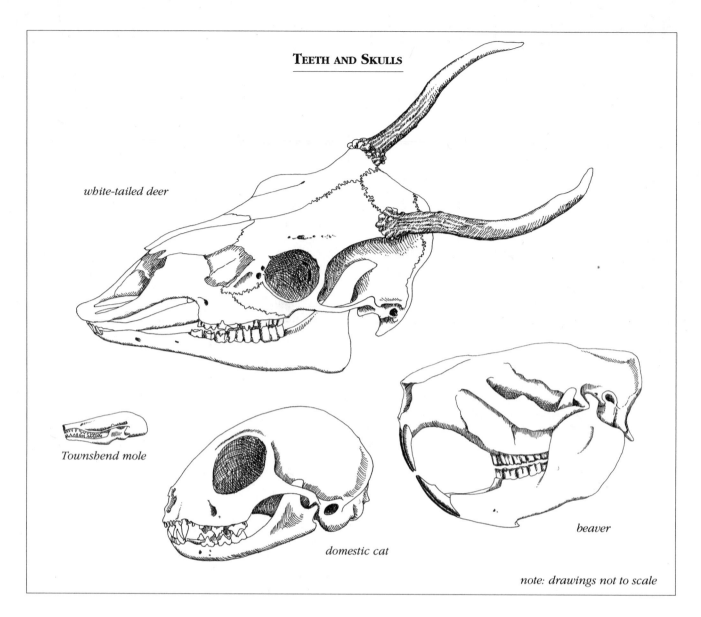

white-tailed deer

Townshend mole

domestic cat

beaver

note: drawings not to scale

all are resourceful, adaptable omnivores. Though humans, which are primates, and opossums, which are marsupials, are both omnivores, our teeth are considerably different. Most humans have 32 adult teeth compared to the opossum's impressive mouthful of 50 sharp ones!

Some animals possess teeth that don't seem to match their eating habits. The coyote, equipped with the sharp teeth of a carnivore, is really an opportunistic omnivore, eating everything from animals to apples to garbage. The woodland jumping mouse, with typical rodent teeth, eats insects as much as it does seeds and fruits. And the panda, with its herbivorous diet, has the teeth and short intestinal tract it inherited from a carnivorous ancestor. It has to spend most of its day eating huge quantities of bamboo to get enough calories to live.

The other features of a mammal's skull are also important indicators of its lifestyle. Eye sockets in the front of the skull provide carnivores, like the cat, with focused, binocular vision useful in hunting prey. Eyes located on the sides of the head give a wide field of vision that allows prey species like the rabbit to keep a lookout all around. Long snouts with bony convolutions belong to animals that have a highly developed sense of smell. And a sharp ridge along the top of the skull of large carnivores provides a strong attachment for muscles used in grabbing and ripping.

Careful examination of an animal's teeth and skull may reveal its identity and can provide valuable information about the life and habits of its owner.

Suggested References:

Burt, William H., and Richard P. Grossenheider. *A Field Guide to Mammals: North America North of Mexico.* Peterson Field Guides. Boston: Houghton Mifflin, 1980.

Merrill, Margaret W. *Skeletons That Fit.* New York: Coward, McCann & Geoghegan, 1978.

Roest, Aryan I. *A Key-Guide to Mammal Skulls and Lower Jaws.* Eureka, CA: Mad River Press, 1991.

Searfoss, Glenn. *Skulls and Bones.* Mechanicsburg, PA: Stackpole, 1995.

Teeth and Skulls

FOCUS: Different types of teeth are adapted to grasp, hold, and chew different kinds of food. Examining the kinds of teeth a mammal has and the shape of its jaw and skull helps us determine the type of food it eats.

OPENING QUESTION: *Why do animals have teeth?*

PUPPET SHOW

Objective: To learn about the differences in the teeth of plant-eating and meat-eating animals.

Perform, or have the children perform, the puppet show. Using the puppets as illustrations, review the following terms: herbivore, omnivore, and carnivore.

Materials:
- puppets
- script

TOOTH TOUCH

Objective: To become aware of the different kinds of teeth humans have and how we use them.

Ask the children to feel their own teeth with their tongues. How many different kinds of teeth do they feel? Give each child several different foods to eat. What actions do the different teeth perform in eating? Have the children hold their hands on their cheeks and jaws while they chew. What moves when they chew? How do their tongues move when they bite, chew, and swallow?

Show a mold or model of human teeth and have the children describe the shape of the jaw and the kinds, shape, and number of teeth. Together list the types of foods humans eat and some ways we prepare our food before we eat it.

Materials:
- mold or model of human teeth (ask your dentist or orthodontist)
- apples, popcorn, fruit leather, celery stalks, other foods that are eaten in a variety of ways

TOOTH TYPES

Objective: To compare the kinds of teeth belonging to different types of mammals.

Show the children unlabeled drawings of herbivore, carnivore, omnivore, and insectivore skulls with teeth. Discuss the kinds of teeth (incisors, molars, and canines) each has or doesn't have. Ask the children to try to determine what the animal might eat and what kind of animal it might be. What observations helped them guess the identity of each animal?

Materials:
- skull drawings of:
  - herbivore (deer family, rodent, rabbit)
  - carnivore (dog, cat, or weasel groups)
  - omnivore (pig or human)
  - insectivore (mole, bat, shrew)

SKULL SKILLS

Part I: Skull Study

Objective: To observe differences in skulls, and to relate skull size and tooth type to animal type.

Divide the children into groups, giving each a skull (unlabeled) to inspect. Once they have had time to examine the skulls, each group passes its skull on to the next group. How is the new skull different? Continue passing skulls until each group has back its original skull.

Now have the groups make detailed observations of a skull by answering these questions:

SKULL INVESTIGATION

How many incisors does the skull have?____ Canines?____ Molars?____

Are most of the teeth sharp or flat?

Looking at the teeth, what do you think this animal ate: plants____, insects____, larger prey____, or a variety of things____?

Do you think the skull is that of a carnivore (meat eater), herbivore (plant eater), insectivore (insect eater), or omnivore (eater of many foods)? Tell why.

Examine the skull's size: is it less than 3" long?____ 3" to 5" long?____ Over 5" long?____

Are the eye sockets large or small?

Are the eyes positioned more to the front or to the sides?

Is the nose long or short?

What animal do you think this is? (Remember, skulls look small without the ears and fur!)

What leads you to this conclusion?

Note: To help young children guess the animal, list the possibilities on the board. Older children could get a much longer list to select from.

When the children are done with their investigations, give each group a card with their animal's picture and related information. Skulls will be introduced to others in Part II.

Part II: What's for Dinner?

Objective: To relate the skull and teeth to the foods eaten.

In the same small groups, have the children invent a fanciful menu for their animal, with appetizers, main course, side dishes, desserts, drinks. Example: "Our animal is a mole. She's an insectivore, so when she went to the cafeteria, she chose Broiled Beetles, Moth Muffins, Junebug Juice, and Cockroach Cake."

Afterward, each small group introduces its animal to the others, shares some key points of interest about its skull, and then reads its menu.

Materials:
- skulls, one for each small group of 4-5 children;
- Skull Investigation sheets
- information card for each skull with animal's picture
- paper and pencil for menus

Note: Fish and Wildlife Departments, high schools, or colleges will often lend skulls.

COMPLETE A JAW

Objective: To design a set of teeth to fit a particular animal's diet.

Give each child or pair of children a cardboard skull (they could select from two or three basic shapes) and some plasticine clay. Have them determine what foods their animal (real or imaginary) eats and then let them form clay teeth for the jaw. They may also cut out paper ears and eyes for their animal and tape these to the cardboard skull. The children may name their animals then display the skulls and teeth in the classroom.

Materials:
- cardboard skulls
- plasticine (nonhardening) modeling clay
- paper, scissors, markers, and tape
- drawings of skulls and teeth from Tooth Types for reference

Note: Younger children may have more success just drawing jaws and teeth. Also, they may need a bit more direction, like "create teeth for a large meat eater" or "create teeth for a small animal that eats nuts."

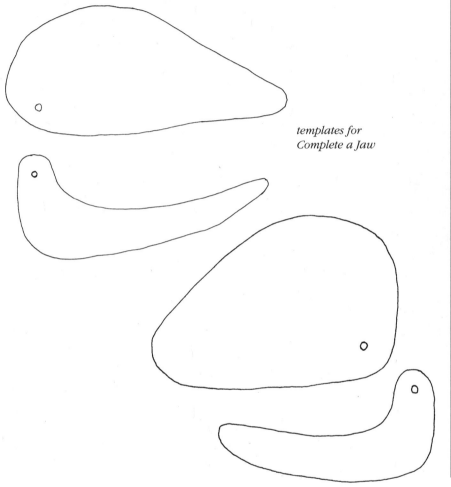

templates for Complete a Jaw

EXTENSIONS

Class Visitors: Fish and Game departments, trappers, taxidermists, naturalists, museums, or illustrators often have collections of skulls or teeth. Invite someone to visit the class with skulls and skeletons.

Teeth Research: Have individuals or small groups research the skulls and teeth of animals that interest them. Some suggestions: anteaters, elephants, wild boars, saber-toothed tigers, whales, hippopotami, bears.

Different Skulls: Make a list of all the kinds of vertebrates the children can think of, for example: fish, turtles, sharks, birds, snakes, dinosaurs, alligators. Do they all have teeth? How do they eat?

Dinner Display: Gather pictures of lots of different animals eating. Can you see them using their teeth in different ways? Sort according to the foods eaten.

Human Teeth: Ask a dentist or an orthodontist to talk with the children about working with teeth.

Teeth and Skulls

Characters: Herbert Hare, Marsha Mouse, Wilma Weasel, Rocky Raccoon

Herbert Hare

Oh, it's so wonderful to munch away in this nice sunny field. *[sniff, sniff]* What's that funny smell? It's not the grass. *[sniff, sniff]*

Marsha Mouse

Don't eat me, don't eat me! **Please** don't eat me! I didn't mean to come into your grassy field. I'll never bother you again if you just **don't** eat me!

Hare

Oh! So, you're what I was smelling. Mouse smell is very different from grass smell.

Mouse

Oh yes, and mouse taste is different, too! Mice aren't as tasty as grass.

Hare

Well, I wouldn't know about that. I don't eat mice.

Mouse

Please, spare me, don't eat...did you say you don't eat mice?

Hare

Have you ever seen a hare's teeth? They're not made for biting into mice or other animals.

Mouse

Then what are they made for?

Hare

They're made just right for chomping on grass, leaves, and twigs.

Mouse

I never knew that. I prefer seeds and nuts, myself. Gee, since you aren't going to eat me, maybe I'll stay in this field a while.

Hare

Oh, you're welcome to. Maybe I'll see you around. Bye, bye.

Mouse

So long. *[Mouse leaves]*

Hare

Gosh, maybe I should have told that mouse about the weasel that comes wandering in this field. She might just like a mouse meal for dinner. I'd better try to find that little mouse again so I can warn her. *[Hare exits; Mouse and Weasel appear on opposite sides of stage]*

Mouse

Now that I don't have to worry about being eaten, I wonder what I can find to munch on in this field. What's that I see over there? Oh, it's a weasel. I'm not afraid of her anymore. She probably can't eat me either. I'll go over and say hello. *[Mouse walks up to Weasel]*

Wilma Weasel

Hey, what are you doing walking over here? How come you're not afraid of me? I could eat you, you know.

Mouse

You couldn't eat me even if you wanted to. Herbert Hare just told me how teeth are made for munching on plants, not animals.

Weasel

His teeth are made for munching on plants. **My** teeth are made for biting into animals – and mice are my favorite.

Mouse

You mmmmean you ccccould eeeat mmmme? YIKES! *[Mouse runs away; Raccoon enters]*

Rocky Raccoon

Hi, Wilma. Well, I don't know what you said to that little mouse, but she sure was in a hurry to get away!

Weasel

Oh hi, Rocky. Well, that mouse is just lucky I'm not hungry right now. She walked right up to me without a care in the world. Imagine thinking my teeth aren't fit for eating an animal! Why, they're **only** fit for eating meat.

Raccoon

Don't you get tired of mice for breakfast, lunch, and supper, Wilma? That's a pretty boring menu! I like to eat some fruit and nuts with my eggs and crayfish.

Weasel

Fruit? Nuts? Yuck! I just like meat. I guess I am a picky eater. Not like you, Rocky. You'll munch on anything, animal **or** vegetable.

Raccoon

You bet I will. In fact, I'm off to the stream right now to find lunch. I saw a bunch of fat frogs down there yesterday. Mmmm Mmm. Care to join me, Wilma?

Weasel

I'd love to, Rocky. A nice cool drink of water would really hit the spot. Then maybe I'll try to find that mouse again. For dinner, this time. *[Weasel and Raccoon exit; Mouse and Hare enter]*

Hare

[out of breath] Oh, here you are! I've been searching for you everywhere. There's something I have to tell you.

Mouse

Yeah, well, there's something **I** have to ask **you.** Why didn't you tell me your teeth were different from a weasel's?

Hare

That's what I was going to tell you, but most animals know that already.

Mouse

How was I supposed to know that?

Hare

Because weasels are **carnivores** and hares are **herbivores,** that's how.

Mouse

Carnivores, herbivores. What are you talking about?

Hare

Carnivores, like weasels, eat other animals. So their teeth are sharp and pointy, made to bite through skin. Herbivores, like us, just eat plants. Our teeth are wide and flat, perfect for chewing up leaves and grasses. And then, there are the omnivores…they'll eat most anything.

Mouse

Oh, I get it now. All I have to do is look at an animal's teeth, and then I'll know if it can eat me or not.

Hare

Well, it might be a bit safer if you just ask me. I know my herbivores and carnivores pretty well, without looking at their teeth.

Mouse

Okay, I'll do that. But right now I think I'll get out of here. There's a certain weasel carnivore somewhere around that might want to stick her sharp pointy teeth into me.

Hare

That's for sure…see you later!

Mouse & Hare

[at same time] Bye! *[both exit]*

Beaks, Feet, and Feathers BACKGROUND

FANTASTIC FLYING MACHINES

What kind of creature has eyes that take up nearly half its head? What animal uses its tail as a brake, rudder, stabilizer, and mate attracter? And who has the warmest, lightest-weight body covering of all? Birds! Birds are remarkable creatures. Some are adapted to fly great distances; others are adapted to dive in the ocean, hunt from the air, or stay warm in extremely cold weather. The greatest challenge for all birds, however, is to find adequate food to give them energy for these and other demands. Specially designed beaks, feet, legs, eyes, and wings help them to meet this challenge.

Feathers, unique to birds, contribute not only to a bird's warmth but also to its ability to fly. Each feather is composed of **keratin**, the same substance that forms the basis of hair and scales. Feathers emerge from tiny growth pits in the skin called follicles, much like mammalian hair follicles. The feathers grow in tracts, with spaces of bare skin in between these feathered areas. Small songbirds have somewhere between 3,500 and 5,000 feathers; waterbirds may have as many as 25,000 feathers.

Close examination of a feather reveals these parts: the shaft, which is the central hollow tube that gives the feather its rigidity; barbs, the parallel strands that attach on either side of the shaft and create the feather's flat surface, or vane; and barbules, which run along the barbs and connect them together with tiny hooks on one side and bumps on the other. When birds preen, they are "zipping up" their feathers, keeping them in good order for flight. A gland above the base of the tail contains oil that the bird spreads on the feathers while preening. Waterfowl especially use a great deal of oil on their feathers. Feathers wear out, of course, so most birds go through an annual or twice-annual molt in which they lose old feathers and grow new ones.

There are three main types of feathers. The flight feathers of the wings and tail have strong shafts running the entire length with flat webs on two opposite sides, presenting a lightweight, yet solid, surface for flight. The down feathers have very short shafts with many noninterlocking barbules to create dead air spaces for good insulation. The contour feathers are smooth body feathers that streamline the bird, and, as may the flight feathers, they carry the colors and patterns that are distinctive of the species.

BIRD FEATHERS

contour

flight

down

Another feature that can help us distinguish between bird species is the bird's beak, or bill. Bird beaks are modified in various ways for obtaining specific types of food. There are wide scoop-like bills that ducks and geese use to shovel for submerged plant and animal food. Most shorebirds have long, slender bills to probe the soft mud and sand for tiny **invertebrates**. The slender bills of chickadees, titmice, and nuthatches are very good at reaching insects in bark crevices, while the broad, short bills of flycatchers scoop insects out of the air during flight. Seed-eating birds have stout bills. The crossbills use their odd scissors-like bills to hold the scales of cones apart while pulling out the seeds with their tongues. Hummingbird beaks and tongues can suck in nectar and are also used to capture small insects inside flowers. Hawks, eagles, and owls all have sharp, curved beaks that, when used in conjunction with the **talons** on their feet, are well adapted to tearing the flesh of their **prey**.

Birds use their feet in a variety of different ways: to walk, hop, run, perch, swim, and to catch or grip their food. Actually, the term *feet* is misleading because a bird actually stands on its toes and uses them to accomplish all those tasks. Most birds have four toes, with the first toe (the hallux) commonly facing backward and the other three forward, which works well for perching. Many other birds, like the owls, parrots, osprey, and woodpeckers, have two toes forward and two back. A bird's feet may also hold the bird in place while it sleeps. When the bird relaxes in sleep, its body slumps down on its feet. In this position, a tendon passing behind the heel is pulled tight. This draws the three forward toes and the hind toe toward each other, clamping the bird onto the twig. When the bird rises to a standing position, the clamp is released.

Swimming birds have a variety of foot designs to aid their propulsion through the water. Most common are webbed feet, like those of a mallard duck whose front three toes are connected by a full web of skin. Birds that dive deep have legs set farther back on the body, and even larger feet, with all the toes webbed.

The feet of **raptors** – birds of prey like hawks, eagles, and owls – have long, sharply pointed, curved talons designed for catching, piercing, and killing their living prey. Ospreys have rough pads on their feet that allow them to grasp slippery fish with greater ease.

What most of us consider the leg of a bird is, anatomically, its foot, with an elongated instep slanting upwards and backwards to the heel. Feathers usually hide a bird's knee, which bends forward as ours does. The length and sturdiness of birds' legs, as we call them, vary according to their feeding habits. A duck uses its short, powerful legs to paddle its webbed feet while the heron's long legs are perfectly designed for standing and wading while it fishes. The ostrich's legs are long and strong, hence its amazing ability to run across open plains.

BIRD BEAKS

sharp-shinned hawk

ruby-throated hummingbird

red crossbill

white-breasted nuthatch

evening grosbeak

Beaks, Feet, and Feathers / 39

Wing size and shape also vary, making it possible for birds to feed and nest in different types of habitats. Birds like partridges and sparrows, which nest and feed in grass-covered or brush-covered areas, can rise almost vertically from the ground with their short, broad wings. Birds that pursue their food on the wing usually have long, narrow, angled wings. Vultures, eagles, and many hawks have long, broad wings that make it possible for them to soar on spread wings when the air currents are right.

Birds' tails are another vitally important tool of flight, as brakes, rudders, and balancers. However, many birds have found other uses for them. Woodpeckers use their tails as stiff props while hammering on trees. A male grouse, turkey, or peacock will strut and display its showy tail for a watching female.

Most birds have extremely keen vision, the best of all the **vertebrates**. Their eyes are large, held by a ring of bone, with the eyelid (except in owls) closing from below, like a reptile's. They also have a special **nictitating** membrane, a clear "third eyelid" that cleans and protects their eyes. Hawks and eagles particularly are able to sight their prey at great distances. Binocular vision and rapidly focusing eyesight enable these birds to accurately pursue their living prey.

Placement of the eyes differs according to feeding habits and the bird's vulnerability as prey. Most birds have eyes on the sides of their heads, allowing broad peripheral vision. Birds of prey have forward-facing eyes for better depth perception used in hunting.

Each bird is uniquely adapted to survive in its own particular environment. Noticing these different adaptations can give clues to the feeding and behavior patterns of the different species and adds a new dimension to watching birds.

Suggested References:

Ehrlich, Paul R., David S. Dobkin, and Darryl Wheye. *The Birder's Handbook: A Field Guide to the Natural History of North American Birds.* New York: Simon and Schuster, 1988.

National Geographic Society. *Field Guide to the Birds of North America.* 3rd edition. Washington, DC: National Geographic Society, 1999.

Proctor, Noble S., and Patrick J. Lynch. *Manual of Ornithology: Avian Structure and Function.* New Haven: Yale University Press, 1993.

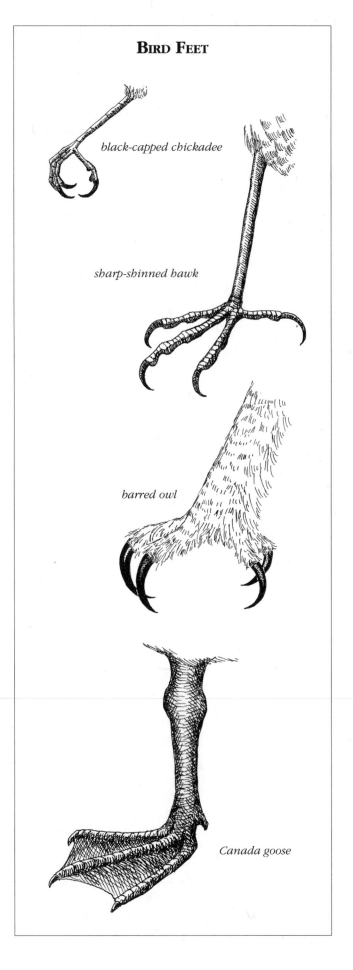

BIRD FEET

black-capped chickadee

sharp-shinned hawk

barred owl

Canada goose

Beaks, Feet, and Feathers

FOCUS: Birds' exceptional adaptations enable them to fly, keep warm, and procure food.

OPENING QUESTION: *What special physical adaptations do birds have?*

FOCUS ON FEATHERS

Objective: To examine the structure and uses of feathers.

Give each child a feather or two to examine. If possible, include flight, contour, and down feathers.

a) As the children describe the parts of a feather, the leader draws them on the blackboard to illustrate a typical feather. Make sure to note the hollow shaft. What differences do the children notice between their feathers? What functions do feathers perform?

b) Have the children gently pull apart the web of the feather. Use a hand lens to see the tiny barbules that project from the barbs. What function might these serve? Have the children try to zip the feather together by pinching and drawing their fingers along the separated barbs from the shaft to the outer edge. Discuss how birds oil and groom themselves.

PUPPET SHOW

Objective: To learn how their different adaptations allow birds to live in various habitats and feed on various food sources.

Perform, or have the children perform, the puppet show. Afterward, hold up each puppet and have the children describe the special adaptations each bird exhibits. Show additional pictures that illustrate the variety of bird beaks, feet, legs, wings, and tails. How might these adaptations help the different birds survive?

MIX AND MATCH

Objective: To explore how a bird's beak and feet are designed to help the bird get its food.

Give each pair of children a picture of a bird. Ask them to look at the shape of the bird's beak, feet, and legs, and, with their partner, talk about how the bird might use these to obtain food and eat it. Spread out pictures of possible bird foods. Ask the children to place their bird picture next to an appropriate food. More than one bird may eat the same kind of food, and one kind of bird may eat several kinds of food. How do the different beaks and feet enable the birds to obtain and eat their food?

Some possible combinations are:

Owl – mouse; Hawk – snake; Eagle – fish; Sparrow – seed;
Robin – worm; Crow – corn; Swallow – flying insect; Gull – mussel;
Heron – crayfish; Loon – fish; Finch – seed; Bluebird – insect;
Pelican – fish

Materials:
• feathers
• hand lenses
• chalkboard, or large paper and markers

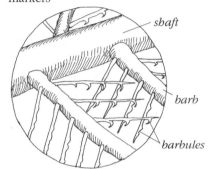

Materials:
• puppets
• script
• props
• pictures of different kinds of bird adaptations

Materials:
• bird pictures
• pictures of various bird foods

PICK A BEAK

Objective: To investigate beak differences by comparing how well different utensils work to obtain different foods.

Set out an assortment of foods in different containers. Brainstorm ideas for what kind of bird food each might represent. (For example, rice grains on a plate could be crawling insects, and cooked spaghetti in a cup might be earthworms.) Give each child a different "beak" utensil and ask the children what foods they might eat, given the designs of their "beaks." Have them try to find the foods they can pick up using only their "beaks." Discuss the advantages and disadvantages of specialized beaks.

MAKE A BIRD

Objective: To review some of the special adaptations that make birds successful.

Divide older children into groups of two or three; youngest ones may do this in small groups with a leader. Give each group a card with one of these adaptation combinations:

~ You paddle around in water with your feet, and catch fish with your beak.

~ You fly after your food and must be able to change directions quickly, and you catch flying insects in the air with your beak.

~ You hammer holes in trees with your beak, and you use your stiff tail to help support you on tree trunks.

~ You eat bugs under bark with your beak, and you climb up and down trees with your feet.

~ You can spot a mouse a half mile away, and you spend a lot of time soaring in the sky.

~ You can see very well in the dark, and you catch live mice with your feet.

~ You catch frogs with your beak, and hunt while wading in deep water.

~ You crack seeds with your beak, and you attract a mate with your beautiful tail.

~ You sip nectar from flowers with your beak, and you have brilliant feathers so other birds can see you.

The children then make body parts (large enough to wear) suitable for the assigned tasks, plus they construct any needed props (like a paper fish). After dressing one child (or the group leader) as the bird, groups take turns introducing their birds and acting out the adaptations.

Materials:
- utensils and tools, such as various forks, spoons, tweezers, pliers, skewers, tongs, straws, nut picks, nutcrackers
- foods in various containers, such as sunflower seeds in a tall jar, rice grains on a plate, jelly beans under shredded paper, grapes floating in water, cereal in a bowl, a whole apple, orange juice in a thin-necked bottle, corn on the cob, nuts in shells, oatmeal on sandpaper

Materials:
- bird adaptation cards
- scissors
- tape
- string
- construction paper
- large sheets of newsprint
- large paper grocery bags
- crayons
- markers

ACTIVITY STATIONS:
1) Mix and Match
2) Pick a Beak

EXTENSIONS

Bird Feeders: Make bird feeders from plastic jugs with the top half of one side cut out. Hang these outside and watch the birds that come to each. Use a field guide to identify them. Keep track of preferred foods, beak design, and the way the birds eat their food. In cold weather, hang a suet feeder and observe the birds that come to it.

Unusual and Endangered Bird Research Project: Have the children research unusual and endangered birds. They can make colored drawings and prepare short written or oral reports about their birds. A few suggestions: spotted owl, ivory-billed woodpecker, bower bird, ostrich, penguin, flamingo, lyre bird, albatross, loon, kiwi, roadrunner, puffin.

Bird Search: Take the children on a bird walk, visiting as many kinds of habitats as possible (like a pond, a field, a beach, a bird feeder). Observe the birds seen outdoors, and for each bird, note the following:

The size of the bird. Is it bigger or smaller than a robin? A chicken?

The color and pattern of its feathers. The shape of its wings and tail.

The shape of its beak. Is it doing anything with its beak? What?

The size and shape of its feet. Does it use its feet when eating?

Can you identify the bird?

Beaks, Feet, and Feathers

PUPPET SHOW

Characters: Cappy Chickadee, Marsha Mouse, Mama Chickadee, Duck, Cardinal
Props: small green leaves to stick on Duck's beak

Marsha Mouse

Wow, Cappy Chickadee, there are so many seeds at this feeder! I could get my fill if we didn't have to keep a lookout for that hungry cat. We'll have to run if we see him, okay?

Cappy Chickadee

You run, Marsha. I can go faster if I fly!

Mouse

Oh, right...I wish I could fly, too. How do you do that?

Cappy

Just lucky, I guess. Lucky to have wings and feathers. Plus, with these hollow bones of mine, I'm lots lighter than you.

Mouse

Uh oh! If you don't want to be a **light snack,** you should fly **now**...there's the cat! See you later!

Cappy

Okay, Marsha – bye! *[both leave, then Cappy returns]* Phew! That darn cat really ruffles my feathers. *[Mama Chickadee enters]* Hi, Ma! I'm sure glad we birds can fly.

Mama Chickadee

Well, if you want to keep it up, you'd better preen your feathers. You look a fright!

Cappy

I was a **flight,** escaping the cat. Poor little Marsha Mouse could only run away. Can all birds fly, Ma?

Mama Chickadee

Well, around **here** they all fly, but even so, you wouldn't believe how different some birds are. Why don't you spend the day visiting some other birds and see for yourself? You could start down by the pond.

Cappy

Okay, Ma, see you later. Call me if you find anything good to eat – I'll be listening. *[both exit; Duck appears, Cappy flies in and hops up to him]*

Cappy

Whoa! You can swim! What **are** you? Are you a bird?

Duck

I certainly am, little chickadee. I'm a duck – haven't you ever seen one before? I can swim with my big feet, and I can fly, too.

Cappy

Golly. What do you eat? Your beak is funny-looking, nothing like my slender bill. I don't see how you can pick any little caterpillars from under bark with that thing.

Duck

Excuse me? Caterpillars? Not in the pond, you silly bird – but I do love snails, and I eat lots of plants and seeds that I find under water, like this! *[tips head down, so only tail is visible and comes up with a mouthful of green stuff]* Now you can see what this beak is for!

Cappy

Wow, that stuff must be yummy. Do you think I could do that?

Duck

No, I don't think you should try it. I keep my feathers all oiled up so I don't get wet and cold. Besides, those skinny legs of yours don't look strong enough for swimming. You do fine flitting around in the trees and perching on branches with your little toes.

Cappy

I guess you're right. Well, my mother told me that we birds are very different from each other, but I had no idea.

Duck

You want to see different? I know who you could go see next.

Cappy

Who's that?

Duck

Kenny Kestrel – he's right over there on the telephone wire.

Cappy

Oh, no, Duck! I already know all about his sharp claws and keen eyes – my mother told me all about him. He **eats** little birds like me in the winter when he can't get grasshoppers. I'd better be going before he spots me. Goodbye, Duck. *[flies off]*

Duck

Oh, uh, sorry about that. *[calling after Cappy]* Goodbye, little chickadee. *[Duck swims away]*

Cappy

[Cappy returns] Whew, imagine that duck thinking I'd want to talk to a hawk. I guess he's too big to be scared of the kestrels.

Cardinal

[Cardinal enters] Hey, Cappy, what are you doing here?

Cappy

Oh, hi, Cardinal. I'm just trying to steer clear of Kestrel. It must be hard for you to hide; you're so red. How come you're so bright and your mate isn't?

Cardinal

Well, this is my turf, and I don't like to share it. I'm red so the other male cardinals know I'm around. My mate does all the egg-sitting, so she needs to be camouflaged.

Cappy

Well, what about that beak of yours? It's so big. Not like Duck's, but...Have you ever been swimming?

Cardinal

No, kid, I haven't ever been swimming, except to take a bath in a puddle. My beak's big like this to crack seeds with, so I don't have to hold them down and hammer on them like you. *[Chickadee-dee-dee is heard from offstage]* Hey, isn't that your mother? You'd better get back to your family.

Cappy

I guess so! I wonder what my flock has found to eat. It sounds like something good! Bye now! *[both exit; Cappy re-enters with Mama Chickadee]*

Mama Chickadee

Why, there you are, Cappy. We're just heading over to the Miller's bird feeder. They **just** filled it up.

Cappy

With sunflower seeds, I hope. I do love sunflower seeds! You know, Ma, now I know how lucky I am to have this bill and these feet. Otherwise, I might have to eat algae or snails or grasshoppers.

Mama Chickadee

Yes, we are lucky, Cappy. Lucky that all birds don't like the same foods we do, so there's plenty for all to enjoy. Now, let's go get dinner.

Cappy

OK, Ma, I'll follow you! *[both exit]*

A BIRD'S LEG AND FOOT BONES

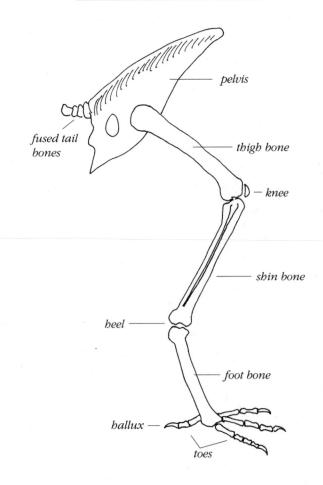

fused tail bones

pelvis

thigh bone

knee

shin bone

heel

foot bone

hallux

toes

Owls

SILENT PREDATORS OF THE NIGHT

The hollow hoots or screeching calls of an owl at night can seem mysterious, even spooky, and the silent, shadowy passing of an owl in flight has startled even the bravest nighttime hiker. Because of owls' **nocturnal** habits, many people are unfamiliar with the life histories of these **birds of prey**. Yet North America has a wide variety of owls, from the diminutive pygmy, elf, and saw-whet owls (smaller than a robin) to the imposing great horned owl (larger than a red-tailed hawk). Although they share many features with other birds, owls have certain distinct characteristics that equip them well for their predatory role in nature. A look at owls from the feet up reveals an array of effective adaptations.

Owls have sharp, curved, vice-like **talons**, or claws, composed of hard **keratin** surrounding a living, growing core. Talons grow at the same rate they are worn down by normal use, and as they grow, they get sharper. Well designed for catching and killing, talons are used by different owls on **prey** ranging in size from grasshoppers to fish, snakes, mice, birds, and skunks. Powered by a tendon that runs from the foot along the back of the leg to where it attaches to a muscle above the ankle, talons lock onto the owl's prey. The firm grip of the tendons means the owl need not rely on muscular strength to carry off its prey, although the prey's weight may hinder the owl's ability to fly.

The unusual structure of their flight feathers allows owls to sneak up on their prey. Whereas most birds' primary wing feathers are smooth-edged, the leading edge of owls' flight feathers is comb-like, minimizing noise by disturbing the flow of air around the wings. In addition, owl flight feathers have velvety surfaces and fringes on the back edges that also help muffle noise. These adaptations make owls the only group of birds that can fly virtually silently, a distinct advantage for catching wary prey.

Like other **raptors**, owls have hooked beaks for tearing their food into bite-sized pieces. Beaks consist of keratinized skin covering bone. Owls often consume smaller prey whole, but larger prey is held in the talons and ripped apart with the beak, the upper and lower edges of which meet like shears. A strong, complex set of muscles controls the neck and beak and aids in this shredding process.

Because much prey is consumed whole, owls swallow a large amount of indigestible material, mainly fur, bones, and feathers. The smallest bones are digested, providing calcium, an essential part of an owl's diet. The indigestible material cannot pass through the owl's small intestine, so owls periodically regurgitate this material as an oblong pellet. Since bones, feathers, and fur are coughed up without having been digested, they usually remain intact. Through careful study of the contents of a pellet, one can determine the diet of an owl.

Although many owls, especially barn owls, use their keen ears to detect prey, owls' highly developed sense of sight also helps them hunt at night. Owls' eyes (like those of all birds) are much larger in proportion to their body than are human eyes. Though most owls are thought to be color blind – seeing only shades of black, white, and gray – they have exceptionally keen sight in the dark. Owls have a very large number of light receptor cells (rods) in their retinas, with very few of the cone cells that distinguish color. One experiment with barn and long-eared owls showed that they could find dead mice under an illumination of 0.000,000,73 foot-candles (equivalent to the light of a standard candle 1,170 feet away)! It would take from ten to one hundred times that amount of light for a human to see. In addition, owls are highly territorial and hunt in the same area for long periods of time. This helps them learn and remember the location of objects, even if they can't be seen well.

Although most owls are nocturnal, they are sometimes active by day. To adjust to brighter light, they can close their pupils to almost pinhole size. To further reduce glare, owls can lower their top eyelid, a feature found only in nocturnal birds. (Other birds close their eyes with the bottom eyelid.) Some species of owls, such as the hawk owl, barred owl, and great gray owl, often hunt in the daytime or at dawn and dusk.

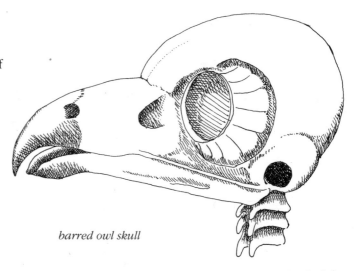

barred owl skull

As hunters, owls must be able to determine the exact distance to their prey. While most birds have eyes on the sides of their heads, thus providing a very wide field of monocular vision, owls' eyes are positioned toward the front of the head, giving them a wide area of binocular vision. This allows them to focus both eyes at once on an object for more accurate depth perception. Because of their large size and conical shape, owls' eyes are fixed in their sockets, unable to move. To compensate, owls have extremely flexible necks with 14 cervical vertebrae. They can turn their heads as much as 270 degrees in either direction, allowing for more than a full circle of vision. Owls can also tilt their heads through a wide range of angles to help them focus.

Owls' ears are equally amazing. Often asymmetrical in shape, size, and placement, their ears can precisely pinpoint the direction and location of a sound, and the parabolic dish shape of the facial disk aids in hearing. Delicately constructed feathers conceal the large ear openings without interfering with sound reception. Like other birds, owls use sound to announce their territories and to attract mates, each species with its own distinctive call. Additionally, many researchers believe that owls rely on their sense of hearing to hunt, using vision to guide them when flying and when making the final strike. In one experiment to determine a barn owl's hearing ability, researchers placed live mice in a totally dark room with an owl. The owl successfully caught its prey in 13 of 17 strikes. When paper, instead of mice, was rustled across the floor (to take away cues associated with odor and body heat), the strikes were still successful. An owl's keen sense of hearing enables it to catch mice under leaf litter or snow, locating them by sound alone.

Owls often hunt and nest in towns as well as in rural and wild areas. Yet, although they may live nearby, we rarely have an opportunity to observe owls because of their nocturnal nature. However, if you learn their calls as well as their habits and habitats, you may be rewarded on your next evening walk with a glimpse of one of these powerful hunters of the night.

Suggested References:

Burton, John A., ed. *Owls of the World: Their Evolution, Structure and Ecology.* New York: E. P. Dutton, 1973.

Ehrlich, Paul R., David S. Dobkin, and Darryl Wheye. *The Birder's Handbook: A Field Guide to the Natural History of North American Birds.* New York: Simon and Schuster, 1988.

Johnsgard, Paul A. *North American Owls: Biology and Natural History.* Washington, DC: Smithsonian Institution Press, 1988.

Sparks, John, and Tony Soper. *Owls.* Newton Abbot, Devon, UK: David and Charles, 1995.

great horned owl

Owls

Owl Moon - read at snack (handwritten)

ACTIVITIES

FOCUS: Owls are uniquely adapted for their lives as nocturnal predators.

OPENING QUESTION: *What special adaptations make owls different from other birds?*

PICTURE PARADE *Kelly*

Objective: To identify some special adaptations that help owls to be successful nighttime hunters.

Show the children pictures of owls and have them point out special features: large forward-facing eyes, facial disk, hooked beak, talons, soft feathers. Discuss how each could help the owl find, catch, or eat its prey. Compare with pictures of songbirds. Discuss other owl features, such as the comb-like leading edge of the first wing feather, large asymmetrical ear openings, flexible neck, and large wings.

Materials:
- an assortment of pictures of owls (include, if possible, pictures of their feet, faces, feathers, and skeletons)
- songbird pictures

CALLING ALL OWLS *Kelly 2*

Objective: To recognize the calls of different owls and use these in identifying the birds.

Play a recording of owl calls, concentrating on local species. Show the children a picture of each owl as its call is played. Stop the tape after each call and have the children try to make the same sound as the owl. Older children may try to identify owl calls and match these with their pictures during another playing of the recording.

Materials:
- recording of owl calls
- pictures of the owls
- tape recorder

(handwritten notes:)
7 calls correspond to the first 8 slides
1.) great horned owl — Who's awake? Me too
2.) short eared owl - barks yappy dog sounds
3.) Eastern screech owl — trilling ghost sounds tremulous
4.) barn owl - human habitats finger nails on blackboard — rusty hinges
5.) barred owl - for you all? Madam, who cooks, who cooks for you? very guttural.
6.) long eared owl — yipping, quacking, cat hoops
7.) saw-whet owl — whistling back up alarm
scrambled short eared, saw whet

OWL EARS *Alice 3*

Objective: To practice locating sounds as a hunting owl would.

Have the children close their eyes and pretend to be owls searching for prey. Make a soft sound and tell the children to listen for this noise and then point toward it when they hear it. First have the children try to do this without moving their heads, and then have them listen while cupping their hands behind their ears and turning their heads toward the sound. The leader makes the noise from different locations while children point to the place they think it's coming from. Ask the children how they were able to locate the sound. What physical adaptations allow owls to accurately locate their prey?

Materials (optional):
- blindfolds
- paper for mouse to rustle

OWL AND MOUSE

Objective: To experience hunting by sound as owls do.

Form a circle with 10-15 children. Choose two children, one to be an owl, the other a mouse. All the others in the circle become trees. Place the owl and mouse inside the circle, with eyes closed, and gently spin them around a few times. With the circle as the outside boundary, the owl must follow and "catch" (tag) the mouse. Instruct the trees to be very quiet or, if the owls are catching the mice too quickly, to say "sshhh…" and sway in the wind. Have the mouse rustle paper or rub hands together occasionally. When the owl catches the mouse, they each choose a replacement from the circle. To make the game more challenging, make the circle bigger.

PELLET INVESTIGATION

Objective: To discover what and how owls eat and examine the contents of owl pellets.

As a group, create a list of foods owls eat. Explain that owls usually swallow their small prey whole, coughing up the indigestible parts in the form of a pellet. What parts do the children think they might find in such pellets? Give an owl pellet to each group of three to five children. Have the children look at its shape and compare it with other pellets. Did it come from a large or small owl? Have the children slowly pull the pellet apart, separating it into piles of bones, teeth, fur, and feathers. Can the children identify the animal(s) inside? Display the bones by gluing them to a piece of poster board. Arrange by type of bone or in the shape of the animal's skeleton. Label any bones that can be identified. Include a sample of fur or feathers.

Materials:
- owl pellets, one for each group of three to five children
- white poster board
- glue
- hand lenses
- charts of mouse and vole skeletons

Note: Owl pellets can be obtained from educational and scientific supply companies.

PUPPET SHOW

Objective: To review the special adaptations of owls.

Perform, or have the children perform, the puppet show. Afterward, discuss what the owl could do that the blue jay couldn't do and why.

Materials:
- puppets
- props
- script

EXTENSIONS

Further Study: Have the children pick different owls to research. They should collect pictures and learn about the range, calls, prey, and hunting and nesting habits of their owls. Are any owls rare or endangered in your region?

Mythical Bird: Owls are characters in the myths and legends of many cultures. Read aloud some of these stories and talk about what the owl has meant to people in different times and places.

It's a Hoot: Using an owl recording and a tape recorder, have the children practice and then record their owl-imitating hoots. Then have the children guess each other's owl calls. Other birds and mammals make noises at night. How many can they learn?

Raptor Rehab: Owls and hawks are sometimes found injured or orphaned. There are a number of good books and articles about their rehabilitation and release. Use some of these sources to learn more about the efforts to help raptors and, if possible, visit a raptor rehabilitation center.

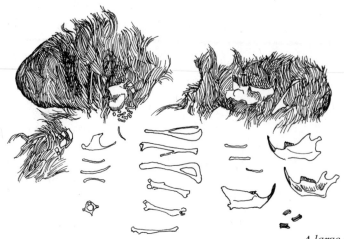

A large owl pellet may contain bones from more than one animal.

Owls

Characters: Ollie Owl, Barry Blue Jay

Props: a pine cone on a stick; paper to flap for wing noise

Ollie Owl

Golly, I can't wait! My good friend, Barry Blue Jay, is coming over for dinner. Here he comes now. *[Blue Jay enters]*

Barry Blue Jay

Hey, Ollie Owl! I haven't seen you for ages. I'm so glad you invited me over. I'm as hungry as can be! What's for dinner anyway?

Owl

Well, rather than have food sitting around, I thought we'd catch dinner and eat it fresh.

Blue Jay

Catch it? What were you planning for our dinner?

Owl

Oh, whatever comes by – skunk, rabbit, mouse.

Blue Jay

Rabbit? Skunk? Yikes! I usually eat insects and berries and seeds, myself. I wouldn't even know how to catch a mouse.

Owl

No problem! Just sit here and watch carefully. You're sure to see something yummy run by.

Blue Jay

But Ollie, it's too dark! My eyes aren't made for seeing at night like yours are. And look, I've got one eye on this side of my head *[turns]* and one eye on this side.

Owl

[looking at Blue Jay's eyes] My goodness, so you do, and they're so small! I've got big eyes, and they're both right in the front of my face. It's much easier to hunt when you can look at one spot with both eyes. Maybe I could help you, Barry, if I just pushed your eyes a little closer to the front.

Blue Jay

[backs up quickly] That's all right, Ollie! I like my eyes just the way they are. I'll try hunting with them like this.

Owl

There go some mice. Quick, let's get them! *[both fly off stage; Owl returns]* Mm-mm, now that's good eating.

Blue Jay

[from below] Here I come, Ollie! You should see the huge mouse I caught. *[comes up with a pine cone]*

Owl

That's no mouse; that's a pine cone!

Blue Jay

Oh gosh, you're right. *[drops cone]* It sure is hard to see down in those dark woods.

Owl

Don't just use your eyes – use your ears, too. I can hear exactly where those mice are. Oh, I hear something else down there. Let's go.

Blue Jay

But I didn't hear anything.

Owl

Don't worry about it. Just follow me. *[both swoop down; Owl comes back up]* Yum. Another delicious mouse. I'm stuffed. *[Blue Jay re-enters]*

Blue Jay

You're stuffed? I'm starved!

Owl

You mean, you still didn't catch anything?

Blue Jay

As soon as I got close, they scampered off somewhere.

Owl

That's because you're too noisy. You've got to fly quietly like me. *[flaps wings silently]*

Blue Jay

Hey, you don't make any noise at all. Let me try that. *[flaps wings; backstage, flap sheet of paper]* How was that?

Owl

Well, I sure heard you. I'm afraid any mouse would have, too.

Blue Jay

Oh, Ollie, it's hopeless. Even if I could see and hear as well as you, I still couldn't be quiet enough to catch anything. I don't suppose you have any leftovers for me?

Owl

No, I'm sorry, Barry. I eat those little mice in one gulp.

Blue Jay

One gulp? Hair, bones, and all?

Owl

That's right. It's head first, down they go! I'm sure you can do it, Barry; it just takes practice.

Blue Jay

I don't know. I think I'd better just stick to eating insects and fruit. Listen, I'm going to head home. I've got a midnight snack of sunflower seeds hidden under some bark. Thanks for inviting me over, Ollie.

Owl

Any time. We'll have to get together again soon.

Blue Jay

Oh, sure – just not for dinner! Bye now.

[both exit]

Thorns and Threats

PLANTS' AND ANIMALS' STRATEGIC DEFENSES

What might prevent you from touching a plant or animal? Sharp prickles, disgusting smells, strong poisons? Examples of all of these abound in the natural world. Every species relies on certain defenses for survival. These adaptations, many of which are common to both plants and animals, succeed if they allow the organism to reach maturity and reproduce.

Projections that can scratch or puncture are one familiar defensive weapon found in both plants and animals. The prickly husks on beechnuts and horse chestnuts protect the developing nuts until they ripen. Thorns and prickles, such as those on blackberry stems or hawthorns, deter grazing by animals. Pasture junipers, with their sharp needles, are rarely eaten.

Similarly, many mammals have formidable projections. Quills allow the slow-moving porcupine to travel more or less unmolested. These modified hairs – covered with hundreds of minute, diamond-shaped, overlapping scales – have barbed tips. When threatened, the porcupine erects the quills with special muscles and either waits for the **predator** to attack or turns its back to the attacker and swats with its tail. Although the porcupine cannot throw its quills, these loosely attached projections do release easily on contact and become embedded in the pursuer, a lasting reminder to would-be predators. New quills begin to grow within a few days to replace the lost ones.

Other threatening projections include horns and antlers. These are mostly used by male members of a species to display vigor or when fighting for dominance. However, horns and antlers can be useful for defense, as in the circle formed by adult musk oxen around the young, with horns lowered against wolves. Tusks (like those of the elephant and the wild boar) are modified teeth that are used not only in food gathering and fighting, but also in defending against attack.

Some animals use chemical, rather than structural, defenses to repel potential predators. To ward off attack, the blister beetle, often seen in late summer, can secrete an orange irritating liquid when disturbed. American toads can secrete a substance that is very irritating to mucous membranes. Dogs, for instance, froth at the mouth after picking up a toad. Many other frogs and salamanders taste bad or numb the mouth.

Some chemicals produced by animals go beyond being irritating to being poisonous. Snakes are probably the most feared for their poison, but that fear is often unwarranted. The majority of snakes are not poisonous. Those that are use their venom primarily in hunting, to paralyze or kill the **prey**, and only as a defense when cornered. Before striking, rattlesnakes hiss, rattle, and threaten to try to scare away the enemy. Spiders are also feared by many, a response far more appropriate in insects than humans. Spiders use the venom in their fangs to paralyze their prey. When used defensively against larger animals, such as humans, spider bites may cause pain or itching to those sensitive to the venom, but rarely do they cause death. The short-tailed shrew, a tiny mammal, is also capable of paralyzing its prey for caching, but on larger victims, the shrew's bite is less poisonous. One man's account of a shrew bite described swelling, burning, and shooting pains.

Poisonous or distasteful chemicals are the most common defense in the plant world. Many of the flowering plants manufacture poisons to keep from being eaten by insects and mammals. These poisons usually also taste bitter, leaving the **herbivore** with a vivid memory of its negative experience.

In addition to poisons, plants have developed other proficient chemical means of warning "don't touch!" Nettles are a good example. The hair tips covering the stems and leaves of the nettle break off easily when touched, injecting a combination of chemicals into the passerby. A swelling, stinging rash follows almost immediately. Poison ivy produces an oil that causes an irritating, itching rash in most humans, but the reaction appears long after the contact, so the oil is probably a defense against **invertebrate** herbivores.

porcupine in threat posture

Another defense used to good advantage by plants and animals is offending the senses. The pungent odors and strong tastes of certain plants protect them. In a pasture, the animals will carefully graze around everything that tastes unpleasant, from buttercups and milkweed to mullein and thistles. Likewise, the skunk's strong, choking, long-lasting scent effectively deters most predators, with the exception of the great horned owl that (like most birds) has a very poor sense of smell. The skunk gives many warning signals when approached too closely, stamping and threatening before releasing the offending chemical. Only a foolish animal ignores these clues more than once.

Body posture and threatening behaviors can also offer defense. Body posture includes the pretense of looking fierce or dead so that a predator will give up. An opossum drops as though dead, lying still, with its mouth open and tongue hanging out, fooling predators that prefer live prey. Owls look huge and fierce as they crouch with feathers raised and wings turned forward, spread wide in a threat display that's often accompanied by hissing. A threatened cat assumes a similar posture: back arched, body turned sideways, tail and fur erect, hissing and spitting loudly with an open mouth full of sharp teeth. Though these behaviors aren't possible for plants, there are a few, like the sensitive plant, that fold up their leaves when touched, leaving the tough and prickly stem showing and not much else.

Tough armor may successfully prevent an attack. Most woody plants, especially trees, are protected by hard bark and tough wood. Nuts are seeds encased in very hard coats that keep them safe from most predators. Animals, too, have evolved a variety of protective coats or shelters: armadillos, turtles, and snails, for example. Fish scales are both tough and slippery. The hard outer body covering of **arthropods** (insects, spiders, crabs, and so on) serves both as a skeletal framework and as protection.

Finally, there is a rather remarkable defense used to advantage by plants and animals: a "take it, I don't need it" adaptation. Because of the way they grow, the grasses and many other plants are able to survive quite a lot of pruning by hungry herbivores without major damage. In the animal kingdom there are similar examples of expendable body parts. Spiders and crayfish can replace a lost leg with a new smaller leg at the next molting. Many lizards can grow new tails. Some ground-feeding birds can relax the muscle of each tail feather, so those grabbed by a predator simply fall out. In time, new ones grow back.

Despite the great variety of effective defenses used by plants, animals, and other organisms, some creatures continue to circumvent these defenses, thus maintaining a balance in the natural world. Rodent teeth are able to nibble through the thick shells of most nuts. Fishers attack porcupines by swatting at their unprotected faces or harassing them out of tree tops. Monarch caterpillars aren't hurt by the toxic leaves of milkweed plants. Instead, the caterpillars store the ingested toxins for their own defense. Almost every animal and plant defense has been challenged by a predator or two that have evolved new behaviors and adaptations to compensate.

The variety of thorns, threats, and other novel defense adaptations bear witness to the many ways to survive in a hungry world.

Suggested References:

Ingrouille, Martin. *Diversity and Evolution of Land Plants.* London: Chapman and Hall, 1992.

Martin, Alexander C. *Weeds.* New York: Golden Press, 1987.

Tinbergen, Niko. *Animal Behavior.* New York: Time-Life Books, 1965.

Rezendes, Paul. *Tracking and the Art of Seeing: How to Read Animal Tracks and Sign.* 2nd edition. New York: Harper Collins, 1999.

Went, Frits W. *The Plants.* New York: Time-Life Books, 1963.

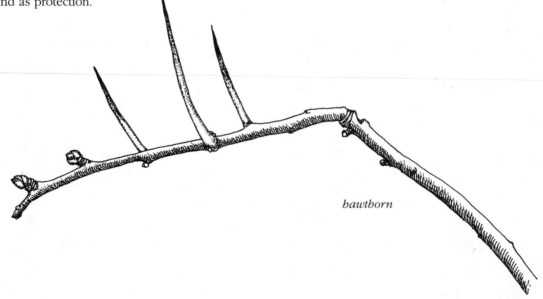

hawthorn

Thorns and Threats

Focus: In order to protect themselves, plants and animals have developed many different defenses.

Opening Question: *How do some plants and animals defend themselves?*

Puppet Show

Objective: To learn some defense strategies used by animals.

Perform, or have the children perform, the puppet show. Afterward, review the defense strategies used by the various puppet characters. Ask the children to try to name a plant that uses a defense similar to that of each animal character.

Materials:
- puppets
- script

Defense Puzzles

Objective: To identify different defense strategies used by plants and animals.

Ahead of time, make up simple Defense Strategy puzzles for each of the categories listed below by gluing labeled pictures of the examples from each category onto an 8 1/2" x 11" piece of paper. Cut the paper between the two or three pictures in such a way that the pieces will fit together again. Make up one puzzle per category.

Give each child a puzzle piece. Ask the children to find the partner(s) whose piece fits with theirs. When they've completed the puzzle, ask puzzle partners to determine the strategy shared by the organisms on their puzzle pieces. Have each group present their defense strategy to the class while the leader records the different categories on a blackboard. Can the children think of any more defenses to add to the list?

Then, examine samples of a variety of protective adaptations used by plants and animals.

Materials:
- Defense Strategy puzzle pieces (for older children, include written description of the defense strategy on the back)
- touchable samples of defense strategies, for example: turtle shell, snail or clam shell, porcupine quills, thorns, cactus, nutshells, strong-smelling plants, fuzzy leaves

Defense Strategies for Puzzles
Prickly
Porcupine – loose, barbed quills stick in enemy
Cactus – spines repel enemies
Beechnut – nut covered with sharp bristles until completely ripe

Smells Bad
Skunk – when cornered, squirts foul-smelling substance
Black cherry tree – twigs and leaves contain foul-smelling chemicals

Stings
Yellow jacket – sharp stinger is attached to poison sac
Nettles – sharp, brittle hairs inject irritating chemical when broken off

Hard Shell
Clam or mussel – closes hard shell when disturbed
Turtle – retreats into hard shell when disturbed
Butternut – nut is encased in hard shell and inedible husk

Looks Scary

Underwing moth – camouflaged until it flies, then flashes large "eye spots" on underwings

Owl – when threatened, spreads feathers and wings to look as large as possible

Puffer fish – when threatened, inflates to several times its normal size

Sounds Scary

Rattlesnake – before striking, rattles tail

Cat – hisses when frightened

Thorns

Blackberry – sharp, hooked thorns discourage enemies from eating plants

Hawthorn – long thorns on twigs discourage enemies from eating plants

Rose – sharp thorns discourage enemies from eating plants

Slimy and Slippery

Slug – slimy coating repels enemies

Eel – slippery coating aids escape

Earthworm – slippery coating aids escape

Tastes Bad or Poisonous

Toad – when eaten, tastes bad and causes predator to get sick

Milkweed – the leaves contain poison and a bad-tasting sticky sap

EAT ME NOT

Objective: To observe the variety of ways plants make themselves inedible or distasteful.

Materials:
- paper plates
- tape
- scissors
- pencils

Ask the children to describe as many plant defenses as they can; record these on the chalkboard (prickly, stinky, hairy, hard, leathery, rough, slippery, distasteful, and so on). Divide the children into small groups. Explain that they must find as many different plant defenses as they can outside. Give each group a paper plate, scissors, and tape. Ask them to tape small samples (postage stamp size) of the plants found to the paper plates and label the defense. Report findings.

Note: Make sure the children know to avoid local poisonous plants!

PANTOMIME PARADE

Objective: To act out the ways some defenses might work.

Materials (optional):
- cards describing different scenarios

Whisper to pairs of children or pass out cards listing the following scenarios:

You are picking a prickly rose...You smell a skunk...You are eating an onion...You are a frightened turtle...You are a cat afraid of a dog...You are a goat using your horns...You are walking through stinging nettles...You are a dog who gets porcupine quills in its nose...You are trying to pick up a slippery fish...You are stung by a bee...You are an opossum playing dead...You stepped on a yellow jacket nest...

Watch the pantomimes, and then let the group guess what's happening in each.

Monster Mouthful

Materials:
- paper
- pencils
- crayons
- markers

Objective: To review defense adaptations by encouraging imagination to outdo nature.

Discuss the difference between defense and offense, listing on the board all the different defense strategies the children can name. Have each child draw an imaginary plant or animal that has super defenses. (Younger children may wish to work in pairs.) Afterward, allow children to introduce their Monster Mouthfuls.

Extensions

Monster Mouthful Predators: Challenge the children to design a critter that has adaptations that allow it to eat their Monster Mouthful creations.

Monster Tales: Have the children write stories about the lives and habitats of their monsters (and their predators).

Plant Defenses: Have the children research local plants with good defenses that might actually hurt people or animals, and then have the children make a chart illustrating them. Agricultural extension agents or poison control centers are good resources.

Animal Defenses: Ask the children to observe their pet animals, farm animals, or animals in nature programs to see what defenses they use. Do they fight very often? Do they use threats more often than they fight? Do they use their defenses more with members of other species or with their own?

American chestnut

Thorns and Threats

Characters: Willie Worm, Benjy Bear, Penelope Porcupine, Todd Turtle, Bonnie Bee, Teddy Toad

Willie Worm

> *[crying]* Boo hoo.

Benjy Bear

> What's the matter, Willie Worm? You sure look miserable.

Worm

> I am miserable. I'm just a poor defenseless worm, and everyone wants to eat me.

Bear

> You do look pretty tasty.

Worm

> That's just what I mean. And I don't have any way to defend myself. Boo hoo. I feel very sad.

Bear

> You're making me sad too, Willie. Survival depends on a good defense, you know. Couldn't you grow big and strong like me? Did you eat all of your breakfast today?

Worm

> I did. But it won't work. I'm not meant to be big.

Bear

> I think you need a defense, Willie, and I'm going to go find you one. You wait right here in this hole, and I'll go find out what I can learn from some other animals. *[Worm exits]* I wonder who I could ask first. *[Porcupine appears behind Bear and he backs into her]* Yikes! Ouch!

Penelope Porcupine

> **Excuse me!** I don't like folks getting too close to me, and I let them know it.

Bear

> I get the point, Penelope Porcupine! Sorry. I didn't mean to bump into you, and I won't do it again.

Porcupine

> Not a mistake most folks make more than once. You know what happens when somebody upsets me? **Swat** with my tail, and the animal's nose ends up looking like a pincushion.

Bear

> That's it! Great idea!

Porcupine

> What's a great idea? You need a pincushion?

Bear

> No, I'm trying to find a defense for Willie Worm, and I'm going to tell him about how you defend yourself. Maybe he can grow quills.

Porcupine

> I don't know about that, but I wouldn't eat him anyway, unless he grows bark. Thin, soft bark, that is. That's my kind of meal, and I'm hungry now. Bye, Benjy! *[exits]* *[Turtle enters; Bear walks along and steps on it]*

Todd Turtle

> Watch out, clumsy! Your big foot would have flattened me if I didn't have this hard shell.

Bear

> Sorry...Hey, that's another good one!

Turtle

> What do you mean, another good one? You like to go around stepping on creatures that are smaller than you? You big bully!

Bear

> No, no, Todd Turtle. You don't understand. I'm looking for the best defense I can find, and you've got a real tough one.

Turtle

> What is this, some kind of a competition? Personally, I shrink from competition – shrink right inside my shell where even bears can't bug me.

Bear

> No, it's not a competition. I'm just trying to find a good defense for Willie Worm because he's so small and defenseless.

Turtle

> Willie Worm? Is he around here?

Bear

> Yeah, he's right over...I mean, no! He's nowhere around.

Turtle

> Well, I'd sure eat him if I could catch him! See you later. *[exits]*

Bear

I'm getting kind of hungry too. I think I smell some honey! *[Bee enters]*

Bonnie Bee

You can smell it, Benjy, but you can't get at it.

Bear

What? Who said that?

Bee

I did. You may be a big bear, and I'm a little bee, but my sisters and I can defend our honey from anyone. A few stings on that shiny black nose of yours and you'd be sorry!

Bear

I believe you. I've been stung once this week, and that's enough. You bees really know how to defend yourselves. Hey, you've just given me a great idea to tell Willie Worm. I'm trying to find a good defense for him. So long, Bonnie Bee. *[Bee exits; rustling noise offstage]* Hmm, what's that rustling in the leaves? Maybe a tasty treat?

Teddy Toad

No – o – o, Benjy. Toads are not tasty treats.

Bear

You sure don't look very appetizing, but that doesn't mean you don't taste good.

Toad

My looks have nothing to do with my secret weapon.

Bear

Secret weapon? This could be another good idea for Willie! What's your secret weapon?

Toad

My skin, that's what. I ooze a terrible-tasting fluid out through my skin that makes animals like you spit me out pretty quick. Now if you'll excuse me, it's awfully bright out here, and I don't want my skin to dry out. *[exits]*

Bear

Gosh, that's another good one. Now let's see if I can find Willie and tell him about all the great defenses he could use. Willie! Willie! Come out, it's me, Benjy Bear.

Worm

[shaking] Are you sure it's safe out there? I'm afraid to come out.

Bear

Sure, it's safe. No one here but me.

Worm

Well, OK, but don't come too close. Did you find me a defense?

Bear

I sure did. I met some animals with great defenses. There was Penelope Porcupine with her sharp quills, and Todd Turtle with his hard shell, and Bonnie Bee with her stinger, and Teddy Toad with his poisonous skin.

Worm

But I can't grow quills, or a shell, or a stinger, and my skin's **very** delicate.

Bear

Shucks, that's too bad. Here, let me pick you up so I can get a close look at you. *[picks up Worm and drops it]* Hey, I can't hold on to you. You're too slippery. Let me try again. *[drops Worm]* Whoops. You're all covered with slime, and I can't hold on to you.

Worm

I'm sorry; I can't help it. It's just the way I am.

Bear

You know what, Willie? You have a **great** defense!

Worm

I do? What do you mean?

Bear

You're so wiggly and slimy, it's hard for your enemies to hold on to you when they grab you.

Worm

You're right! I may not have thorns or threats, but I'm pretty good at worming my way out of a ticklish situation. Thanks for reminding me, Benjy. I feel much better.

Bear

I'm glad, Willie. See you later. I'm going off to look for something to eat – something I can hang on to, that is!

Frogs and Polliwogs

Miraculous Transformation

There is something irresistible about a frog. Not a thing of conventional beauty with its bulging eyes, gaping mouth, and slippery skin, it nevertheless intrigues children. A visit to a pond is almost always punctuated by shrieks of "There's a frog!" – followed by immediate lunges to catch it or cautious efforts to sneak up on it. Usually, neither approach is successful, and the frog slips nonchalantly away or leaps suddenly into the water.

Frogs and toads are **amphibians**, animals that live part of their lives in water and part on land (*amphi bios* – double life). Ancient amphibians were the first **vertebrates** to emerge from water and crawl out on land, probably during the Devonian Age, when much of the earth was swampy and warm with intermittent droughts. Modern amphibians are still tied to water; most species breed in water and spend a significant portion of their lives in damp or wet environments.

The nearly 4,000 different species of frogs and toads make up the order Anura, the tailless amphibians. Though similar in many ways, there are notable physical differences between frogs and toads. Most frogs have smooth, moist skin, and their long hind legs make them excellent jumpers and swimmers. The inch-long spring peeper can hop up to 17 inches per leap. (If we had comparable ability, then a four-foot-tall human could jump 68 feet!) The larger green frog is also a good jumper, able to go 10 times its own length in one bound. At the end of their long hind legs, aquatic frog species have webbed feet to help in swimming. Toads generally have relatively dry, rough skin and can live in drier environments than frogs. Toads lose less water through their skin than frogs do, and they are equipped with a patch of special skin on the lower belly through which they can absorb water quickly. To rehydrate, a toad just has to find a puddle to sit in.

As cold-blooded creatures, unable to generate their own heat, frogs must hibernate (**brumate**) during winter. They become inactive during the cold months of the year, hidden from sight in the mud at the bottom of ponds or under the ground. During this time, their metabolism slows dramatically. The little oxygen they require for survival is absorbed through the skin. Some species, such as wood frogs, chorus frogs, gray treefrogs

young gray treefrog

and spring peepers, can actually withstand being frozen solid. When temperatures get close to freezing, these mostly terrestrial frogs quickly lose large amounts of water, and their bodies produce an antifreeze that protects their body tissues, enabling the frogs to survive a northern winter in the leaf litter.

Frogs emerge from brumation during the spring, and some species begin courtship right away. Most species return to water to mate and lay their eggs, even those that live a largely terrestrial life as adults. Frogs breed from early spring through summer, each species according to its own particular timetable. Spring peepers and wood frogs are usually the first to appear, while bullfrogs wait until later in the summer.

Males arrive at the breeding ponds before the females and begin calling. Each species of male has its own distinctive call that serves to establish its territory and attract females. Females are usually silent. A frog makes its calls by pushing air back and forth from mouth to lungs, vibrating its vocal cords. Vocal sacs in the throat region inflate, acting as air reservoirs. Frogs usually begin calling at dusk and continue through the night or until a mate is located.

To mate, the male crawls on the back of the female and grasps her around the body behind the forelegs. As the female deposits her eggs in the water, the male releases his sperm over them, fertilizing them externally. After laying and fertilizing their eggs, most frogs abandon them.

Egg-laying strategies vary. Female wood frogs lay large globular masses of eggs. The spring peeper lays her tiny concealed eggs one at a time. Green frogs and bullfrogs lay masses of thousands of free-floating eggs. The American toad lays bead-like strings of eggs. Amphibian eggs are covered by a gelatinous substance that swells to form a thick protective coat. The number of eggs laid in a season

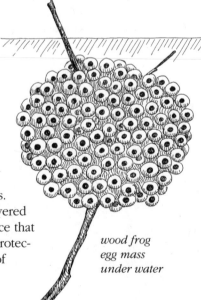

wood frog egg mass under water

varies from hundreds for spring peepers to over 20,000 for bullfrogs. Incubation time is also different from species to species, depending in part on the water temperature and the amount of sunshine available to warm the generally dark-colored eggs and speed their development.

The tiny tadpoles (also called polliwogs) that hatch from the eggs are very different in physical form and behavior from the adult frogs or toads they will become. Tadpoles have round bodies, vertically flattened tails, and gills – adaptations for a strictly aquatic existence. Tadpoles are mostly **herbivores**, with mouths adapted for scraping algae off submerged objects and nibbling vegetation. The time period from tadpole to adult depends on the species and on the weather. Some require as little as a few days to make the transformation; for others it takes two or three years.

The tadpole phase is the most critical time in the life of frogs and toads. Of the large numbers that hatch, very few reach maturity. They become **prey** for numerous aquatic animals, such as predaceous diving beetles, crayfish, and dragonfly **nymphs**. They are also at the mercy of the environment. Tadpoles living in vernal pools, or temporary ponds, must reach adulthood before these bodies of water dry up during the summer or fall. In addition, because of their thin, permeable skin, amphibians are very susceptible to pollution, such as acid rain, pesticides, and other chemicals.

Over time, the tadpole begins to grow hind legs. Then its front legs emerge, blocking the gill openings. Simultaneously, lungs develop to replace the gills. The eyes migrate somewhat toward the top of the head and bulge out, and the mouth changes from a tiny opening with rasping edges to a wide gape capable of catching and holding live prey. The tail, which provides a major source of nourishment during this transition period, is gradually absorbed as the tadpole becomes an adult.

An adult frog is carnivorous, eating primarily insects. Its tongue is adapted to hunting; it is long and sticky and attached at the front of the mouth so it can dart out to capture passing prey. Large species, like the bullfrog, are also known to eat mice, small birds, and other frogs. The toad in the garden is probably feasting on slugs at night.

To make the transition necessary for their double life, frogs are equipped with many effective and miraculous adaptations. By moving effortlessly between land and water, they remind us that the natural landscape is a seamless whole.

American toad

Suggested References:

Conant, Roger, and Joseph Collins. *Reptiles and Amphibians of Eastern and Central North America*. Peterson Field Guide Series. Boston: Houghton Mifflin, 1991.

Reid, George, and Herbert Zim. *Golden Guide to Pond Life*. New York: Golden Press, 1967.

Stebbins, Robert C., and Nathan W. Cohen. *A Natural History of Amphibians*. Princeton, NJ: Princeton University Press, 1995.

Tyning, Tom. *A Guide to Amphibians and Reptiles*. Stokes Nature Guide Series. Boston: Little, Brown, 1990.

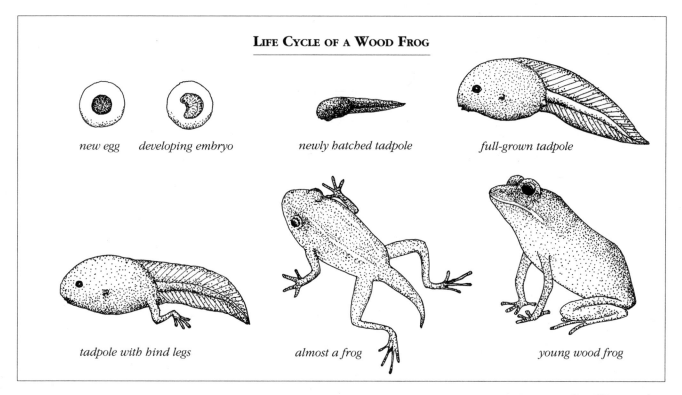

LIFE CYCLE OF A WOOD FROG

new egg *developing embryo* *newly hatched tadpole* *full-grown tadpole*

tadpole with hind legs *almost a frog* *young wood frog*

Frogs and Polliwogs

FOCUS: Frogs and toads can live in a wide range of habitats and temperatures because of their unique physical adaptations and specialized life cycles.

OPENING QUESTION: *What are some of the special adaptations of frogs and toads?*

PICTURE PARADE

Objective: To learn some of the special adaptations of frogs, toads, and tadpoles.

Discuss the characteristics of amphibians. Show pictures of different kinds of frogs and toads and the places they inhabit. Be sure to include photos of local species. (You may also want to play a tape of their calls.) While looking at the pictures, have the children help you list frogs' special adaptations. Then show a chart of frogs in their various stages of metamorphosis and describe the process to the children.

Materials:
- pictures of frogs and toads
- chart of frog life cycle
- (optional) tape of frog songs and a tape player

MERRY METAMORPHOSIS

Objective: To role-play the growth and development of frogs from egg to adult.

Read through the Merry Metamorphosis script with the children listening. (Older students can do the reading themselves.) Assign one child to be the diving beetle. Then have the rest role-play the egg, tadpole, and adult stages of a frog as you read the script aloud a second time. Afterward, ask the children why they think frogs lay hundreds or thousands of eggs at a time.

Materials:
- Merry Metamorphosis script

FROG OR TADPOLE?

Objective: To identify some of the differences between frogs and tadpoles.

Divide the children into two groups, one representing frogs, the other tadpoles. Tell them you're going to hold up pictures or words that show features of frogs, tadpoles, or neither. If the correct group claims the feature by raising their hands, they get the card. Otherwise, through group discussion, decide which group should have the card. Play until all appropriate cards have been correctly claimed and the "neither" cards are weeded out. Afterward, give each group a couple of minutes to discuss how these adaptations help their critter survive in its particular habitat and then share their conclusions with the whole group.

Frog Features: four legs; big eyes on top of head; eats insects, worms, and sometimes other frogs; big wide mouth; sticky tongue; long hind legs; lungs; hops on land; sings.

Tadpole Features: tiny round scraper mouth; no legs; tail; gills; lives only in water; tiny eyes on side of head; no voice; eats plants.

Neither: furry; has wings; has eight legs; covered with feathers; breathes fire; has curly horns.

Materials:
- signs (and pictures) with characteristics of frogs, tadpoles, and "neither"

MOVING MEALS

Objective: To become aware of frogs' ability to spot and catch moving prey.

Explain that frogs, like most predators, eat live prey, expertly catching moving insects within their reach. Discuss how they do this. (good eyesight, especially for movement; long, sticky tongue; quick jumping) One child is appointed "Frog." The rest of the children are "insects" and form a line facing the frog. The frog, standing with her back to the insects and about thirty feet from them, calls out "1-2-3, food for me" and turns. Any insects seen moving are "eaten" (pointed to) and must freeze. The object is for insects to get to safety by crossing an imaginary line even with the frog. How could an insect avoid being eaten by a frog?

FROG JUMPS

Objective: To compare a frog's ability to jump with our own.

Ask the children to line up and see how far each can jump from a standing start. Measure their jumps. Can they jump as far as they are tall? A spring peeper is about one inch long and can jump 17 inches. Have children calculate how far they would need to be able to jump to match a peeper's ability. What adaptations enable frogs to jump so well?

Materials:
- measuring tape
- paper
- pencils

SPRING SERENADE

Objective: To demonstrate that each species of frog has a different courtship call that the male sings to attract the females of his species.

Briefly discuss the reasons for frog calls and for the different songs of different species. Then, divide the children into two groups. Give out matching sets of calling cards to each group, one per child. The first group acts out the role of male frogs and gathers at the "pond," a place at a distance from the second group, the female frogs. At a signal, the males start singing their assigned songs. The females, with their eyes closed, first quietly listen for their assigned sound, and then open their eyes and go to their frog singer. One male may attract more than one female; others may attract none, as in nature. To do over, separate the two groups, shuffle the cards, and start over again with the other group now singing.

Materials:
- matching sets of calling cards with one of the following frog calls: quack (wood frog); peep (spring peeper); jug-o-rum (bullfrog); katung (green frog); or others from your area

FROG HUNT

Objective: To look for frogs in any stage of development, and assess frog habitat.

Go to a pond or other wetland. In small groups, search along the edge of the water for eggs, tadpoles, and adult frogs or toads. Specimens may be caught (gently) for closer examination. Record observations on Frog Field Notes sheet.

FROG FIELD NOTES
Make notes on any frogs observed, and sketch the stages seen. Include date; time of day; and sizes and numbers of eggs, tadpoles, and adults. Release specimens carefully by lowering container into the water, then tipping.

Assess the pond for frog and toad habitat requirements. Look for:
- Shallow water (warm water for tadpoles and insects to develop in)
- Water plants (places to anchor eggs, for tadpoles to hide)
- Tall plants around the pond (places for frogs and insects)
- Insects (may be hiding on a cold day)

Rate the pond Good, Fair, or Poor for each requirement. Is there anything that could be done to make this site a better place for frogs?

Materials:
- white dishpans
- nets
- pencils
- Frog Field Notes sheets
- clipboards

SHARING CIRCLE

Objective: To review and personalize what's been learned.

Ask each child to complete the following sentence: "One special thing about frogs (or toads, or tadpoles) is _____." Encourage children to name specific features and adaptations.

EXTENSIONS

Adopt a Pond: Make several repeat visits to a place where frog eggs have been seen. What happens to them? If you see tadpoles, what are they doing? Are any adult frogs living in or around the pond?

Frogs and Relatives Research: There are many amazing kinds of amphibians in the world. Have the children choose a species to research, describe, and illustrate, creating a bulletin board or reference notebook. Visit the many good sites on the worldwide web that deal with amphibian research and monitoring. Some are specially designed for teachers and students.

Amphibian Issues: Declining, disappearing, and deformed amphibians are of concern all over the world. What can the children learn about the causes? What kinds of things are being done to help conserve amphibians? Are there local efforts that children could learn about? Invite someone involved in these efforts to visit the classroom.

Raising Tadpoles: Ask your state wildlife agency if and when it is okay to raise tadpoles in the classroom. (Some states require that you get an educational permit.) If you do this, be sure to release them to their original home!

adult wood frog
two years old (may live more than three years)

Frogs and Polliwogs

If you look closely in a pond
By chance you just may see
A mighty mass of frog eggs
Floating in the water, free.
*[Group huddled together moving as one unit
in different directions.]*

Moving with the water
As it ripples here and there,
How nice to be a frog egg,
Not a worry or a care!

But then – oh no – what's that you see
Swimming all about?
A big fat diving beetle
So you quickly yell, "Watch out!"
*[One person as the diving beetle swims
toward group.]*

For that old diving beetle
Is looking for a lunch
And would really be quite happy
With some frog eggs in a bunch.

But then, by luck, the wind blows strong
Away the eggs do float,
And that hungry beetle now must seek
Another frog egg boat.
*[Group moves away. Diving beetle
person leaves.]*

As days go on those frog eggs
Will start to look quite strange –
If you look closely in that pond
You're sure to see the change.

Coming from that mass of eggs
Are not a bunch of frogs,
But rather little tadpoles
Also known as polliwogs.
[Group begins to separate.]

They have no legs to hop with
Or lungs to breathe the air,
For they stay under water
Eating plants that grow down there.
*[Each child now moving about with feet
together and hands at their sides.]*

They're sort of round up in the front,
Their backs have wagging tails.
They don't look a bit like frogs
But more like tiny whales!

If you look closely in that pond
And watch them every day,
You'll see the tadpoles changing
In yet a different way:

On each side of the wagging tail
Legs will start to grow –
The tadpoles start to use them
For swimming to and fro.
*[Children now shake both legs and begin moving
about with legs apart, hands still at sides.]*

Later front legs will appear
And that's not even all,
For as they do, you'll notice
The tails will get quite small.
*[Children can now wiggle arms from elbows down,
upper arm still touching sides.]*

They gather at the pond edge
With big eyes bulging out.
They breathe with lungs and start to eat
The bugs that swim about.
*[Children take deep breath and then
pretend to eat.]*

With long hind legs to hop with
And a tail that's gone for good,
Those tiny little frog eggs
Now look just like they should.

If you look closely in that pond
To see your polliwogs,
You won't even find them –
You'll meet some baby frogs!
[Children hop, hop, hop.]

Habitats

An animal's habitat is the community in which it lives. A habitat can be likened to a home in a child's mind – a place in which food, water, shelter, and familiar plants and animals can be found. Forest, field, pond, and stream are all unique habitats, their different physical conditions making each one home to a particular combination of living organisms.

Throughout this theme, children will compare a variety of different habitats and become familiar with some of their inhabitants. The forest floor is home to shrews and millipedes, while grasses and crickets are found in open fields. Tadpoles live in warm, shallow ponds, but brook trout prefer cold, fast-moving streams. Children will discover what makes each habitat suitable for certain organisms but not for others. They will explore how habitats change over time and how these changes affect what species can live there. And, while studying each habitat, children will examine the interactions among the inhabitants. Predators and prey, producers and decomposers, all are important links in the flow of energy from sun to plant to animal and back to plant again.

What would it feel like to be a millipede in the leaf litter or a cricket in a grassy lawn? Children find out by lying down on the forest floor or sitting quietly in a field. The activities in these chapters invite children to experience each habitat intimately while exploring its physical characteristics and becoming acquainted with its inhabitants.

woolly bear caterpillar

Life in a Field

EXPOSED TO THE ELEMENTS

Welcome to a field. Close your eyes and feel the heat of the sun and the coolness of the wind. Listen to the sounds of insects hidden in the grass. Here is a habitat distinctly different from forest or stream, desert or tundra. A close-up look at a field reveals a complex mix of plant and animal inhabitants that face the unique challenges of this exposed environment.

A field is home to many different kinds of plants and animals, providing them with the necessities of life: food, water, shelter from **predators** and the elements, and space in which to live and reproduce. Some insects and smaller mammals may spend their entire lives in a single field, while larger animals may frequent them for hunting or grazing.

At first glance, a field may seem to contain nothing but grass, yet closer examination reveals a wonderful blend of diverse plants, providing food and shelter for numerous small animals. Just as a forest has distinct layers, so does a field, only on a smaller scale. An upper layer is made of the flowering stems of tall grasses and leafy stalks of plants like goldenrod. Shorter grasses and a variety of plants like wild mustard and daisies make up the middle layer, while low-growing plants like strawberry and dandelions are found close to the ground. Beneath all of these is a tangled layer of dried plant material. It forms a thick mat and provides a place for mice and voles to tunnel in safety. And the soil upon which all this life depends, itself hosts myriad organisms in every teaspoonful.

True grasslands, such as those found in the North American plains, exist only in places where the rainfall is too low to allow trees to grow and too great to support desert plants. In most other areas, fields exist because they are maintained by grazing or mowing. In these transient fields, the grasses and flowers would gradually be replaced by woody shrubs and then trees if natural succession were allowed to occur. In most places, children will investigate and explore fields that have been mowed within five years. These fields will likely host a number of familiar plants: goldenrod, clover, yarrow, daisies, buttercups, Queen Anne's lace, black-eyed Susan, and, of course, grasses.

black-eyed Susan

A field is one of the more exposed habitats, susceptible to extremes of heat and cold, drought and storms. This quality of being exposed to the elements means that those that inhabit a field must either be specially equipped to do so or must be able to leave the field when necessary. Plants that can withstand the drying winds and hot rays of the summer sun adapt to these rigorous conditions through an amazing variety of shapes and textures. Some, like milkweed, have stiff leathery leaves that do not dry out. Others, like clover and goldenrod, have toothed or divided leaves, which allow for greater air circulation and cooling by the wind. Plants like yarrow or Queen Anne's lace have lacy leaves, allowing very little surface to be directly exposed to the hot sun.

Texture, too, protects leaves and stems from drying out. Examine a black-eyed Susan leaf through a hand lens; its bristly white hairs reflect the sun's rays, illustrating another effective method of preventing excess heating. Noticing the myriad designs of plants in a field helps to highlight the diversity of plants sharing this one habitat. It is not necessary to know the names of all the plants to see and feel the differences among them.

The more diverse the plant life in a field, the wider the variety of animals that can live or feed there. Easiest to find are the insects, such as grasshoppers, ants, crickets, leaf hoppers, bees, butterflies, beetles, and their relatives the spiders and daddy longlegs. **Amphibians** – toads, frogs, and salamanders – are scarce, unless there is a damp or protected area nearby because they lose too much body moisture in these sunny, dry areas. Snakes, on the other hand, protected by a scaly skin, find sufficient shelter under the grasses or in rodent holes. A field is a good place for these **reptiles** to hunt for mice and other food. Most mammals cannot fill all their habitat requirements in a field, but a great many depend on the field for food. Plant eaters such as mice, woodchucks, voles, rabbits, and deer spend much of their lives in open fields, and they in turn are a food source for predators that visit but rarely inhabit a field: foxes, coyotes, hawks, and owls.

young white-footed mouse

American kestrel

the food web, the less likely that a change will disturb the whole system. If one species of plant disappears, most animals can shift to another food source. Thus, although the makeup of the community may change, it is unlikely to collapse completely.

A field is a habitat bustling with activity. It is a fascinating place to study because of the diversity of life to be found there and the complex relationships among those that live in or visit it. A field is also a beautiful place to sit quietly, listening and watching the life around you.

Suggested References:

Allen, Durward L. *The Life of Prairies and Plains.* New York: McGraw-Hill, 1967.

Molles, Manuel C., Jr. *Ecology: Concepts and Applications.* New York: McGraw-Hill, 1999.

Newcomb, Lawrence. *Newcomb's Wildflower Guide.* Boston: Little, Brown, 1977.

Smith, Robert Leo. *Elements of Ecology.* New York: Harper Collins, 1992.

Stokes, Donald. *A Guide to Observing Insect Lives.* Boston: Little, Brown, 1983.

The interdependency of plants and animals found in a field illustrates some concepts basic to all successful habitats. Survival of plants and animals depends on their having enough energy to carry on the business of living. The transfer of energy from one organism to the next can be illustrated by a **food chain**. The original source of energy is the sun. Plants receive energy from the sun, which allows them to produce their own food. This process, called **photosynthesis**, enables a plant to manufacture sugars and starches from carbon dioxide and water in the presence of **chlorophyll** and sunlight. Plants are thus called primary **producers**. Dependent upon plants for food are the vegetarian animals, or **herbivores**, such as rabbits and mice. In the context of a food chain, they are called **primary consumers**. Then there are meat-eating animals or **carnivores**, like foxes and hawks, which are **secondary consumers**. Because most animals have multiple food sources, they are linked in many possible food chains, which in any given habitat are interconnected in a **food web**.

When one portion of a field is disturbed, there is a reaction throughout. Fire or herbicides may destroy the plants in a field, leaving the mice and grasshoppers to die without food, which in turn starves the foxes and hawks that prey upon them. However, the more complex

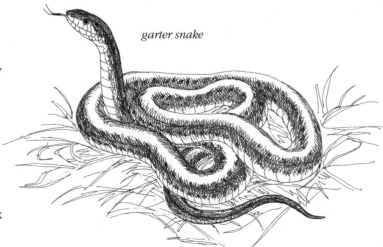

garter snake

Life in a Field

FOCUS: A habitat provides plants and animals with food, water, shelter, and a place to raise young. Fields are home to many interdependent species of plants and animals.

OPENING QUESTION: *What basic needs must a habitat meet?*

PUPPET SHOW

Objective: To identify the basic components of a habitat.

Perform, or have the children perform, the puppet show. Afterward, review the four basic components of a habitat. Did the various puppet characters have the same habitat requirements? How were they different?

Materials:
- puppets
- script

WHO AM I?

Objective: To become familiar with animals that live in or visit a field.

Ahead of time, make signs for each of the following animals. Include the name or a picture of the animal on each card, and list the food it eats.

Bee — flower nectar and pollen
Butterfly — flower nectar
Deer — grass, leaves, twigs
Rabbit — grass, twigs, berries
Grasshopper — grass, leaves
Mouse — seeds, berries, roots
Woodchuck — grass, flowers
Cricket — leaves, fruit, seeds
Earthworm — dried plant material
Caterpillar — leaves
Bluebird — grasshoppers, crickets, caterpillars, spiders
Spider — bees, other insects
Robin — earthworms, berries, caterpillars, spiders
Mole — insects, earthworms, roots
Skunk — fruit, bird eggs, grasshoppers, crickets, spiders
Snake — frogs, mice, insects
Owl — mice, rabbits, skunks
Hawk — mice, insects, birds, snakes
Fox — mice, rabbits, insects, berries
Coyote — deer, rabbits, mice, woodchucks, insects, fruit

Materials:
- signs with the picture or name of each animal and its food

First, introduce the animals by holding up the signs and briefly discussing what each eats. Are they carnivores or herbivores?

For younger children (grades K-2): Have one child tape a sign to the leader's back without letting the leader see it. The leader asks yes-or-no questions of the group until he or she can guess the animal. Useful questions might be: Am I bigger than a lunchbox? Can I fly? Do I have two legs? Do I have wings? Do I have scales? Do I eat meat? Repeat as often as time (or interest) allows.

For older children (grades 3-6): Begin as above, with leader demonstrating the activity using one of the animal signs. Then, tape a sign to each child's back. At "go," the children should find a partner, look at each other's signs, and ask each other yes-or-no questions to try to figure out what animal is on their sign. This could also be done in small groups with each child taking a turn asking questions.

For both: Afterward, display all signs. Ask the children if these animals spend all of their time in fields. Which ones might we see — where and when? Which ones use other habitats as well?

FIELD FOOD WEB

Objective: To see how animals and plants in a field are interconnected in a food web.

Begin by asking three children to help you demonstrate a food chain. Give one a "Grass" headband, and the others the "Rabbit" and "Fox" cards from the Who Am I activity. What would happen to the rabbit if there wasn't any grass to eat? Explain that energy flows through a food chain and that food chains such as this are part of more complex food webs.

Now, divide the children into three groups representing plants, plant-eaters (herbivores) and meat-eaters (carnivores). Give all the plants green headbands. Use the picture cards from Who Am I to identify the herbivores and carnivores. To form the field food web, the leader places the "sun" on the floor in an open space. Have the plants sit in a circle around the sun and discuss how the sun provides energy for plants to make food. Have each plant hold one of the strings from the sun. Now give each of the herbivores two pieces of yarn. Have them find one or two plants they would eat and connect themselves to those plants with the yarn. Finally, give each of the carnivores one long piece of yarn. Ask each of them to hold on to the center of the yarn and pass the ends to the two herbivores they'd most like to have for supper. What would happen to the animals if all of the plants died? What about if just one kind of plant died? Which creatures could move to another habitat?

SILENT WATCH

Objective: To silently become part of the life in a field.

Go out to the field, younger children in small groups with a leader, older children singly or in pairs. Spread out so there is a feeling of being alone. Sit very still, not moving, so that life in the field resumes its activities. (This can last two or three minutes for younger children, longer for older ones.) What did the children see? Hear? Smell? Enjoy?

FIELD FINDS

Objective: To observe the diversity of plant life in a field and note the presence of many small animals that inhabit that field.

Note: Be sure to alert children to avoid any poisonous plants growing in the area.

Divide the children into small groups and give each group a Field Finds card. Each group should visit a separate section of unmowed field, looking for and checking off items on the Field Finds card. How might life in a field be difficult for plants? For animals? What will the field be like in a few years if no one mows it?

Materials:
- a "sun" made of a yellow paper plate with five to seven three-foot pieces of string attached to the edge
- five to seven green paper headbands labeled with one of the following: leaves, twigs, berries, flowers, seeds, grass, roots
- animal cards from Who Am I activity
- 2 ten-foot pieces of yarn per child

Materials:
- a watch or stopwatch

Materials:
- Field Finds cards
- pencils
- clipboards
- hand lenses

Find a plant that is:
As tall as your waist
As tall as your knee
Shorter than your ankle
Prickly

Find a leaf that has:
Smooth edges
Edges with tiny teeth

Find a leaf that is:
Fuzzy or hairy
Thick and leathery
Long and thin
As wide as your foot
As long as your little finger
Lacy

Find animals such as:
A crawling insect
A hopping insect
A flying insect
A spider
An insect on a flower
An insect resting on a plant

Find a flower that is:
As big as your fist
Smaller than your thumbnail

Find animal sign like:
A spider web
Worm castings
A hole in the ground
A mouse runway
A partly-chewed leaf
A cocoon
An insect gall

Can you find any of the plants pictured here?

yarrow

white clover

common plantain

timothy grass

SHARING CIRCLE

Objective: To personalize what has been learned by imagining what it is like to live in a field.

Have the children complete this sentence: "If I lived in a field, I'd want to be a _____ because _____."

EXTENSIONS

Inventory: The children should choose a favorite spot in the field. Mark it with stakes or string. Check it once a week to see which plants are in flower, which have gone to seed, which have died, and what creatures are active. Use a flower guide to identify the plants. Keep a record or a journal of the study plot.

Imagine Yourself as an Insect: Suggest that the children each write a story pretending they are small insects or animals in a big field. What do they see above them? How do they get around? What do they do each day? At night? In winter or in bad weather? Are they afraid of anything?

Look at the Roots: Dig up a clover or other leguminous (pea family) plant, such as alfalfa, vetch, or trefoil. Use a hand lens to observe the bumps on the roots (nitrogen-fixing nodules). Research their natural history. How would these nodules help the plant they are on? How would they help the field habitat? Dig up a grass plant and a Queen Anne's lace plant. How are the roots different? Which could survive a drought better? Which could hold the soil best against erosion by wind or water? How are the leaves suited to life in a field?

Life in a Field

Characters: Marsha Mouse, Mother Mouse, Maria Monarch, Cindy Squirrel, Cappy Chickadee, Bernie Beaver

Marsha Mouse

Hey, Mom, what's a habitat?

Mother Mouse

Well, dear, a habitat is like a home. It's where animals or plants live. Why do you ask?

Marsha

Because someone told me I'd have to find a good habitat when I grow up. But how can I find one if I don't even know what it is?

Mother

That's true. It would be hard. I know...Why don't you go talk to some of the other animals around here? Ask them what habitat they live in and why. Now there are four important things a good habitat needs to provide. When you get back, let's see if you can tell me all four. *[Mom exits]*

Marsha

Oh goody. This should be fun! And here comes Cindy Squirrel. I'll ask her first. *[Cindy Squirrel enters]*

Cindy Squirrel

Hi, Marsha!

Marsha

Hi, Cindy. I'm trying to find out what a habitat is. Where do you live?

Squirrel

I live in the forest.

Marsha

That's right. You're so good at climbing trees.

Squirrel

Well, yes, but I live in the forest because that's where I can find my favorite foods.

Marsha

I know – acorns!

Squirrel

Right – acorns and hickory nuts and pine seeds and beech nuts and...Hey, all this talk about food is making me hungry. See you later. *[exits]*

Marsha

Bye, Cindy, and thanks! Now I know one of the four things that a habitat has to provide: food! Hey, there's Maria Monarch flitting around in that field. I'll ask her next. Hey, Maria!

Maria Monarch

Oh hi, Marsha. What's on your mind?

Marsha

I'm trying to learn about habitats. I know a habitat provides food, but I'm not sure what else it provides for animals and plants.

Monarch

Oh, not just food, Marsha. Every living thing needs water, too. Lucky me, I get my food and water at the very same time – by drinking the sweet nectar from flowers. Like this: slurrrrrrp. MmmmMmm, that's good. Off I go! *[exits]*

Marsha

Food and water, of course! That's two out of four. Now who can I ask? Oh look, there's Cappy Chickadee. *[Cappy appears, singing chick-a-dee-dee-dee, then disappears]* Hey, where did he go? *[Cappy appears again, singing]* Wait, Cappy. I just saw you sitting on that tree stump a minute ago and then you disappeared! How did you do that?

Cappy Chickadee

Why hello, Marsha. Chick-a-dee-dee-dee. I didn't disappear. I just slipped into a hole, right here in this rotting stump. *[while singing, disappears, then quickly reappears]*

Marsha

Oh, is there a tunnel in there?

Chickadee

Chick-a-dee-dee-dee. There sure is. And, it leads right to our nest. We pecked out a nice cavity in the rotting wood and laid our eggs deep inside where they'll be safe and warm.

Marsha

What a good place for a nest. Hey, that must be something else that a habitat has to provide – a place to raise young.

Chickadee

You're right about that. *[cheeping noise of baby birds]* And by the sound of those young 'uns, I'd better get right to work finding them some food! Bye, Marsha. *[Chickadee exits, singing]*

Marsha

Bye, Cappy. And thanks! Now I know three things that a habitat needs to supply. I wonder what else there could be? I'll visit the pond and see who I can find there. *[loud clap]* Yikes! What was that noise? *[Beaver enters]*

Bernie Beaver

Oh, that was just me, Marsha, smacking my tail on the water. That's how I tell the other beavers to dive for cover. I thought I saw a predator, but it was only you.

Marsha

Gee, I didn't mean to frighten you, Bernie.

Beaver

Oh that's okay, Marsha. Better to be safe than sorry, I always say.

Marsha

Well, Bernie, I'm glad you're here because I want to ask you about habitats.

Beaver

Well, you're asking the right fella, 'cause I live in the best possible habitat – the pond! Got everything I need right here or close by.

Marsha

Well, you've definitely got lots of water, and I suppose there's plenty of fish to eat, too.

Beaver

Oh no, no, no, Marsha. Beavers don't eat fish. We cut down nearby trees then drag them into the pond where we can safely eat the tender bark off the branches.

Marsha

Safely? You mean the pond protects you?

Beaver

Sure it does. We dam up the pond to make the water nice and deep, then we build a lodge with an underwater entrance. No coyote or fox can catch us in the lodge, and they don't chase us in the water either. Plus, it's a nice snug place to spend the winter.

Marsha

Gee, now I know the fourth thing that a habitat needs to provide – protection – from both predators and bad weather. Thanks, Bernie. I've got to run home and tell my mom what I've learned. *[Beaver and Marsha exit; Mother Mouse and Marsha re-enter]*

Mother Mouse

Welcome home, dear. So, did you discover four important things about habitats?

Marsha

You bet I did! I found out that a good habitat has food, water, shelter, and a place to raise a family.

Mother

That's exactly right, Marsha. Great job. Why, now you can start thinking about what you need in your own habitat.

Marsha

Oh, I already know just what I want. I'll find a nice sunny field with lots of seeds to eat and plenty of dew to lick off the grass in the mornings. And I'll make lots of tunnels in the matted grasses close to the ground. They'll protect me from predators. And I'll make my babies a nest lined with the softest thistle down. It'll be wonderful! Well, I'd better start looking right away. After all, I'm almost grown up now, aren't I, Mom?

Mother

Yes, you're such a big little mouse, Marsha. Off you go!

Marsha

Bye, everyone. See you in the field! *[both exit]*

Forest Floor

HOME OF THE HIDDEN WORKERS

The forest floor is a habitat bursting with invisible energy. It is home to billions of unsung heroes, the decomposers, who break down many tons of plant and animal debris over the course of a year. Through decomposition, important minerals and nutrients are returned to the soil where they can once again be taken up by plants.

Thanks to the towering canopy of trees, the floor of the forest remains cool, dark, and moist, free from winds, and protected from extremes and sudden changes in temperature. It is a remarkably rich habitat, for each year a vast amount of dead leaves, needles, seeds, and other plant material settles upon the forest floor. Here, this material becomes part of the layered mat known as the leaf litter.

Examining the layers of leaf litter reveals much about the process and stages of **decomposition** occurring on the forest floor. On the surface, leaves appear intact and may be freshly fallen or dried and brown. Beneath this layer, the leaves are damper and often perforated with many tiny holes, the work of diminutive mites and insects. These holes allow bacteria and **fungi** to enter the inner tissues of the leaf and begin the process of decay. Further down in the leaf litter, look for skeletonized leaves in which all the soft tissue has been removed, leaving only the stem and an intricate network of veins. Digging deeper reveals leaves that are darker-colored with a slimy feel from the microorganisms that coat them. White strands of fungi can also be found in these deeper layers. Beneath this is the **humus** layer, where the plant material has been broken down into indistinguishable fragments. Finally the leaf litter is transformed back into its raw materials, the nutrients so important to life, here stored in the forest soils.

What type of life thrives in this specialized habitat? In a teaspoon of soil there may be billions of bacteria, millions of protozoa, and thousands of algae and fungi. These organisms derive their energy from decaying plant and animal material. Most of these are so small they can only be seen with a microscope, but the larger fungi are easily seen with the naked eye. The main body of a fungus consists of masses of underground **hyphae**, which look like white threads in the soil or under the bark of dead or dying trees. One ounce of forest soil may contain up to two miles of these fungal threads. Mushrooms are merely the fruiting bodies of fungi that appear when the fungus is ready to produce spores. Fungi break down dead plant matter by secreting enzymes into the plant tissue and then absorbing the nutrients as they are released.

A whole host of **arthropods**, animals with jointed legs and hard **exoskeletons**, live in the top few inches of the forest floor. These include insects, millipedes, and mites. Springtails are wingless insects that live in the leaf litter, in crevices in bark, and in the soil. Their food consists of tiny specks of plant material. They are most

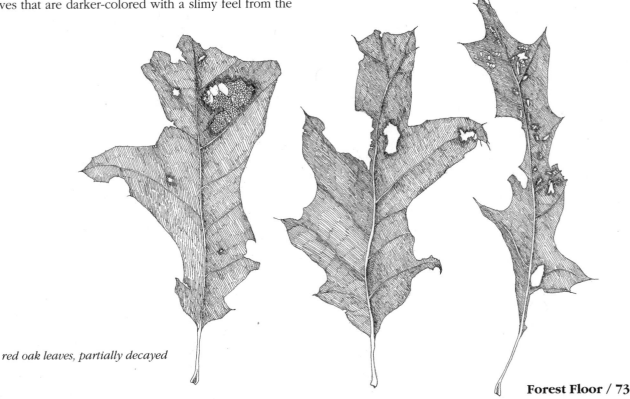

red oak leaves, partially decayed

notable for their special method of locomotion – their folded-under tail straightens with a force that allows them to spring two feet up in the air. Look, too, for the long tube-shaped millipede with its hard, segmented outer shell and numerous legs. Millipedes feed on dead leaves and decaying wood, breaking them down into smaller fragments in the process. Mites are also important **decomposers**, feeding on plant and animal material. Although each mite is no bigger than the head of a pin, they are numerous enough to make up five percent of the live weight of forest soils.

A larger inhabitant of the leaf litter is the brightly colored red eft. It is the juvenile stage of the red-spotted newt, a primarily aquatic salamander. For several years, it takes to the woods and lives on the forest floor before returning to the pond as an adult. Its brilliant orange color warns potential **predators** to stay away, for its skin contains poisonous substances.

red eft

Digging into the soil beneath the litter layer reveals evidence of burrowing animals. The most important of these are the earthworms. Rather than pushing soil aside as they burrow, they actually eat their way through it, grinding up the soil and plant matter in their gizzards. Worms consume their own weight in leaves and other plants every 24 hours. Their tunnels allow air and water to penetrate the soil and create openings for insects, bacteria, and fungi to enter. As they snack their way through the soil, they also enrich it with their **castings**, which contain substantial amounts of nitrogen, phosphorus, and potassium. Additionally, worms are a great food source for many other forest floor creatures, including moles, shrews, and various **reptiles**, and **amphibians**.

A worm's body is made of a series of rings, or segments. The front end of an earthworm is closest to the **clitellum**, or band, an enlarged section about the width of five or more segments, located partway along the body. This front end is equipped with a mouth and other vital organs. It continues to live even if the worm gets severed, and can regenerate a new hind end. Earthworms breathe through their leathery skin, which must remain moist or they suffocate. To travel, a worm first stretches itself out and then pulls its back end forward. Tiny hooked bristles called **setae** keep its body from slipping back during this accordion-like movement. The setae, four pairs on all but the first segment, are very hard to see, but you can feel them when you let a worm slide through your moistened hand.

Another common burrower is the secretive shrew. The shrew is perfectly constructed for burrowing in the soil, with its nose made of bones to serve as a plow and its body covered with velvety fur. This tiny mammal is known as one of the fiercest animals around because of its willingness to attack almost anything. The shrew has a very high metabolic rate and must eat every hour, making hibernation impossible. A shrew's menu may include snails, insects, worms, mice, snakes, birds, and other small animals.

Most people enjoy the serenity of a walk through the forest, but few of us realize that beneath our feet is another world with its own unique community of inhabitants. In this seemingly static environment, the process of decomposition is churning away, returning vast quantities of plant and animal debris to their elemental components. Although we cannot see this vital process in action, we can see evidence of it all around us when we take a close-up look at the forest floor.

Suggested References:
Andrews, William, ed. *A Guide to the Study of Terrestrial Ecology.* Englewood Cliffs, NJ: Prentice-Hall, 1974.
Appelhof, Mary, Mary Frances Fenton, and Barbara Loss Harris. *Worms Eat Our Garbage.* Kalamazoo: Flower Press, 1993.
Hunken, Jorie. *Ecology for All Ages.* Old Saybrook, CT: Globe Pequot, 1994.
Kricher, John C., and Gordon Morrison. *Ecology of Eastern Forests.* Peterson Field Guides. Boston: Houghton Mifflin, 1988.
McCormick, Jack. *The Life of the Forest.* New York: McGraw-Hill, 1967.
McLaughlin, Molly. *Earthworms, Dirt, and Rotten Leaves: An Exploration in Ecology.* New York: Atheneum, 1986.

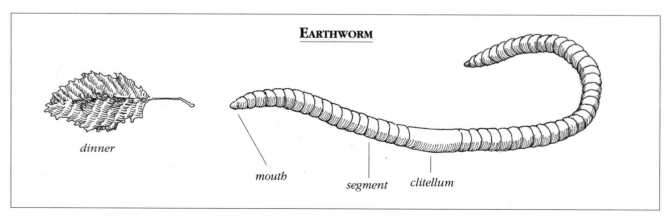

EARTHWORM

dinner

mouth

segment

clitellum

Forest Floor

FOCUS: The forest floor is a habitat teeming with life, much of which is involved in the decomposition of leaf litter and the recycling of nutrients.

OPENING QUESTION: *What lives on the forest floor?*

PUPPET SHOW

Objective: To learn about some of the inhabitants of the forest floor.

Perform, or have the children perform, the puppet show. Afterward, review the different forest floor inhabitants that Ellie Eft meets. What important jobs do the Fungus and Earthworm have?

Materials:
• puppets
• script
• props

LIE DOWN AND LOOK

Objective: To experience what life is like on the forest floor.

After entering the woods, have the children lie down on their backs. Ask questions such as:

How far away are the treetops?

Do you see shapes in the spaces between the leaves?

Are the leaves closest to you moving?

Are the leaves farthest up moving?

Is the ground under you soft or hard?

Have the children turn over onto their stomachs.

What is covering the forest floor?

Dig a "nose hole" and sniff. Is the smell familiar?

Is it wetter or drier, warmer or colder under the leaf litter?

What's there?

Have children poke around a bit with their fingers to see the decayed leaves, insects, and roots.

Materials:
• plastic bags to lie on if the ground is very wet

DIGGING DEEPER

Objective: To explore the layers of the leaf litter and to see decomposition in action.

Divide the children into small groups or pairs. Give each a loop of yarn. Tell them they will be investigating the top few inches of the forest floor. Have them place the yarn on the forest floor and very carefully begin taking off the layers within the loop. Have groups make a record of their findings by placing samples of each layer on the cardboard, and then drawing or making notes on any creatures seen. Where is decomposition furthest along? Did they notice any white root-like threads? What animals did they find in the leaf litter? Replace the leaf litter.

Materials:
• 20-24" strands of yarn, looped, one for each group
• pencils
• clipboards
• sheets of cardboard
• (optional) insect field guides

FOREST FORAY

Objective: To notice different features of the forest environment.

Outside, have the children work in small groups to find the items listed below (or assign each team a few of the tasks) and record their examples.

> **FOREST FORAY CARD**
> Find:
> Something soft and squishy
> Something rough
> Something smooth and slippery
> A hole that might be a home for an animal
> The spot in the forest that gets the most sunlight
> The spot in the forest that gets the least sunlight
> The driest place
> The wettest place
> Leaves in four different stages of decay
> The oldest thing in the forest
> The youngest thing in the forest
> Three pieces of evidence that insects have been around
> Something that doesn't belong in the forest. Remove it if it's trash.

Afterward, gather the children in a circle to share findings.

Materials:
- Forest Foray cards
- clipboards
- pencils
- (optional) trash bags

Springtails are tiny.

MEET AN EARTHWORM

Objective: To examine the anatomy of earthworms.

Give each pair of children an earthworm on a paper plate covered with a moist paper towel and a hand lens. Allow some time for the children to observe the worm and ask questions. Which end is the front of the worm? How can they tell? Does the worm have a top and a bottom side? Encourage the children to pick up the worm. Can they feel the bristles (setae)? Can they find the band (clitellum)? Using a diagram of worm anatomy, see if the children can identify the parts of the worm. Give the children some worm castings to examine. In what ways do worms help the fields and forests? (fertilize, decompose, aerate, feed other animals, mix soil layers)

Materials:
- earthworms, one for every two children
- hand lenses
- paper plates or shoe boxes
- paper towels
- water mister (keep worms moist, not wet)
- worm anatomy diagram
- worm castings

WORM PREFERENCES

Objective: To test earthworm preferences for different habitat conditions.

Allow children to work in pairs or small groups to perform one or more of the following investigations. Be sure to leave 5-10 minutes for the worms to respond to the choices. Have the children record their results.

Light or Dark? Have the children place a moist paper towel on a paper plate and make a low cover with a piece of paper or cardboard over one half of the plate. Place a worm so that it is between both covered and uncovered halves. Put the plate under a bright light. What does the worm do?

Wet or Dry? Have the children place a moist paper towel next to a dry paper towel on a paper plate, leaving about 1/4" of space between the two towels. Place a worm between the towels, so that it just touches both paper towels. Which way does the worm go?

Materials:
- worms
- paper plates
- cardboard or construction paper
- clamp or gooseneck lamps
- paper towels
- water
- maple leaves and pine needles

Forest Type? Have the children cover a paper plate with a moist paper towel then place a handful of decomposing maple leaves on one half of the plate and pine needles on the other half, leaving 1/4" between the piles. Does the worm seem to prefer one forest type?

Afterward, carefully return the worms to a soil-filled container. Ask the children to share the results of their investigations. What do these findings tell us about the habitat needs of worms?

MYSTERY BAGS

Objective: To identify a variety of forest floor components through the sense of touch.

Put several different objects commonly found on the forest floor into separate bags, and pass the bags, one at a time, halfway around a small circle of children. Each child in that half of the circle feels the object and gives an adjective describing it. The other half of the circle tries to guess the bag's contents. The last child gets to reveal the object. Repeat so that everyone gets the chance to feel and to guess. You may want to end by hiding a candy jelly worm in the mystery bag – then give each child one as a special snack.

Materials:
- cloth bags
- forest floor objects such as wood, nut, acorn, moss, leaves, bark, stick, pine needles, rock, fungus, bone, cone, snail shell
- (optional) candy jelly worms

EXTENSIONS

Elf Houses: While exploring a forest floor, have the children construct little houses out of forest floor materials (rocks, sticks, shed bark, fallen leaves) for an imaginary, clothespin-sized forest dweller. Children can go on a house tour to visit each other's ELF houses.

Make a Worm Habitat: Create a worm jar in a clear container (e.g., a large glass jar, plastic peanut butter jar, or a soda bottle with the neck cut off). At the bottom, place a layer of small stones. On top of that, place a layer of topsoil, then food for the worms – partly decayed leaves, a little finely chopped vegetable waste, a tablespoon of oatmeal or cornmeal. Place a layer of damp, fallen leaves on the top. Place the worms in the jar. Make daily or weekly observations of worm activity.

Give the worms a variety of food, a tablespoon at a time, when needed. The worm jar should be covered with a cloth and kept dark so that it doesn't dry out, but can still get air. Release the worms before winter or keep them until spring.

Bring the Forest to the Classroom: Use dishpans or buckets to bring sections of forest floor into the classroom. Carefully examine the layers of decomposition, and put any creatures found in jars or bug boxes. Examine with hand lenses before releasing.

Forest Fantasy: Have the children write and illustrate a story about the life of a (real or imaginary) tiny forest creature.

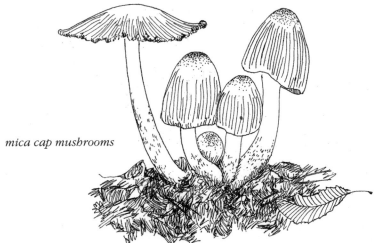

mica cap mushrooms

Forest Floor

Characters: Ellie Eft, Papa Newt, Dead Leaf, Wendy Worm, Freddy Fungus (mushroom puppet with white threads attached at the base), Mildred Millipede, Sammy Shrew

Props: maracas or a jar of beads to shake

Ellie Eft

Papa Newt, Papa Newt, look at me! My skin has turned orange. What's happening?

Papa Newt

Oh, my little girl, Ellie Eft, is growing up. Why, it's time for you to head off to the forest floor for an adventure of your own.

Eft

But I like living here in the pond.

Newt

Oh, don't worry. You'll be back in a few years when you're all grown up.

Eft

OK, Papa Newt! So, where is this forest and how do I get there?

Newt

Well, you'll have to cross this field, and it can get hot, dry, and windy by midday. You'd better leave now while it's still cool.

Eft

Goodbye, Papa! I'll be back when I'm a big newt like you.

Newt

Goodbye, Ellie, and good luck! *[both exit]*

Eft

[Eft re-enters; Dead Leaf appears mid-stage]
Ahhhh! At last. It sure was hot out in that field. Here the trees block the sun, and the forest floor feels cool, moist, and comfy. I can even burrow under this rotten old leaf for safety.

Dead Leaf

Did you call me a rotten old leaf?!

Eft

Oh, excuse me, I didn't mean to…

Leaf

Well, I **am** a rotten old leaf and proud of it, too! Why if we leaves didn't rot, we'd pile up so high that you couldn't see the forest for the leaves! Thanks to the Decomposers, that's not going to happen.

Eft

Decomposers? Who are they? Do they write songs?

Leaf

No, no, no! Composers write songs. Decomposers break us leaves and things down into smaller and smaller bits, until we're nothing but nutrients in the soil. Then plants use us to make new leaves.

Eft

So where are these Decomposers? I don't see anyone else around.

Leaf

Why, they're everywhere! Some are too small to see, like bacteria, but others are so big they can make a whole leaf disappear. Oh! The ground is shaking. Company's coming for dinner. So long. *[exits; Worm enters]*

Wendy Worm

MmmmMmm. That sure was a tasty morsel.

Eft

Why, Wendy Worm, you ate the leaf!

Worm

Sure did. Ground it right up in my gizzard. It's well on its way to becoming soil. A simple matter of in one end and out the other.

Eft

Well, I know just what you mean. You look like a great meal to **me!**

Worm

I would make a good meal, but first you have to catch me, nya, nya-nya, nya, nya!
[Worm dives behind stage]

Eft

That's easy. That worm went down right here. *[exits in same spot; returns draped in white threads]* Oh! I lost him! Got tangled up in these white roots. Why, they're everywhere down there.

Freddy Fungus

[enters, its threads draped on Eft] Those are me! I mean, I am they. I mean, I'm Freddy the Fungus and those are my fungal threads. I'm very attached to them!

Eft

Oh, sorry. I didn't know you mushrooms had roots.

Fungus

They aren't really roots. They help me break down my food, so I guess they're a bit like stomachs.

Eft

[sounding worried] Food? Um, exactly what sort of food? Nnnnot little orange efts, I hope?

Fungus

No, not efts – just dead leaves, roots, and bark. Now if you'll let go of me, I'll just thread my way right over to that rotten twig. MMMmmm. See you later, Ellie. *[exits]*

Eft

Goodbye, Freddy! Gee, I'd like to find something to eat, too. There seem to be lots of insects here on the forest floor. *[shake maracas to a 4/4 beat]* Maybe that's one now.

Mildred Millipede

[sound repeats, then Millipede appears] Hup, two, three, four, five, six, seven, eight, nine, ten, eleven, twelve. Company, halt!

Eft

You're no insect with all those legs! Who are you?

Millipede

We're Mildred Millipede on the march. We're the leaf patrol, on the lookout for dead leaves under attack.

Eft

The leaves are under attack?

Millipede

Of course. They're being attacked by bacteria and fungi. Then, once they're broken down a bit, we eat them. Now, if you'll excuse us, we have work to do. Forward, march! Hup, two, three, four… *[maracas sound as Millipede exits]*

Eft

I guess I'll stay out of her way – wouldn't want to impede a millipede! Hey, the ground is shaking again – wonder if it's another worm. I'll get it this time. *[Shrew appears]* Hey, you're not a worm!

Sammy Shrew

A shrewd observation.

Eft

What a long nose you have! And what short, velvety fur, and beady eyes, and sharp teeth…

Shrew

All the better to bore tunnels with, slip through easily, spy my prey, and then bite it!

Eft

Yikes! I know who you are. You're a shrew! *[sobbing]* A forest floor predator if I ever saw one. I bet you eat little creatures like me.

Shrew

Yuck! Not this time. I just ate. Besides, that orange skin of yours tells me you're nasty tasting. I'm not hungry enough to want to eat you. But maybe I'll catch you later. *[exits]*

Eft

Phew, that was a close call. Life on the forest floor is quite the adventure! You never know what's around the next tree or under the next leaf. But I'm off to find out. *[exits]*

Rotting Logs

TEMPORARY HOMES ON THE FOREST FLOOR

Trees, both living and dead, play important roles in the forest ecosystem. A rotting log provides an ever-changing habitat for creatures big, small, and microscopic. And as the log decays, it helps to maintain forest health by releasing essential nutrients back to the soil for the next generation of plants.

A rotting log passes through a series of stages as it progresses from tree to soil. During each stage, the log hosts a particular community of plants, animals, and other organisms that consume the log or each other. In so doing, they gradually change the structure of the log, creating a habitat no longer suitable for themselves, but increasingly appropriate for another group of inhabitants. As some inhabitants move out, others move in, creating a continuous parade of temporary log dwellers. Those leaving the log seek a newer, less decomposed log to inhabit, while those moving in come from a log that is now too rotten to meet their habitat needs.

Each rotting log resident feeds off the resources of this special habitat while also contributing to the larger forest ecosystem. **Decomposers** such as **fungi** and bacteria break the wood down into chemical compounds that become available for other organisms. **Scavengers** like millipedes and earthworms eat plant and animal remains and become **prey** for other creatures. Small **predators**, such as the daddy longlegs, hunt and eat insects, spiders, and snails while large predators, like the pileated woodpecker, drill oblong holes in search of carpenter ants. Once the log is well decomposed, plants make an entrance, creating new food sources for **herbivores** and providing shelter for small animals.

This succession of living things in decaying wood is specific to each kind of tree. Even the progression of organisms in different trees of the same species can vary. External conditions, including moisture, temperature, and climate, help determine who lives in the log and the rate at which it decomposes.

A healthy, living tree is well protected from decay. Its wood is composed mainly of cellulose, a tough carbohydrate that is indigestible to most organisms, and its waterproof bark contains a natural fungicide that resists rot. But disease, weather, fire, old age, and animal use can weaken the tree and eventually lead to its death.

Decomposition begins while a tree is still standing. Bark beetles burrow under the bark to lay their eggs, creating openings in the tree's armor. Their tunnels allow moisture, air, and the **spores** of fungi to enter the tree. Tiny strands of the fungi, called **hyphae,** penetrate the wood cells and live off the tree's stored starches and sugars. Some fungi produce chemicals that actually dissolve the cellulose. Mushrooms and other fungal fruiting bodies grow on the outside of the tree, signaling its internal destruction. A whole host of wood-boring beetles may also be hard at work excavating the inside of the tree. **Frass** is the

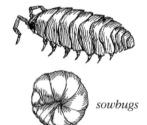

sowbugs

sawdust-like material they leave behind during their work. Dark-colored frass was passed through their digestive tract while light-colored wood chips were merely chewed and deposited in and around their tunnels. Carpenter ants may move in at this stage, excavating elaborate galleries within the tree.

A standing dead tree, or snag, is also attractive to many larger, more noticeable forest creatures. Woodpeckers come in search of insect meals, and animals such as flickers, raccoons, and flying squirrels take up residence in the tree's cavities. The fisher, a large member of the weasel family, often tears apart snags looking for squirrels. This animal activity creates more openings for moisture and fungi to enter, weakening the tree until eventually it falls.

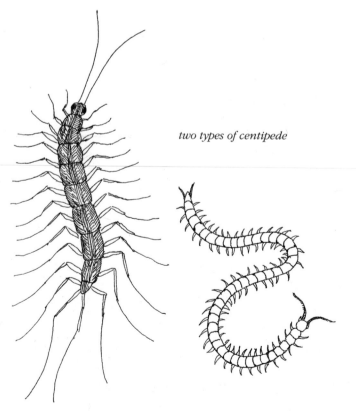

two types of centipede

Once on the ground, the fallen log accumulates moisture, and the decomposition process accelerates, as both fungi and bacteria work best in moist conditions. More mush- rooms appear on the log's

wireworm (click beetle larva)

surface, indicating increasing decay. Mushrooms provide food for many bacteria, insect larvae and adults, slugs, and snails. Most of the bark is now gone, and the outer wood becomes "punky," or soft and moist. Fungi-eating beetles move in, along with such scavengers as sowbugs (also called pillbugs or roly-polies). Predators like salamanders and centipedes may be found inside the log or between the log and the damp ground on which it rests. Chipmunks establish burrows under the log, and snakes hibernate within the log during the winter. Spiders take up residence in the higher, drier, more solid portions of the log. And the log often becomes carpeted with lush mosses.

As the log's wood slowly turns to **humus**, ferns and other plants begin to take root on its surface. Even the seedlings of certain trees, such as hemlock and yellow birch, gain a foothold in cracks and openings and thrive on the "bones" of the dead tree. These plants thrive on the log's moisture and the nutrients that are now becom- ing available. The physical action of their growing roots further breaks down the log's structure.

Gradually the outer shell collapses and becomes covered with fallen leaves and other plant debris. Most of the wood-boring insects leave for younger rotting

logs, and decay is carried on by earthworms, bacteria, and fungi. Eventually, it is difficult to decide whether the mound is wood or forest floor. The newly formed soil, rich in organic matter and nutrients, sprouts many seedlings.

The death of a tree is hardly the end of the story. In fact, it signals a new beginning, for as it decays, the tree becomes a thriving habitat for successive waves of creatures, each furthering the process of decomposition. Eventually the tree becomes nutrient-rich soil, providing a nursery bed for the seedlings of future trees.

Suggested References:

Hughey, Pat. *Scavengers and Decomposers: The Cleanup Crew.* New York: Atheneum, 1984.

Jackson, James. *The Biography of a Tree.* Middle Village, New York: Jonathan David, 1978.

McCormick, Jack. *The Life of the Forest.* New York: McGraw-Hill, 1966.

Schwartz, George I., and Bernice S. Schwartz. *Life in a Log.* New York: Natural History Press, 1971.

Tresselt, Alvin. *The Dead Tree.* New York: Parent's Magazine Press, 1972.

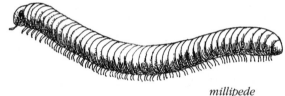

millipede

a rotting log with a click beetle, daddy longlegs, and snail

Rotting Logs

FOCUS: A rotting log serves as a habitat for many plants and animals, which vary according to the log's stage of decomposition.

OPENING QUESTION: *What might live in a rotting log?*

TIME MACHINE

Objective: To observe the stages of a log's decomposition.

As a costumed professor, explain to the children that you have invented a remarkable time machine that will show what happens to a tree when it dies, but you need their help for the machine to work right. Divide the children into four groups: one to whistle like the wind, one to tap their fingers like the rain, one to make chomping noises like chewing insects, and one to chant the passing seasons: "spring, summer, fall, winter." Open the flap of the time machine to reveal a recently cut piece of wood, then close it. Progress through the first four years, moving the "seasons" dial as the groups make their noises. Then open the time machine (in which your hidden operator has switched logs) to see a log in initial stages of decay. Remove log from machine and ask children what they notice about the log. (some decay, bark gone, insect tunnels) Repeat this process for the passage of four more years, after which the secretly replaced log will look punky and soft, and then again for the last four years (total of twelve years), when the log will be chunks of soft, rotten wood and newly formed soil. (You may end with a seedling growing in the rotting wood.) Afterward, inspect and compare the four stages.

PUPPET SHOW

Objective: To learn that a rotting log can provide a home for many different animals.

Perform, or have the children perform, the puppet show. Discuss what will happen to the log and which character will be able to live in it the longest.

LOG LOOK

Objective: To explore a rotting log.

In small groups outside, have children gather around a rotting log and close their eyes. (If activity must be done indoors, place rotting logs, one for every four children, on plastic tarps or newspapers.)

Direct children to:
>   **Listen** — As they tap the log, does it sound hollow or solid, wet or dry?
>   **Smell** — Does it smell wet or dry? Is it like anything they've smelled before?
>   **Feel** — Does it feel hard or soft, wet or dry, rough or smooth?

Now have the children open their eyes and predict what the inside of the log will be like and what they might find there. Then have the children explore the log by gradually pulling it apart. Collect any small creatures in white

Materials:
- four small logs in various stages of decay, from freshly cut to fully rotted
- time machine – large cardboard box with flaps that open in the front and back, painted with a seasons clock that has a movable dial, plus other numbers, gadgets, and dials
- (optional) professor's costume

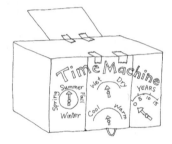

Materials:
- puppets
- script
- props

Materials:
- rotting logs, one for every four children, depending on size of logs
- hand lenses
- paper and pencils
- flashlights
- bug boxes or white dishpans
- (optional) newspaper or plastic tarps, and field guides

dishpans or bug boxes for closer examination. Use hand lenses; flashlights are helpful for seeing into crevices. Ask groups to keep track of their discoveries and questions. Later, each group can share its most interesting findings. Return the log pieces and their inhabitants to the places where they were found. Older children may want to use field guides to identify their insects and other findings.

BARK BEETLE INVESTIGATION

Objective: To examine bark beetle tunnels and learn about the insect's life cycle.

Tell the children that they will be examining evidence of an organism that helps start the process of decay in a dead tree. DON'T tell them anything more about the bark beetle at this time; instead, tell the children that they will be trying to solve a mystery. Explain brainstorming rules: all ideas count, no ideas are wrong, there's no debating now — the purpose is to generate as many ideas as possible and evaluate them later.

Divide the children into small groups, and give each group samples of bark beetle engravings to study. Ask the groups to think up theories about how the patterns were created. Why and how were the tunnels made? Why do the branch tunnels get wider as they get farther from the center? Afterward, groups may share their theories and key observations.

Now, read the Engraver Beetle Story aloud and help the children see the evidence in their engraving samples. Discuss how the work of bark beetles helps the process of decomposition by making places for water and fungi to get into the log.

Materials:
- bark beetle engraving samples
- hand lenses
- pencil and paper
- Engraver Beetle Story

SHARING CIRCLE

Objective: To share feelings and observations about a rotting log.

Sit in a large circle. Ask each child to complete the following sentence:
"One thing that surprised me about rotting logs was _____."

EXTENSIONS

Bark Beetle Medallion: Children can make a medallion by flattening a piece of polymer clay into a disk and pressing it onto some bark beetle tunnels to make an imprint. Use a toothpick to make a hole for necklace string and then bake the disks on a cookie sheet. (Note: Clay and play dough do not work because they stick to the bark.)

Pill Bugs: Help the children put about six pill bugs in a large jar darkened with a removable cylinder of black construction paper. Add several slices of apple and raw potato, which provide food and moisture. Ask the children to watch and record the bugs' activities at different times of the day. Do they seem to prefer light or dark? How much do they eat? Release the pill bugs back to their home within a week.

Terrarium: Set up a terrarium with some soil and a rotting log. Keep it moist, and observe any changes over time. Fungal threads may grow, mushrooms may appear, insects or spiders emerge, eggs hatch, slime molds appear and disappear.

Word Games: Make up a crossword puzzle or a word search puzzle using rotting log plants and animals.

Rotting Logs

Bark beetles are tiny insects that eat the wood of trees. One kind is called the engraver beetle. This beetle eats the inner bark of the tree, between the bark and the wood. The beetles themselves are so tiny, no bigger than a peppercorn, that most people never notice them. But the wonderful mazes of tunnels they engrave in the wood tell us when they've been around. *See if you can find one of the main tunnels (longer and wider than the others) and the branch tunnels, which start small and get wider at their ends.*

Early in the summer, adult engraver beetles fly through the forest, looking for just the right tree. Each kind of engraver beetle likes a particular kind of tree. When they find the right kind, they chew a hole straight through the bark and create a small chamber in the wood. *Look for a short wide branch on one of the main tunnels. It may be at one end or near the middle.* This is the mating chamber. After mating, the male leaves and the female chews out a main tunnel, making notches all along it. *Look closely to find the tiny notches.* One egg is laid in each notch. In some species there will be two females that tunnel in opposite directions from the same mating chamber.

Later in the summer, the eggs hatch into beetle larvae. They are hungry, so they start to chew through the wood, away from the notches in the main tunnel. *Notice that the ends of the tunnels are wider than the beginnings, for as the larvae grow, their tunnels get larger, too. Notice how the tunnels rarely cross.* When one tunnel makes a bend, the neighboring tunnels do, too. Living in the complete darkness of their solitary tunnels, it is surprising that they don't bump into each other. It is thought that the beetle larvae can feel the vibrations as neighboring larvae chew in their tunnels, and this helps them to stay away from each other.

When winter comes, the larvae stop tunneling and rest through the cold months. In the spring, they form pupae and change into adult engraver beetles with wings. *Look for the chambers where they spent the winter and turned into pupae.* Finally ready to leave their tunnels, the engraver beetles chew their way straight out through the bark. Then they fly away to find a tree where they can start their own families.

The next time you are in the woods, look closely under the loose bark of dead trees for the beautiful carvings that tell you engraver beetles have been at work.

bark beetle magnified

actual size

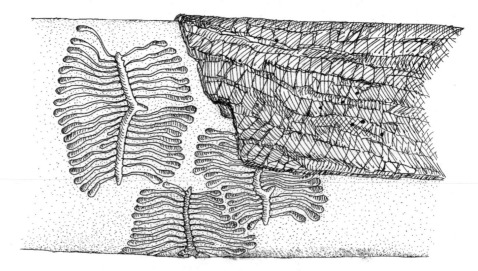

*bark beetle tracks on ash branch,
and entrance and exit holes through bark*

Rotting Logs

Characters: Rocky Raccoon, Benjy Bear, Charlotte Spider, Wendy Worm
Props: A real or constructed rotting log; a card with the word "directions" written on it and taped to a stick

Rocky Raccoon

Benjy Bear, I've been looking for you! As secretary of the forest, you must have a list of all the available homes around here.

Benjy Bear

I sure do. Are you needing a new home, Rocky Raccoon?

Raccoon

Yes, I am. Why, my old place is so leaky that I've been soaked for the last three days. Now, I don't need anything too fancy – no wall-to-wall moss or anything – just a roomy hole that's fairly dry.

Bear

I have just the place for you, Rocky. It has a soft, comfortable floor and thick, well-insulated walls. It'll keep you warm in the winter and cool in the summer. Just follow these simple directions, and you'll have no trouble finding it. *[hands him the "directions"]*

Raccoon

Thanks a lot, Benjy. *[walks off; Spider appears]*

Bear

Why, hello, Charlotte Spider. What brings you out this way?

Charlotte Spider

Well, Benjy, I'm in need of a home. Do you have anything for me?

Bear

Just what kind of a place are you looking for, Charlotte?

Spider

Well, I just love old homes. The older the better, with lots of little cracks and crevices for me to crawl into and hide. And I'll need a safe place for my egg sac and a spot where I'll be protected enough to spend the winter.

Bear

Why, I've got the perfect place for you, Charlotte. This home will help protect you from predators and be a great place for you to find food. These directions will show you how to get there. *[hands her the same "directions"]*

Spider

Thanks, Benjy. I'll go find it right now. *[Charlotte exits; Worm appears]*

Wendy Worm

Benjy, can you help me? I'm having a terrible time finding a home.

Bear

Don't worry, Wendy Worm. Just leave it to me. Now, what kind of a home do you want?

Worm

We worms go for something cool, damp, and quiet. I'd like a soft place with lots of rotting things for me to eat.

Bear

Well, it just so happens I know of a place with nice, rich, damp soil and lots of rotting leaves to eat. Here are the directions to get there. *[hands her same "directions"]*

Worm

Thank you, Benjy. I knew I could count on you. *[both exit; log comes up; Raccoon enters]*

Raccoon

That rotting log over there must be the new home Benjy was talking about. *[walks over and inspects log]* It's nice and hollow down at this end, and the roof seems good and solid. It looks perfect! But I'd better check it out. I'll just take a little snooze inside to see how well it works. *[Raccoon exits behind log; Spider enters, humming the song "Itsy Bitsy Spider"]*

Spider

Hey, this looks like the place. This rotting log must be my new home. *[Worm enters from opposite side]*

Worm

I'm here. This must be the place Benjy was talking about. Oh, hi, Charlotte.

Spider

Hi, Wendy. Welcome to my new home! Can I show you around?

Worm

Your new home? Benjy told me it would be **my** new home!

Spider

Well, there must be some mistake. We can't both live in the same place! Why, spiders need lots of small cracks to hide in and places to catch prey.

Worm

And I'm a worm. I'd be a prisoner inside solid walls. Dry wood and bark are too rough on my sensitive skin. I need rotting leaves to eat and moist dirt to make it easy to move around.

Spider

Then, Benjy must have made a mistake.

Worm

Well, I'm going to go find him. He must have given one of us the wrong directions.
[Worm exits; Raccoon appears from behind the log]

Raccoon

Hey, what's all the fuss about. Can't you see I'm trying to take a nap in my new home?

Spider

Your new home? When did you move in? How did you find this log?

Raccoon

Benjy Bear told me about it. He said it would be better than my old leaky log, and he was right. It's perfect.

Spider

Well, he sent me here too, and he also told Wendy Worm this log would be perfect for her! He must have been very confused!

Raccoon

I don't know. That Benjy is one smart bear.

Spider

Do you think he thought we could all find homes in the same log?

Raccoon

Well, there's a nice big hollow space at this end that's just perfect for me.

Spider

And the middle of the log has great places for me to crawl around in and plenty of juicy insects to eat, too.

Raccoon

Unfortunately, the other end of this log is so rotten it's almost turned into soil. That wouldn't make a very good home for anyone.

Spider

Oh, but it would! It'd be perfect for Wendy Worm. It's cool and damp, and there are lots of rotting leaves for her to eat.

Raccoon

Well then, I guess this rotting log can be home sweet home for all of us.

Spider

I'd better go find Wendy and tell her she can live here, too.

Raccoon

Good idea. But, before you go, there's just one more thing.

Spider

Yes, Rocky?

Raccoon

Welcome home!

Spider

Welcome home, Rocky. *[both exit]*

Animals in Winter

MANY WAYS TO COPE WITH WINTER

Geese honking overhead and chipmunks darting underground with cheeks full of seeds signal winter's inevitable arrival. When the season of cold, ice, and snow sets in, animals find it more difficult to get around, find food and water, and stay warm. Nature has prepared these creatures to meet the challenges of winter in different ways. They may migrate, sometimes traveling thousands of miles to reach a winter habitat; they may become dormant or hibernate, seeking safe shelter in which to spend the coldest parts of the season; or they may remain active, foraging as best they can.

Migration is a seasonal, mass movement of animals to find adequate food supplies or to reproduce. The most obvious migrators are the numerous birds that fly south during the fall. Warblers and swallows leave to find active insects, and water birds seek open water with accessible food. Some ocean fish also migrate to warmer southern waters. Although few land mammals migrate, some do make impressive yearly treks. Certain caribou herds travel hundreds of miles from their northern calving grounds to the southern part of their ranges. Many whales migrate, as do some bat species. Perhaps the most amazing migrator is an insect, the monarch butterfly, which may fly thousands of miles to reach a particular winter roost in southern California or Mexico.

Some creatures need only move a short distance to find more favorable conditions. Certain insects and spiders that have summered in bushes and trees descend to weeds and grass roots for the winter. Others snuggle down into cracks and crevices in trees, rocks, or buildings. Ground-dwelling insects, snails, earthworms, and salamanders, as well as tree and wood frogs, hide under leaf litter and in rotten logs or descend into the ground. Many land turtles and toads burrow into the soil below the **frost line**. Certain aquatic creatures – such as fish, turtles, insects, and salamanders – move to deeper water or creep down into the mud. Snakes may huddle with other snakes in shelters below the ground or among rocks.

Once sheltered for winter, cold-blooded animals that cannot maintain an adequate body temperature enter an inactive state. Their body processes, including breathing, slow down substantially. Many of these creatures (such as toads and some insects) also manufacture a special antifreeze that helps prevent the animal's tissues from rupturing.

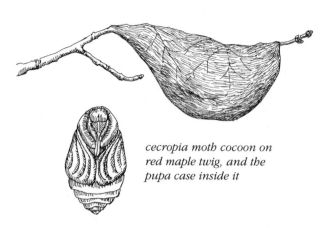

cecropia moth cocoon on red maple twig, and the pupa case inside it

Insects overwinter as eggs, **larvae**, **nymphs**, **pupae**, or adults, depending on the species. For many insects, the egg or pupal stage is most suited to winter since each is immobile, with a protective covering to withstand the cold.

Because mammals and birds are warm-blooded, they maintain a particular body temperature regardless of changes in the weather. The greatest winter challenge for these animals is to find enough to eat. Some mammals that feed on green plants, grains, or insects are not able to find sufficient food in winter. Often these animals eat heavily in the summer to store extra layers of fat, which serve as nourishment and insulation whether they become dormant, hibernate, or remain active through the winter.

Certain mammals enter periods of winter **dormancy**, during which their body processes slow way down and they go to "sleep." Body temperature drops slightly, causing a corresponding drop in metabolism, which conserves energy. During mild spells, dormant animals may wake up and venture out in search of food and

eastern chipmunk

water, or they may eat stored food they gathered during the fall. Skunks, raccoons, and chipmunks are common dormant mammals.

Although hibernating mammals like the woodchuck, little brown bat, and jumping mouse look merely asleep, they actually enter a specialized state that is different from dormancy. **Hibernation** does not happen suddenly. Periods of drowsiness and wakefulness alternate until profound lethargy takes over. A study of woodchucks showed a variation of three days to a month between full activity and hibernation. During hibernation, the animal is motionless, often for days at a time, and it breathes very slowly, as little as once per minute. Its body temperature plummets and its heart beats slowly and irregularly. A woodchuck's summer temperature is around 96.8°F while its hibernating temperature may be as low as 37.4°F. Its heart averages four to five beats per minute during hibernation as opposed to 160 beats per minute when active.

Biologists disagree over whether the black bear becomes dormant or truly hibernates. Whatever its winter state, the bear is a miracle of survival during this most severe season. It does not eat, drink, urinate, or defecate throughout its winter sleep. Although its temperature drops only about 10°, its metabolic rate decreases by half. During this winter fast, the female carries out a pregnancy, gives birth, and nurses her young, using only the energy she's stored as fat.

Another way some animals cope with winter is by remaining active. This group includes **predators** – such as weasels, foxes, bobcats, and coyotes – and **prey** species that eat bark, twigs, nuts, or seeds – like porcupines, deer, snowshoe hares, and squirrels. Birds that can find adequate food also may remain in northern areas, such as owls that eat small mammals, woodpeckers that find a year-round supply of insects hidden within

trees, and a variety of seed-eating birds. Active cold-blooded animals overwinter in places where they can keep up their body temperatures and find food. Fish may be seen swimming under the ice in deeper lakes. Honeybees remain in their hive, feeding on stored flower pollen and honey. By eating a high-energy diet and constantly vibrating their wings, honeybees can keep the whole hive as warm as 95°F in the middle of winter.

Just as dormant and hibernating animals seek special winter shelters, many active animals move to winter habitats to ensure adequate food, water, and protection. For example, during severe weather in northern climates, white-tailed deer gather under medium to large short-needled evergreens in "**yards**" where the thick tree cover provides protection from winds, and the shallower snow makes movement easier. In the fall, moose move from wetlands to forestlands in search of winter browse. Squirrels prefer cavities in trees rather than leaf nests in branches for their winter shelters. Birds may hide in holes or seek evergreen roosts as protection from bitter winds, icy temperatures, and hungry predators.

Winter tests the survival abilities of northern animals. A keen observer sees busy fall preparations and winter evidence of survival on the edge. This harsh season clearly illustrates how nature copes with adversity.

Suggested References:

Buck, Margaret W. *Where They Go in Winter*. New York: Abingdon Press, 1968.

Halfpenny, James C., and Roy Douglas Ozanne. *Winter: An Ecological Handbook*. Boulder: Johnson Books, 1989.

Hunter, M. L., J. Albright, and J. Arbuckle, eds. *The Amphibians and Reptiles of Maine*. Maine Agricultural Experiment Station Bulletin 838, 1992.

Marchand, Peter. *Life in the Cold*. Hanover, NH: University Press of New England, 1987.

Rezendes, Paul. *Tracking and the Art of Seeing*. 2nd edition. New York: Harper Collins, 1999.

Stokes, Donald. *Guide to Nature in Winter*. Boston: Little, Brown, 1976.

chickadee

Animals in Winter

FOCUS: Animals have differing habitat requirements during the winter, depending on their levels of activity and the availability of food.

OPENING QUESTION: *Why is winter hard for animals, and how do they cope?*

PUPPET SHOW

Objective: To identify some common ways animals cope with winter.

Perform, or have the children perform, the puppet show. Afterward, use the puppet characters to discuss various winter survival strategies. How do people prepare for and cope with the rigors of winter?

Materials:
• puppets
• props
• script

WINTER TALES

Objective: To explore the winter survival strategies of a variety of animals.

Divide the class into small groups. Give each group a Winter Scenario card that identifies an animal and its winter survival strategy. Provide time for the groups to plan and practice skits that show what their animal does in winter. Sound effects and props may be used. After each skit, other children try to guess the identity of the animal and its winter survival method.

Materials:
• Winter Scenario cards

WINTER SCENARIOS

• **Monarch Butterflies:** We fly to Mexico, away from all the cold and snow. There we gather together in clusters and hang from trees, while we bask in the warm sun.
• **Canada Geese:** Listen for our honking in the fall and look up to see us flying in V-formation, winging our way south to winter near the seacoast.
• **Honeybees:** In winter we cluster around our queen inside the hive. Those of us on the outside of the cluster warm the hive by vibrating our wings at high speeds. Then we trade places with other bees in the center and eat the honey we have stored to feed us all winter.
• **Beavers:** We keep busy, felling trees and storing branches underwater near our lodge. Once the pond freezes over, we stay warm and dry inside our lodge, munching on these stored branches.
• **Woodchucks:** We eat our fill all summer long and then dig a cozy burrow in which to spend the winter. We curl up in a tight ball, tail over nose, and take it S-L-O-W, hibernating the winter away.
• **Wood Frogs:** We hop about in the leaf litter catching insects to eat. When it gets cold, we burrow under the leaf litter to spend the winter in the deep freeze, waiting for spring rains to thaw us out.
• **Ermine:** We bound across the snow and dive into tunnels to catch mice, a favorite winter meal.
• **Woodpeckers:** Finding our favorite meal is the same, fall, winter, spring, or summer. Listen for us tapping on tree trunks and dining on any insects that we find under the bark.
• **Snowshoe Hares:** The snows of winter don't keep us inside. In our new white coats we are hard to see except for our wiggly noses. With our big furry feet we can hop on top of the snow, nibbling on evergreens and buds.
• **Bears:** We're fond of sweets, especially honey fresh from the hive, but we'll eat almost anything as we lumber through the woods preparing for our winter rest. Snuggled warm in a cozy den we sleep soundly, living off our plentiful stored fat.

Animal Bingo

Materials:
- Animal Bingo cards
- Animal Bingo Clues
- edible Bingo chips
  (animal crackers or other snacks)

Objective: To identify wintering habitats for a variety of common animals.

Give each child a Bingo card and Bingo chips. Read Animal Bingo clues describing a variety of winter behaviors and winter habitats. Children share ideas and guess which animal each clue describes. When correct, they cover up the appropriate animal's picture with a Bingo chip. Play until someone gets Bingo, or until all the animals are covered up.

Afterward, review the variety of habitats mentioned in the clues including: hemlock groves, treetops, brush pile, rotting log/leaf litter, burrow under forest floor, tree stump, hive, sunny south, snow-covered field, tall pine tree, stream, mud under pond, cave, rocky hollow.

Animal Signs

Materials:
- Animal Signs cards
- pencils
- clipboards

Objective: To discover signs of animal activity outdoors.

Divide the children into small groups. Give each group an Animal Signs card plus a pencil for taking notes.

Animal Signs Card

Look for the following:

Tracks of at least 3 different animals.

Five potential food sources. Who might eat them?

Three signs of animals having eaten.

Homes or shelters for 3 different animals.

Stop, look, and listen – what other signs of animal activity do you notice?

Pretend you are an animal – find a good place to take shelter from a storm.

Winter Mural

Materials:
- large winter mural depicting a variety of winter habitats (holes in trees, underground tunnels, pond with lodges, rotting log, stone wall, or rocky outcrop)
- pictures of common local animals

Objective: To identify and review appropriate wintering habitats for a variety of common animals.

Display a large mural of a winter scene and hand out pictures of animals to pairs of children. One at a time, ask each pair to place their animal in an appropriate winter habitat, explaining why that particular location was chosen.

SHARING CIRCLE

Objective: To share new knowledge about what animals do in winter.

Have the children choose an animal they'd like to be and complete the sentence, "I'd like to be a _____ in winter because _____."

EXTENSIONS

Naturalist's Calendar: Keep a calendar of animals and animal sounds, tracks, and sign noticed by the children. As winter turns to spring, are more and/or different animals seen?

Favorite Animal: Have the children choose a favorite animal to research: how it spends the winter, what foods it eats, how it keeps warm, where it lives, when it has young. They could draw its picture and write a report or give an oral presentation, or write the facts on a sheet of paper and make everyone guess what animal it is.

Helping Winter Habitats: Invite a forester, naturalist, or game warden to talk about winter habitats. How can people help improve winter habitats for animals? How could the school area be a better place for animals in winter? Are there any things that the children could do as a class? Make plans to do so.

ANIMAL BINGO — CLUES *(See next page for sample Bingo card)*

White-tailed Deer: I nibble buds and twigs all winter long, and when it's really cold I "yard" up with my companions under the hemlock groves.

Gray Squirrel: In winter I get plenty of exercise hopping from branch to branch in my treetop home, or digging up acorns and hickory nuts that I buried in the forest floor. When I rest, I curl up in a snug hole in a tree or in a soft, leafy nest.

Snowshoe Hare: Winter finds me snug and warm under the tangled branches of young evergreens, and I'm camouflaged in my soft fur coat, so I can venture out to feed on twigs whenever I feel hungry.

Woolly Bear: In winter I lie under the bark of a rotting log or a warm blanket of leaf litter, curled up in a tight woolly ball.

Eastern Chipmunk: I fill my burrow up with a pile of seeds so I can sleep and snack all winter long in my home beneath the forest floor.

Striped Skunk: A rotting log or old tree stump is a good place for me to make my winter den, for I can sleep through the coldest days, snug and warm in my black and white pajamas.

Blue Jay: Winter winds may sway my home high in the treetops, but I have a good view of all that goes on below, and you can be sure I'll set up a noisy alarm if an enemy comes in sight.

Honeybee: In the fall, my nest mates and I make our home safe and warm for winter by chinking all the cracks and storing plenty of sweet golden food to eat.

Robin: My winter home is warm and sunny, but to get there I must fly up high, buffeted by the chilling winds, for mile upon weary mile.

Red Fox: Straight as an arrow my neat tracks cross a snow-covered field, but they may show signs of a scuffle where I stop to catch a tasty snack in the snow. When it's time for a rest, I may curl up in a sunny spot next to a tree or big rock.

Great Horned Owl: Winter or summer, spring or fall, I'm hard to see while nestled against the trunk of a pine tree high above the ground, but on a still winter night you may know me by my ghostly hooting call.

Brook Trout: A bubbling stream is the place for me, in summer or in winter, too. But when the water's nearly frozen, I may dive deeper to find flowing water or look for warmer springs feeding into my stream.

Green Frog: When pond and stream are frozen, a winter's day will find me buried deep beneath the mud, as cold as ice, just waiting for warmer days to thaw me out so I can swim and hop about.

Wood Turtle: I hibernate in winter, buried deep in the mud beneath the stream where I live in summer, safely resting within my armored shell.

Little Brown Bat: My home in winter is topsy turvy, but that's the way I like it as I hibernate hanging upside down in a crowded cave with family and friends packed close around me.

Garter Snake: I rarely make social calls, except in winter when I slither into rocky hollows with others of my kind where we hibernate through the cold winter months.

ANIMAL BINGO CARD

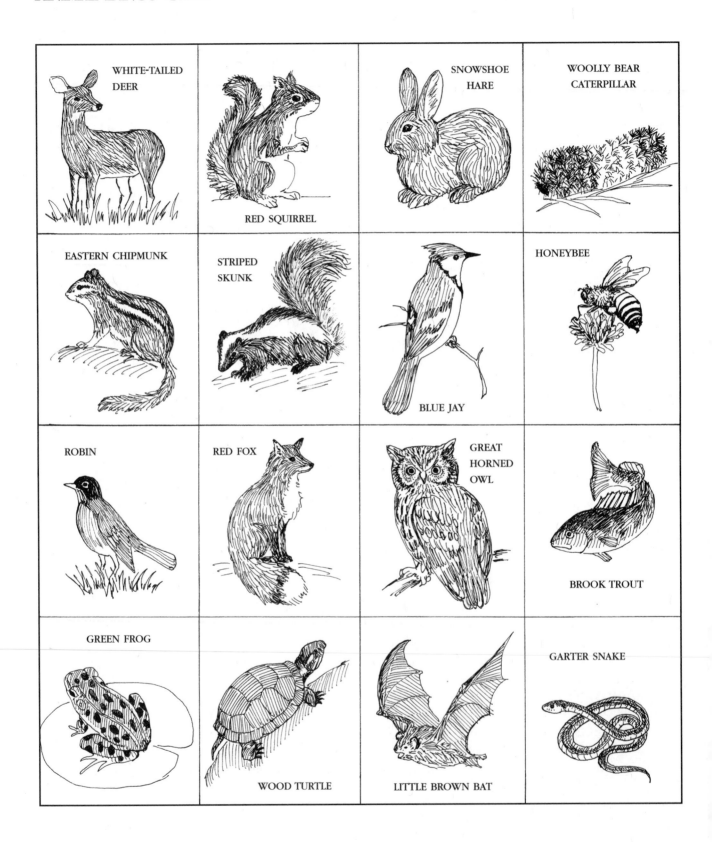

WHITE-TAILED DEER

RED SQUIRREL

SNOWSHOE HARE

WOOLLY BEAR CATERPILLAR

EASTERN CHIPMUNK

STRIPED SKUNK

BLUE JAY

HONEYBEE

ROBIN

RED FOX

GREAT HORNED OWL

BROOK TROUT

GREEN FROG

WOOD TURTLE

LITTLE BROWN BAT

GARTER SNAKE

Animals in Winter

Characters: Herbie Hare (with a white coat and a brown coat), Robby Robin, Woody Woodchuck, Betty the Mourning Cloak Butterfly, Rocky Raccoon, Charlie Chipmunk, Marsha Mouse

Props: cotton balls, white cloth, tiny suitcase

Herbie Hare

[in summer coat] Brrr, it's chilly this morning. I'll just keep hopping about. *[Robin appears carrying a suitcase]* Hi, Robby Robin. Where are you going with that suitcase?

Robby Robin

Well, Herbie, can't you see winter is coming? I'm migrating.

Hare

Migrating? Does that mean traveling?

Robin

Yes, it does. I'm flying south. Anyone with any sense migrates to a warmer place to spend the winter. Take my advice, Herbie, pack your bags before the snow comes. *[exits]*

Hare

Bye, Robby. *[Woodchuck appears]* Hi, Woody Woodchuck. Robby Robin just told me everyone's going south for the winter. Are you going, too?

Woody Woodchuck

Oh, not me, Herbie. Pass the berries, please. I stay right here. *[munches]* Pass me some leaves, thank you. *[more munching]* You see, *[munching]* I hibernate.

Hare

You hibernate? Does that mean you eat a lot?

Woodchuck

That's part of it. First I eat a whole lot and get nice and fat. Mind passing a little more mint please? Thank you. *[munching]* Then I get sleepy and I fall...*[snores]*

Hare

Woody! Woody! Wake up! Is that all hibernating is? Just eating and sleeping?

Woodchuck

No, no, it's much more than sleeping. My heartbeat slows way down, and so does my breathing. Why, even my temperature drops!

Hare

Yuck! Who wants to spend the winter as an ice cube!

Woodchuck

Well, I'm not completely frozen, not like Teddy Toad. He's got some special toad antifreeze that protects his body even if he freezes solid. But hibernating's all right by me. When the snow flies, I'll be curled up in my burrow. You won't see me until spring, so...I'll just say good *[yawns]*...g'night. *[snores; exits]*

Hare

I can see Woody's not going to be much fun this winter. I wonder if everyone hibernates or migrates. I guess anyone who can fly just heads south.

Betty Butterfly

[from offstage] Nope, that's not true!

Hare

Who said that? *[Butterfly enters]* Oh, it's you, Betty Butterfly. I didn't even see you hiding under that bark.

Butterfly

I'm not hiding. I've just found myself a good place to spend the winter.

Hare

Shouldn't you be migrating to someplace warmer?

Butterfly

Well, monarch butterflies migrate, but not us mourning cloaks! We stay right here. So do most other insects. When the days get shorter, our bodies slow down. We don't eat, and we hardly breathe. We'll get busy again when it warms up. Bye, Herbie. *[exits]*

Hare

Bye, Betty. See you in the spring...Gee, seems like winter's going to be awfully lonely around here. *[hops off]*
[White cloth appears, covering the stage; Hare re-enters, now in white coat]

Hare

Hey, what's this white stuff all over the place? *[hops around]* Look, I can make pictures with my feet just by hopping around. Oh goodie, here comes Rocky Raccoon. Hello, Rocky! What's all this white stuff? *[Raccoon appears, hunting around for something]* And what are you looking for?

Rocky Raccoon

Well, first of all, this white stuff is snow, my boy, and it means winter is upon us. And that is why I'm looking for the cozy nook I had last year. It was just over here, an old log with a snug hollow where snow can't get in.

Hare

Don't tell me – you're looking for a place to hibernate?

Raccoon

Heavens, no. I'm not a hibernator. I am what they call dormant. This means I sleep a lot, but I often wake up on a winter morn and stroll about. On colder days I stay home. Ah, here it is, the perfect nook! And so my friend, I'm off for a little nap. *[exits]*

Hare

Bye, Rocky. Sleep well.

Charlie Chipmunk

[pops up] Hi ya, Herbie. It's me, Charlie Chipmunk. I'm dormant too, but I'm a lot smarter than Rocky. When he gets hungry, he has to get up and go looking for lunch. Not me. I just fill up my hole with seeds and sleep right on top of them. Whenever I wake up hungry, I just roll over and grab a bite to eat. You oughta try it. Catch ya later. *[exits]*

Hare

Bye, Charlie, have a nice sleep. Maybe I should become dormant too and sleep through the winter. I'll try counting buds to get to sleep. One, two, three, four, five, six...this isn't working. I don't even want to sleep! Who cares if it's cold out? I just want to hop and skip and jump. *[Mouse pops up]*

Marsha Mouse

Well, then, join the party, Herbie, because I'm active all winter, too.

Hare

You are, Marsha Mouse?

Mouse

Yes, and so are vole and squirrel and chickadee and deer...

Hare

Hey, that's cool.

Mouse

And beaver and muskrat are too, though they're usually stuck under ice...

Hare

I'll have lots of company!

Mouse

But you have to watch out, because fox and coyote are active, too. And so are bobcat and fisher and otter and mink and weasel and, oooh...I'm scaring myself.

Hare

Well, I'm glad that I don't have to hibernate or go away, but I might have known that all those predators would hang around, too. Hey, I wonder if this new white coat I've been growing will help me to hide from them?

Mouse

Sure, Herbie, you look just like a snowball. A snowball with a twitchy nose!

Hare

A snowball. That gives me an idea – take that! *[throws cotton ball at her]*

Mouse

Missed me! *[throws cotton ball back]* Gotcha! *[they both throw more snowballs, including some at audience]*

Hare

Hey, this winter's gonna be fun after all! Bye, everyone! *[both exit]*

Snug in the Snow

SNOW IS A WELCOME BLANKET FOR MANY

A shelter of snow may not seem like the coziest place in which to spend a cold winter day, but to the animals that must survive northern winters, snow is a fact of life, and differences in snow cover can mean either survival or death. Snow serves as insulation, much like the fiberglass in the walls of a house or the feathers in a down vest.

When snow lands, the flakes face every which way, creating air spaces around them. These air spaces trap warmer air radiating up from the earth. The depth of the snowfall and its consistency determine how much air the snow can trap. Fresh snow insulates better than old snow, for once on the ground, the snow crystals break down or bond together and the air spaces between them become smaller. Eventually the snow becomes a latticework of ice and air spaces. As the snow ages, a region of brittle, loosely arranged crystals forms at its base, creating a layer in which small animals are able to move about with ease.

Beneath a thick cover of insulating snow is a stable environment, the **subnivean** layer, in which temperatures stay within the range of 20°F to 30°F. (In contrast, winter air temperatures above the snow can fluctuate widely and sometimes plunge to 30° or 40° below zero.) The subnivean layer is an environment of near darkness; only eight percent of available light can penetrate through a foot of snow, and only one percent can penetrate through two feet.

Many small mammals spend the winter in this dark, sheltered zone next to the ground. It provides crucial cover for small mammals that would freeze to death if exposed to the harsh elements for too long. Because their body surfaces are so great in proportion to their body volume, they lose heat rapidly in cold winter temperatures. Smaller animals are also less able to carry a coat thick enough to withstand continual exposure to cold.

Two of the small, subnivean mammals are the shrew and the vole. Shrews have long pointed snouts, tiny black eyes, and short legs. Their small ears are almost hidden beneath their soft fur. Voracious **predators** with an extremely high metabolism, shrews must eat almost continuously to prevent starvation. Their diet consists largely of insects and earthworms, though they will eat small mice or shrews as well as berries and nuts. Voles, like mice, are rodents that primarily eat vegetation. In the winter they often nibble the bark at the base of trees. They resemble – and are often mistaken for – mice but have smaller eyes and ears and a short tail. As the snow melts in the spring, their extensive snow tunnel networks near the ground are revealed.

The under-the-snow layer provides a good habitat for these and other animals. Many hide seeds and nuts for winter use. Weeds and grasses, flattened by the snow, add their seed heads to the larder. Bark of shrubs, small trees, and surface roots provide additional fare. And snow provides excellent cover from predators, made hungrier by winter's cold and scarcity of **prey**.

Most winter activity for these small mammals takes place beneath the snow, but tracks reveal that they often run across the top of the snow hunting for new food or taking the easiest route to a stored cache. Such exposure makes them vulnerable to predators. Even though much of this scurrying occurs under the cover of darkness, many predators, such as owls, hunt at night. With their acute hearing, owls can even locate these animals under the snow and plunge, talons first, into the snow to snatch them.

While some small animals use the snow primarily as a protective cover, others make use of it for hunting, roosting, or storing food. Red squirrels tunnel through the snow to find caches of nuts and seeds and to establish escape routes. Weasels, slimmer than you might think, hunt on the surface but may follow their prey – squirrels, mice, and voles – right into their tunnels. Grouse dive into deep snow or tunnel in to make cozy roosts for the night, providing protection against the cold and predators.

Variations in snow consistency can mean life or death to those who walk on it or live in it. A crust, if thick enough, can support even large animals like coyotes, aiding in winter travel. Yet the same thick crust can trap a sleeping grouse or prevent owls from reaching their tunneling prey. Powdery snow is light enough for large animals to walk through, but smaller creatures or those with short legs often flounder. On the other hand, deep, heavy snow may hinder animals that must walk through it, like deer, fox, coyote, and bobcat, while lighter animals like deer mice, meadow voles, and grouse, are able to travel easily on its surface.

Suggested References:

Kirk, Ruth. *Snow.* New York: William Morrow, 1978.

Marchand, Peter J. *Life in the Cold.* Hanover, NH: University Press of New England, 1996.

Stokes, Donald W. *A Guide to Nature in Winter.* New York: Little, Brown, 1976.

Stokes, Donald W., and Lillian Q. Stokes. *A Guide to Animal Tracking and Behavior.* New York: Little, Brown, 1986.

Webster, David. *Snow Stumpers.* New York: Natural History Press, 1968.

vole

Snug in the Snow

FOCUS: Many animals that are active in winter find warmth, protection, and even food beneath a blanket of snow.

OPENING QUESTION: *How does snow benefit certain animals during the winter?*

PUPPET SHOW

Objective: To learn how snow protects small animals from severe cold weather and predators.

Perform, or have the children perform, the puppet show. Discuss the benefits of snow for the animals involved.

Materials:
- puppets
- script
- white cardboard stage cut with slanting tunnel (2" wide) leading to a small (3") chamber

SNOW CRITTER CONCENTRATION

Objective: To identify some of the animals that spend their time in or under the snow.

Divide the children into two teams. Show them a mural of a winter scene with 10 removable flaps attached. Explain that hidden beneath the flaps are pictures of 5 different animals, 2 of each kind, that may seek shelter under the snow. One at a time, the children take turns trying to find matching pictures by raising two of the flaps. When a pair is found, their pictures are uncovered. Teams take alternate turns guessing until all matches are found and uncovered. Discuss how the snow helps each of the animals.

Materials:
- winter mural with removable flaps covering animal pictures, 2 of each kind. Possibilities include white-footed mouse, weasel, grouse, meadow vole, shrew, red squirrel, beaver.

DIORAMAS

Objective: To create models of the winter homes of animals.

Have the children work in pairs or small groups to construct a diorama depicting life under the snow. Give each group a shoe box. Have them turn the box on its side and decorate the upper side to represent the surface of the snow and the inside of the box to represent the subnivean layer, the habitat under the snow. Provide clay or play dough for making animals that will be placed in the diorama, either under or above the snow.

Materials:
- small boxes
- twigs, dried weeds, and evergreen sprigs
- white paper
- cotton batting and/or Styrofoam to represent snow
- clay or play dough

INSULATION INVESTIGATION

Objective: To compare the insulating properties of different materials.

Divide the children into pairs or small groups, giving each group two film canisters. Prepare hot gelatin by mixing a quart of very hot water with 4 tablespoons of gelatin. Explain that each group must try to insulate one of the canisters from the cold by wrapping it in a material of their choice. The second canister will be a control and will not be wrapped. Display a selection of possible insulating materials and have each group or pair make a selection. Now fill the canisters with gelatin and ask the children to work

Materials:
- thermos of hot water
- gelatin
- plastic film canisters, 2 per group
- insulating materials such as plastic bubble wrap, newspaper, Styrofoam packing "peanuts," plastic and paper bags, hats, mittens, fur, down-filled vest
- rubber bands
- thermometers (optional)

quickly to wrap one of their canisters. Have the groups place both canisters outside, side by side, one wrapped in the insulation chosen and the other unwrapped. Children should feel the control canister every five minutes, and, when it feels cool to the touch, bring both canisters inside. Have them unwrap the insulated canister. Ask the children if they feel a difference in the temperature of their two canisters. Is one cooler? Which one? Did the gelatin in the unwrapped canister harden? Did the gelatin in the insulated canister harden? Have the children explain what they observed.

MOUSE HOUSES

Objective: To discover, firsthand, places where animals have made tunnels under the snow.

In small groups, have the children look for animal holes or tunnels in the snow. They may leave some seeds or nuts in them. The children can check during the next few days to see whether the food they left has been eaten.

SHARING CIRCLE

Objective: To share new discoveries about staying warm in winter.

Gather the children into a circle and have each one complete the sentence: "I'd like to be a _____ in winter, and I'd stay warm because I would _____."

Materials:
• seeds or nuts

white-footed mouse

EXTENSIONS

Layer of Fat: Place 2 cups of vegetable shortening into a plastic zip-closing bag. Place a second bag inside this bag of "fat." Have the children place one hand into the inner plastic bag so that the shortening surrounds their hand. Then ask them to place both hands, one bare and one covered by the bag of shortening, into a dishpan full of cold water. Does the water temperature feel the same to both hands? Why not?

Comparing Temperatures: On a cold day when there is at least a foot of snow, have the children measure and record the temperature on top of the snow and under the snow at ground level. How much difference is there?

Snow Story: Ask the children to pretend they live under the snow and have each write a story about it. Is it dark or light? quiet or noisy? cozy or lonely?

short-tailed weasel

Snug in the Snow

Characters: Marsha Mouse, Marvin Mouse (wearing easily removable ear muffs, tail warmer, and scarf), Wilma Weasel (in white winter coat)

Stage: white cardboard with a tunnel cut from the top slanting down toward a larger, more circular opening near the bottom; a strip of brown at the bottom with weed seeds stuck to it

Prop: baking sheet for rumbling noise

Marsha Mouse

[on top of the snow] Oh, I'm so excited. My favorite cousin, Marvin Mouse, is coming to visit me. He was a little bit scared about coming north in the cold winter, but I told him it would be lots of fun.

Marvin Mouse

[dressed warmly] Hello! Is that you, Marsha Mouse? It's me, Marvin!

Marsha

Marvin?! I can hardly see you under all those clothes. Why are you wearing all that?

Marvin

Well, I wasn't about to come here without my scarf, ear muffs, and tail warmer.

Marsha

But Marvin, you don't need those winter clothes. You'll be plenty warm without them.

Marvin

Oh, I don't think so. I'm used to the warm south. Besides, I bet everyone up here wears warm clothes like this in the winter.

Marsha

Well, **people** do, but we mice don't need those. Marvin, how would you like a really warm coat — one that would trap little pockets of warm air all over your body?

Marvin

Oooh! Now **that** sounds warm and toasty. I'd like that a lot!

Marsha

Well then, Marvin, take off those clothes!

Marvin

What?! Well...OK...if you say so. But don't peek! *[Marvin shakes and his clothes fall off]* Now what?

Marsha

Now, how do you like your fur coat?

Marvin

Why, I guess I **am** wearing a nice warm fur coat! Mmm, I do feel warm and toasty. *[wind noise]* But, brrrr, when the wind blows, I get a chill.

Marsha

I do, too, Marvin. We mice don't have long, thick fur coats like foxes or bears, but we have other ways of staying warm. Come on. I'll show you. Time to burrow under the snow!

Marvin

How is that going to make us warm? Everybody knows snow is freezing cold.

Marsha

You'll see. Come with me. *[both walk down tunnel into chamber under the snow]*

Marvin

Hey, it **is** warm down here. I'm not cold at all.

Marsha

I told you. It may not be as warm as Florida, but my little home under this fluffy blanket of snow is quite comfy.

Marvin

It's very cozy. *[loud rumbling noise]* Uh oh! What was that rumble I just heard? Maybe it's a snow avalanche. Let's get out of here.

Marsha

That was no avalanche, Marvin. It was just my stomach.

Marvin

Your stomach? Wow. You must be very hungry!

Marsha

I'm starving. Let's get something to eat.

Marvin

Oh no, you mean we have to go back outside?

Marsha

No, no. There's food right down here. If we just tunnel through the snow, we'll find delicious seeds to munch on. *[both mice eat the weed seeds]*

Marvin

Yum! You're right. There's lots of good food under here. *[loud sniffing noises heard]* Hey Marsha, did you hear that? *[Weasel slowly crosses stage]*

Marsha

Quick, Marvin, get back! It's Wilma Weasel. Shhh, stay perfectly still and quiet. Oooh, I hope she doesn't smell us. *[both exit]*

Wilma Weasel

[standing at top of tunnel entrance] Ah ha! I smell mouse. MmmMm. Two flavors! Luckily for these two mice, I just ate lunch. But maybe I'll drop in on them later...for dinner. *[Weasel exits; both mice reappear in tunnel]*

Marsha

OK, Marvin, the coast is clear.

Marvin

What was that weasel doing here, Marsha?

Marsha

The same thing we are, looking for a place to stay warm and find food. The only difference is, **we're** her food!

Marvin

Can weasels follow us into these tunnels?

Marsha

You bet they can!

Marvin

Oh dear, now what are we going to do? I heard her say she'd be back for dinner!

Marsha

Don't worry, Marvin. I've got lots of places where we can hide from a hungry weasel and miles of tunnels to take us there. We'll be safe and sound under the snow. You just follow me.

Marvin

I'm right behind you, Marsha. Let's go! *[both exit]*

White-tailed Deer

ELUSIVE BEAUTY OF WOODS AND FIELD

Creatures of both field and forest, white-tailed deer embody the wild beauty of North America. While they inhabit much of the continent, we often see no more than a fleeting glimpse of these elusive mammals. Throughout their range, white-tailed deer occupy a variety of habitats, including hardwood and coniferous forests, wetlands and creek beds, and the borders of meadows and fields.

White-tailed deer move from one type of habitat to another over the course of a year. These seasonal movements correspond to the changing needs of the deer. Fawns are born in early spring in thickets, fields, woodlands, or swamps when the vegetation is just emerging. As soon as a fawn is able to walk, its mother will move it to a new sheltered location – hidden in tall grass or tucked in next to a fallen log or tree trunk – to hide it from **predators**. Camouflaged by its dappled coat, a newborn fawn has little or no scent, giving it added protection against discovery by predators. Fawns grow quickly throughout the spring on a rich diet of doe's milk, which has almost twice the solids of cow's milk and nearly three times the protein and fat.

Each spring, bucks grow new antlers that they shed every winter. Because antlers are made of bone, bucks expend a great deal of energy growing new ones each year. Their diet, mainly grasses, twigs, acorns, and fungi, must be rich in calcium and phosphorus. A young buck can only produce small spikes because the nutrients in its diet are still being used for body growth. A mature buck can grow a larger rack because its body has already reached full size. The quantity and quality of food available in a buck's habitat help determine the size of its rack.

As spring turns to summer, food becomes more abundant. Fawns follow closely at their mothers' heels, grazing during the early hours of the morning and again in the evening. Deer feed upon the young leaves of woody plants such as maple, oak, and willow, and upon broad-leaved plants and grasses in the meadows. Other favorite foods are cultivated crops such as corn, lettuce, clover, and alfalfa. Deer prefer to spend their time on the edges of fields where they can easily bound to safety in the nearby woods. Like cows and other **ruminants** with four-chambered stomachs, they can eat and run, temporarily storing the food in their first stomach. Later, in safety, the deer regurgitate the food, rechew it as cud, and pass it on through the last three stomachs for final digestion.

As **prey** animals, deer are well suited to detect and avoid predators. Their large eyes are quick to notice the slightest movement and allow them to see at twilight and in the dim light of a forest. Deer can run faster than 30 mph, leap more than 25 feet, and clear obstacles as high as eight feet tall. Acute hearing and a phenomenal sense of smell give deer early warnings of trouble. And their uncanny ability to hide comes from having an intimate knowledge of every nook and cranny in their home range. If needed, their sharp hooves can be dangerous and effective weapons.

In late summer, deer seek out ponds, streams, and wetlands for water and succulent plants. Then in the fall, they shift habitats yet again to hardwood forests or mixed stands where they look for foods rich in fats and oils, like acorns, beech and hickory nuts, berries, apples, and the newly fallen leaves of deciduous plants. These foods provide the additional calories deer need as winter approaches.

Deer also prepare for winter by replacing their reddish-brown summer coats with grayer, thicker coats of fur. The winter coat contains hair that is both hollow and crinkled and much longer and denser than the summer fur. Thus, warmer air is trapped both inside the hairs and between them, providing good insulation against winter's cold. The gray color helps deer blend in with the bark of the trees.

Deer tend to associate in larger family groups during the fall and a pecking order develops within and between groups. Dominant individuals will chase others away from the best food sources, and the subordinates – usually the young, the old, and the infirm – will be left with the remains. These individuals will be less prepared for the hardships of winter and are more likely to succumb to starvation. However, grazing in a group provides an advantage, for there are more eyes to see and ears to listen for the approach of predators. When a deer spots danger, up goes the white tail, signaling group members to flee to safety.

Winter presents many hardships to the white-tailed deer. Cold, ice, snow, and a diminished food supply are chief among these. The severe cold of northern winters puts great demands on the deer's ability to stay warm. Deep snow covers up much of their food supply and makes travel difficult. Their small hooves and thin legs easily break through snow or crusts of ice, and deer expend much energy as they flounder in the deep snow. Their food supply is restricted to woody plants that are difficult to digest and increasingly hard to reach as the supply of low branches is used up.

For these reasons, deer often seek shelter in protected areas during the harshest part of the winter. In northern parts of their range, deer congregate in stands of conifers that help to block the bitter winds and heavy snow. These areas, known as deer **yards**, are usually on a south-facing slope or in protected lowlands with relatively shallow snow. The deer regularly tramp down paths throughout the area so travel remains possible within the yard.

Deer **browse** on tough woody foods in winter, such as the needles of hemlock and white cedar, and the twigs and buds of hardwoods like maple, birch, sumac, oak, and aspen. Because they lack upper front teeth for biting, deer use their molars to tear off twigs from branches, leaving a ragged end. This can easily be distinguished from rodent browse, as rodents slice through twigs with their sharp front teeth leaving neat, angled cuts. Deer also eat the outer bark of several kinds of trees, using their lower incisors to scrape upwards along the tree trunk to remove the strips. Some of these foods are fairly nutritious, but others simply fill stomachs, providing little nutrition. An adult deer is said to need a minimum of two and one-half pounds of browse per day. Healthy adult deer can safely lose up to 30 percent of their body weight, but in a severe winter when deer stay within the yard's confines, they may use up the food supply and their body fat and finally face starvation.

Wintering yards are often identified by the conspicuous signs of heavy deer browsing resulting from the animals' confinement in these relatively small areas. Especially here, deer may have a major impact on their environment. When their population numbers are very high in an area, they will browse so heavily that they inhibit the growth of new saplings, leaving little food for future generations of deer.

As winter nears its close, deer seek out the first green shoots emerging around springs and vernal pools.

Bucks regain their strength and prepare for the growth of antlers while does become solitary again, using their energy to feed and prepare for the arrival of their fawns.

The white-tailed deer's adaptability, keen senses, speed, and agility account for its success in many different habitats across the North American continent. Throughout the year, a deer will move from one habitat to another in its square mile of home range, seeking food, water, shelter, and a place to rear its young. Its graceful beauty and elusiveness make it one of the most loved and admired of all wild creatures.

Suggested References:

Godin, Alfred. *Wild Mammals of New England*. Baltimore: Johns Hopkins University Press, 1977.

Hall, L. K. ed. *White-tailed Deer: Ecology and Management*. Harrisburg, PA: Stackpole Books, 1984.

Patent, Dorothy Hinshaw. *Deer and Elk*. New York: Clarion Books, 1994.

Rezendes, Paul. *Tracking and the Art of Seeing*. 2nd edition. New York: Harper Collins, 1999.

Rue, Leonard Lee, III. *The Deer of North America*. New York: Crown, 1978.

young doe

White-tailed Deer

FOCUS: White-tailed deer, like many other animals, occupy different habitats during the year as their requirements for food, water, and shelter change with the seasons.

OPENING QUESTION: *What are the habitats of the white-tailed deer?*

PUPPET SHOW

Objective: To learn about the seasonal changes in habitat of the white-tailed deer.

Perform, or have the children perform, the puppet show. Afterward, review the important habitat requirements of the white-tailed deer in different seasons.

Materials:
• script
• puppets
• props

SHOW AND TELL

Objective: To examine different parts of the deer's body and consider how they are related to habitat and lifestyle.

Working in small groups, have the children examine the deer parts listed below while thinking about the questions supplied in each collection.

COLLECTION 1

Skull: How would you describe the size of a deer's eye sockets? How might the size of deer's eyes help them to see at dusk or in a dark forest? Examine the teeth. Do deer have upper and lower front teeth?

Nipped twigs: Compare the nipped twigs. Try to determine which twig was bitten off by a deer and which by a rabbit. What observations lead you to your conclusion?
Hint: Unlike deer, rabbits have sharp upper and lower front teeth.

COLLECTION 2

Antlers: What do you think these antlers are made of? How might the size of a deer's antlers reflect the habitat in which it lives?

Winter and/or summer pelt: Examine the pelt. What color is it? Are the hairs hollow or solid; crinkly or straight; long or short? How might a different pelt for summer and winter help a deer? Do you think this is a winter pelt or a summer pelt?

COLLECTION 3

Leg: What do you notice about the shape and size of a deer's leg? Of a deer's foot? How might the design of the legs and feet help deer in certain habitats or situations? In what conditions or habitats might a deer have difficulty moving around?

Tape: Unroll the measuring tape to 8' to see how high a deer can jump. How does this help a deer running from a predator in the woods? Unroll the tape to 25' to see how far a deer can jump. How does this help a deer running from a predator in a field?

Have groups rotate among these collections. Afterward, have the children share their thoughts about the questions posed.

Materials:
• a deer skull
• rabbit and deer browse
• antlers
• winter and/or summer deer pelts
• portion of lower leg including the hoof
• 25' measuring tape
• hand lenses

Note: Deer hunters, Fish and Wildlife departments, or private collectors are usually willing to lend deer parts.

DAILY DIET

Objective: To discover how much food a deer needs in winter to stay alive.

Provide a large pile of twigs and slender branches (preferably varieties eaten by deer), a kitchen scale, and several empty grocery bags. Have the children fill and weigh the bags until they reach 2.5 pounds, the amount an average deer needs to eat per day in order to stay healthy through the winter. Do you think it would be hard for deer to find this much food to eat every day through the winter?

Materials:
• twigs and branches
• kitchen scale
• grocery bags

POPULATION PRESSURE

Objective: To understand how food supply limits the size of the deer population.

Divide the children into groups of twos, threes, and fours and give each group eight pretzels. Explain that each group represents a deer herd and the pretzels are the amount of food available in their wintering territory. Each deer needs a minimum of one pretzel per month to remain alive. As you announce the months one by one, beginning with December, children must take and eat a pretzel or, if none remain, collapse of starvation. Did any group live through the winter? Do you think deer share their food? Which deer in a herd would get to eat first? Which starve first?

Materials:
• pretzel sticks, 8 per group

EAT AND RUN

Objective: To experience the benefits of living in a herd for safety from predators.

Divide the group in half. One half of the children will be deer. Give each of these children a strip of white cloth and a paper lunch bag, representing their white tail and stomach respectively. The rest of the children are trees and must choose a place to stand in a selected play area. Designate a safety zone. Scatter popcorn or other grain around the play area for the deer to collect in their paper bag stomachs. Secretly select one of the trees to be a predator. Provide a prop to represent this role, but caution that it should be kept hidden until the predator is ready to pounce.

Tell the children that a predator lurks in the forest. Deer must collect food while keeping a wary eye out for predators. When a deer spots the predator, it must signal to the others by waving its white flag. Upon seeing this signal, deer try to flee to the safety zone without getting tagged by the predator. Let the children switch roles. How does it feel to be the deer? Does living in a group help them to escape from predators? How?

Materials:
• strips of white cloth
• paper lunch bags
• "safety zone" sign
• popcorn or other grain
• headband with ears or other prop representing the predator

twigs browsed by deer

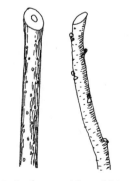 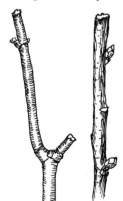

twigs browsed by rabbit

HABITAT MURALS

Objective: To illustrate the seasonal habitat requirements of the white-tailed deer.

Discuss the various habitats where deer might be found during the four seasons of the year. Divide the children into four groups, assigning each group a season, and provide them with a large sheet of mural paper. Ask each group to draw a habitat for deer that includes the important components needed during their assigned season. After the habitat is drawn, children use track stamps to make a trail that tells the story of the deer's daily activity within this habitat. When all are finished, have groups look at each other's murals and follow the trails to locate the different habitat components.

Materials:
- large mural paper
- pencils
- markers, crayons, or paints
- deer track stamps (cut sponge cushion to shape, and glue on wood blocks)
- inkpads

ACTIVITY STATIONS:
1) Daily Diet
2) Population Pressure

EXTENSIONS

Resources: Invite a game warden in to talk about the deer population in your state.

Story Time: Read the story "Lightfoot, Blacktail and Forkhorn" from *The Burgess Animal Book for Children* by Thornton Burgess (New York: Grosset and Dunlop).

Habitats for Sale: Have the children write a classified ad describing the habitat they created in Habitat Murals.

Antler Antics: In small groups, have the children research the various types of animals that have antlers. Provide a variety of materials and let them construct antler models to wear.

white-tailed buck

White-tailed Deer

PUPPET SHOW

Characters: Freddy Fawn, Herbie Hare, Dosie Deer, Woody Woodchuck, Charlie Chipmunk

Props: acorn glued on a puppet stick; signs saying "Summer" and "Fall"

Herbie Hare

Well, hello there! You're a newcomer to the thicket! What's your name?

Freddy Fawn

[loud whisper] Shhh, I'm not supposed to make noise!

Hare

Oh, I see. I'll have to guess then. Let's see… Is it Francis?

Fawn

[whispers] No.

Hare

Ferdinand?

Fawn

[whispers] No!

Hare

Franklin, Fitzwilly, Finnegan? *[Fawn giggles]* Oh, I give up. Can you whisper it?

Fawn

[whispers] It's Freddy.

Hare

I should have guessed! Welcome to the thicket, Freddy Fawn. You'll like it here. It's a nice, safe place to grow up. Now I'm off to find breakfast, and it looks like your breakfast is coming to you! *[exits; Deer enters]*

Dosie Deer

Good morning, Freddy. I'm back.

Fawn

Mommy! *[stands close to her]*

Deer

I found lots of good plants to graze on in the meadow this morning, so you'll have plenty of milk to drink.

Fawn

Can I go out in the meadow with you, Mommy?

Deer

In the summer when you're a little bigger, Freddy. First you have to get stronger so you can run back here into the thicket if there's danger.

Fawn

I am strong. See? I'm steady on my legs. *[wobbles]* Almost.

Deer

Oh, you are getting strong. In no time at all you'll be able to come with me. Now drink and sleep, and soon you'll be a strong little deer. *[both exit]*

[hold up a sign saying "Summer"]

Woody Woodchuck

Hmm, now let's see, how does it go? How much wood could a woodchuck chuckle? No, that's not it. How much good would a… *[Fawn enters]*

Fawn

Hi. I'm Freddy Fawn. Who are you?

Woodchuck

I'm Woody Woodchuckle…I mean chucklewood, I mean…Woodchuck!

Fawn

Hi, Woody Whatever-you-are. This is my first day in the meadow. I'm big enough to graze for my own food. But I have to be on the lookout for danger. You aren't a danger, are you?

Woodchuck

No, no. Woodchucks are strictly vegetarian.

Fawn

Oh, we deer are, too. I'll be coming here to graze all summer with my Mom. Don't worry, I can keep a lookout for predators because I'm tall enough to see over the grass – well, almost.

Woodchuck

Now you be sure and let me know if you see one.

Fawn

OK, and when you see my white tail go up, you run for the forest.

Woodchuck

[chuckles] Not on these short legs of mine. When there's danger around, I'll just pop into my hole, like this. *[exits]*

Fawn

Pop goes the Woodchuck! See you tomorrow, Woody. *[exits]*

[hold up sign saying "Fall"]

Fawn

Good morning, Mom. Where are we going today? There isn't much left to eat in the fields anymore. The grass is all brown.

Deer

That's why it's time for us to move into the forests, Freddy. We'll find hickory nuts, beechnuts, and acorns in the woods. You'll love them.

Fawn

I know I like hay and I know I like corn, so I'm sure I'll like haycorns!

Deer

Not haycorns, silly. Acorns. You'll see when we get there. *[both exit; brief pause then Fawn re-enters]*

Fawn

Mom was right; there's lots of good stuff to eat here in the woods. I've had hickory nuts and beechnuts, but I still haven't found any acorns. I wonder if you find them in holes in the ground like this one. *[Chipmunk pops up]*

Charlie Chipmunk

Hey, watch where you're putting your big nose! This is my den!

Fawn

[jumping back] Oh, sorry. I was just looking for acorns.

Chipmunk

You aren't going to find any acorns under this maple tree! What you need is an oak tree. C'mon, I'll show you. *[acorn appears]* Here, try this.

Fawn

[munching, then acorn exits] These are good! Even better than hay or corn.

Chipmunk

Oh yeah, and there's plenty for all of us – unless they get buried under an early snow.

Fawn

Snow? What's that?

Chipmunk

Oh, come on, kid! It's this cold white stuff that covers the ground in winter and makes it hard to find food and get around. That's why we chipmunks fill our burrows full of food in the fall. Got a mouthful to carry down right now, so fee oo ayter! *[exits]*

Fawn

Gosh, I wonder if I should be storing acorns, too? *[Deer enters]*

Deer

No, Freddy. You don't need to do that. When it starts to snow and get very cold, we'll move to a sheltered area with other deer.

Fawn

That sounds nice. Where will we go?

Deer

To a grove of evergreens. The branches make a good windbreak and they catch the snow so it won't be as deep as in other places. We can curl up under the evergreen trees at night for protection from the winter storms.

Fawn

Does it ever get warm again?

Deer

Oh yes, when spring comes. Then we'll move to the swamps and pools where the first green leaves come up.

Fawn

We live in a lot of different places, don't we, Mom?

Deer

We sure do. We change our habitat with the seasons.

Fawn

I guess we're in the habit of changing habitats!

Deer

I guess we are, and it's not a bad habit at all, if you're a deer.

Fawn

Let's go find some more acorns. I'll lead! I'm big enough to find my own way – *[both exit with Fawn leading, then return with Deer leading]* Almost! *[both exit]*

Streams

THE CHALLENGE OF A MOVING, WATERY WORLD

From the quiet trickle in late summer to the thundering roar after a spring rain, a stream is a dynamic, ever-changing ecosystem. Various habitats within the stream – turbulent riffles, quiet pools, the bottom of rocks, gravel beds – shelter distinct communities of plants and animals that can be easily explored by those willing to wade in.

Among the oldest bodies of water on earth, streams form when volumes of water, unabsorbed by the soil, are pulled by gravity across the surface of the earth. As the water moves, it carves a course for itself, following the path of least resistance.

A stream's current sets it apart from still-water bodies and creates varying conditions along its course. The current acts like a giant mixer, folding air into the water as it tumbles over the rough streambed. Oxygen from the air dissolves into the water and serves as a critical resource for aquatic life. Many stream creatures, such as trout, require high levels of dissolved oxygen to survive.

The speed of the current determines the texture of the streambed and the distribution of organisms. Rapidly moving water picks up small rocks and other particles and carries them downstream, where they are dropped when the current slows. The fastest sections of a stream contain only the largest rocks as all other materials are scoured away, while still pools have muddy bottoms, with the smallest particles settling there. Stonefly **nymphs** cling to rocks in fast-flowing water whereas dragonfly nymphs prefer the mud where the water is quiet.

mayfly nymph

Moving water also picks up soil runoff and decomposing plant and animal matter along the way. These dissolve or become suspended in the water, creating a dilute soup of vital nutrients. Many stream inhabitants take advantage of this natural transport system and catch and eat the materials that are delivered to them. Certain caddisfly **larvae** construct nets that snare small organisms right from the water as it rushes by.

The presence of plants in and around a stream has a profound effect on life within the stream. The plants growing along the banks provide shade and, along with cold springs, help to keep stream water cool. Cold water holds more dissolved oxygen than warm water, and oxygen is critical to the survival of many stream animals. Stream bank plants also hold soil in place, preventing **erosion**.

Plants living in the stream are dependent upon available sunlight, nutrients, dissolved oxygen, and the ability to maintain a foothold in the moving water. Algae and moss often form a slippery film, clinging tightly to rocks submerged in the current. Most rooted plants can only become established in the slower backwaters.

Animal life in the stream is diverse and abundant due to high oxygen levels and the variety of habitats. Of all the stream creatures, the insects are probably the easiest to find and observe. Each species is adapted to carry out the normal functions of life (eating, breathing, moving or holding fast, laying eggs) while living in a rushing, changing world.

In the still pools at the water's edge, you may see an insect flitting along the surface, its shadow frequently more noticeable than the creature itself. This is the water strider, whose hair-fringed legs skate across the stream's surface as it prowls for unlucky insects that have fallen in, or for aquatic insects that venture up to the surface.

Not all stream insects, however, can be seen from the banks. Many of them spend much of their lives submerged and must be sought underwater amid the rocks, gravel, and plants. Where the stream courses around and over rocks and gravel, riffles of disturbed water create a home for the riffle beetle. The larva, called the water penny, is flat and oval and clings closely to the sides of rocks.

underside of a water penny

Turn over a rock and you are likely to find common stonefly and mayfly nymphs. Flattened and with strong hooks on their feet, they dwell within the narrow space between the bottom of a rock and the streambed where the current is very weak. They are so well camouflaged

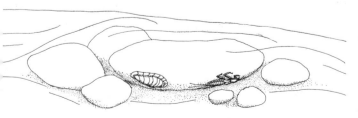

water penny and mayfly nymph on underside of a rock

that you may not notice them until they scurry for cover on the opposite side of the rock. Mayflies have gills along their abdomens and usually three feathery-looking, tail-like appendages. Stoneflies have gills that look like fuzzy tufts at the base of each leg and only two tail-like appendages. Most mayflies and some stoneflies are **herbivorous**, while some of the larger stoneflies prey on insects.

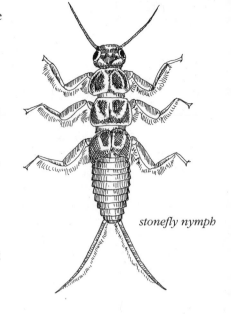
stonefly nymph

black fly larva

As you lift more rocks, you may see some tiny, quarter-inch, dark larvae swaying in the current. Black fly larvae are unmistakable, with their swollen rumps anchored firmly in place. Sieve-like hairs project from the sides of their heads to strain algae, tiny animals, and plant debris from the water. If dislodged, these larvae will creep, spider-like, back up the current on a silken thread.

The gravels of the streambed provide yet another type of habitat, where crane fly larvae may be found in abundance. Cream-colored and maggot-like in appearance, a crane fly larva is one and one-half to two inches long and bears some appendages on its head that look like the fleshy feelers on a star-nosed mole. Adult crane flies resemble giant mosquitoes, but they do not bite.

As you continue to turn over rocks, sift gravel, and explore the stream, you may discover the caddisfly larva within its tube-shaped home made of sand grains or leaf pieces. These larvae construct their homes by weaving an intricate tube of silk threads that is closed at one end. Depending on the species, the silk tube is

caddisfly larva

covered with sand grains, leaf pieces, or small sticks. These elaborate cases supply both protection and ballast in flowing waters. Some cases are cemented to rocks while others are free, allowing their inhabitants to travel with their homes on their backs.

The presence or absence of certain insects can reveal a stream's health. The removal of gravel from the streambed, the addition of pollutants, or the erosion of soil from banks without vegetation can burden a stream ecosystem. Observing the profile of insect populations within the stream can help determine water quality. For example, the presence of stoneflies indicates a healthy stream because they are rare or absent in polluted or oxygen-poor streams.

While insects are important stream inhabitants, they are just one group of animals that depend on these moving waters. A great array of creatures can usually be found, from crayfish to fish, salamander to snail, each living in the stream habitat best suited to its needs and abilities. Some eat plants; others eat decaying organic matter; still others prey on animals. Each plays a role in the stream's **food web**, which extends beyond the stream into the terrestrial environment around it. Many land animals frequent streams to drink water or find a meal. On your next trip to a stream, look for the tracks of these visitors.

A stream is a constantly changing and sometimes perilous ecosystem. Yet despite the hardships, many organisms thrive in its rushing water. As you gaze at the sparkling riffles in a stream, listen to its gurgling song, and feel its chill with your bare toes, consider the great diversity of life it harbors.

Suggested References:

Caduto, Michael. *Pond and Brook: A Guide to Nature Study in Freshwater Environments.* Englewood Cliffs, NJ: Prentice-Hall, 1985.

Edelstein, Karen. *Pond and Stream Safari: A Guide to the Ecology of Aquatic Invertebrates.* Ithaca, NY: Cornell University Press, 1993.

Klots, Elsie B. *The New Field Book of Freshwater Life.* New York: G.P. Putnam's Sons, 1966.

Reid, George K., and Herbert S. Zim. *Pond Life.* New York: Golden Press, 1967.

Stapp, William B., and Mark K. Mitchell. *Field Manual for Water Quality Monitoring: An Environmental Education Program for Schools.* Dexter, MI: Thomson-Shore, 1988.

Streams

FOCUS: A single stream can contain many different habitats that host a variety of plants and animals.

OPENING QUESTION: *What would it be like to live in a stream?*

DREAM A STREAM

Objective: To introduce the various habitats within a stream.

Sitting in a circle, have the children close their eyes and imagine they are sitting near a stream. What does it look like? Sound like? Ask each child to think of two or three words to describe their imagined stream and share these with the group. Sort the children into groups of three or four according to the various stream habitats represented by their different descriptions, such as fast-flowing, slow-moving, sandy bottom, cool and shady, and so on. Give each group a 3' section of mural or newsprint paper upon which to illustrate the group's particular stream habitat. Afterward, join the sections together, reviewing the different types of stream habitats represented. Ask the children what life might be like in the different stream habitats they've drawn. (You may want to use this completed stream for the Stream Mural extension activity.)

Materials:
- pencils, crayons or markers
- roll of newsprint or mural paper

PUPPET SHOW

Objective: To identify some common stream insects and their adaptations to stream life.

Perform, or have the children perform, the puppet show. Using the puppet characters, discuss the insects' adaptations and why those were necessary in a stream habitat.

Materials:
- puppets
- script
- props

STREAM SALLY AND SAM

Objective: To identify some important adaptations necessary for survival in the stream habitat.

Working in small groups, have the children use the materials provided, along with those found in the classroom, to transform an adult or one of their group members into a stream critter. Their newly created stream critter should be given a name and must have special adaptations allowing it to breathe, see clearly, catch food and eat it, move around on the bottom, and keep from getting washed away, all while submerged under water. Ask each group to introduce its critter and explain how it is adapted to life on the stream bottom.

Materials:
- colored paper
- scissors
- tape
- yarn or string
- paper cups
- plastic eating utensils
- pipe cleaners
- balloons
- straws
- crayons
- egg cartons
- cardboard tubes from toilet paper, paper towels, or gift wrap

STUDY A STREAM

Objective: To explore the variety of habitats in a stream and the life each supports.

Using the Stream Study Sheet as a guide, have the children explore the stream in small groups. Provide each group with needed equipment and, if possible, assign different stream habitats for each to investigate. Afterward, bring groups together to share their collected critters and compare results. What differences in the various stream habitats were recorded? Ask the children to share their ideas about what might explain these differences. Be sure to release each critter back in the part of the stream where it was found.

RELEASE CEREMONY

Objective: To return insects to their proper home in the stream.

Gather the children together and gently tip the collecting basin into the appropriate stream habitat, returning the insects back to their homes. Have the children repeat the following poem.

> *Swim away, float away, dive away, leap –*
> *You're free to go, I'm not going to keep*
> *You from living your life.*
> *You deserve to swim free.*
> *Thanks for sharing this time with me.*

Materials:
- Stream Study sheets
- clipboards
- pencils
- white dishpans
- 10' lengths of string
- hand lenses
- small plastic containers
- nets
- thermometers
- field guides
- microscope, if available
- oranges

EXTENSIONS

Fast Floaters: Divide the children into small groups and have each group choose a natural object (cone, twig, leaf, acorn) for a floating race. Place spotters at the finish line to see which object crosses first, then have each group place its object in the water at the same time and place. After the race, discuss why the winner might have won, considering shape, size, stream flow, and obstructions.

Stream Mural: Have the children draw the animals collected at the stream. Cut out the drawings and place them on the Dream a Stream mural in the appropriate location.

Stream History: Have the children interview local residents about the history of a local stream (uses, floods, bridges, pollution).

Diary of a Stream: Visit a stream regularly with the children and have them keep a journal of their observations.

Stream Mapping: Have the children look on a map to see where the stream comes from and where it goes. Try visiting the stream at different points along its course and see how it changes. For older children: Have them use a map to follow the water in their stream to the ocean, noting each body of water it travels through along the way.

Streams

Select a section of the stream and answer the following questions.

THE STREAM

Circle the word that best describes the bottom of the stream:

mud   sand   gravel (2" or less)   rocks   boulders

How quickly is the water moving? (Measure and mark off 10 feet with a string, then drop an orange in and time how long it takes to go this distance. If possible, compare the speed of the water at two sites, one with fast water and one with slower water.)

Number of seconds: Site #1 _____ Site #2_____

Does the area with faster water have bigger or smaller rocks than the area with slower water?

What is the temperature of the stream? (Use a thermometer.) Number of degrees: _____

THE PLANTS

Are plants growing on the edge of the stream? Which of the following?

trees   bushes   ferns   grass   flowers

Is the stream mostly shaded or mostly sunny?

Are plants growing in the stream?

Are plants growing on rocks in the stream? Feel some rocks for slippery algae. Look for moss.

Are there dead or decayed plants on the bottom? leaf litter? twigs or branches? muck?

THE ANIMALS

Look for signs of animals (paths, droppings, prints, browsing) and list them below.

How do you think these animals are using the stream (drinking water, hunting prey, eating plants)?

Are there animals on top of the water? How do they move?

Do you see any animals swimming in the water? fish? insects? others?

Take a rock out of the stream and examine it for animals. Look at them closely with a hand lens. Do they scurry away? How are they attached to the rock?

Collect some stream critters. Put them in a white dish basin containing some stream water. Observe how they move.

Can you identify any of the insects? Which kinds did you find?

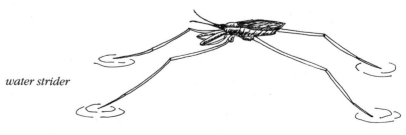

water strider

Streams

Characters: Willy Water Bug, Gregory Grasshopper, Mabel Mayfly Nymph, Walt Water Penny

Prop: coat hanger; sink plunger; a pretend magnifying glass made from a 10" circle of cardboard with a 9" hole in the center, attached to a handle with a circle of clear acetate for the lens.

Willy Water Bug

Oh, the start of another boring day swimming about this little pond! I wish I had some excitement in my life. I'm ready for an adventure. *[Grasshopper enters]*

Gregory Grasshopper

Hi, Willy Water Bug. I heard you complain about being bored. If you want an adventure, why don't you come out and hop in the fields with me?

Water Bug

Oh, thanks, Gregory Grasshopper. That sounds like fun, but remember I'm a water bug, and we water bugs need to live in the water.

Grasshopper

Well, why not try living somewhere besides this boring pond? What about living in a fast-moving stream? All that rushing, bubbling water would be exciting, don't you think?

Water Bug

Hey, that's a great idea, Greg. Is there a fast stream nearby?

Grasshopper

There sure is. It's just across this field. Come on, follow me! *[Water Bug and Grasshopper cross the stage together]* Here's the fastest part of the stream. Jump right in!

Water Bug

Thanks, Gregory! *[Water Bug jumps off and Grasshopper exits]* Oh buggy, now this **is** exciting. Yikes! This water is moving pretty fast. Oh no! I'm getting swept away. Help! Help! *[Mayfly Nymph appears]*

Mabel Mayfly Nymph

Oh my, it looks like you're in trouble. Don't worry, here I come! I'll save you! *[Mayfly pulls Water Bug from the stream]*

Water Bug

Thank you! You saved my life. Who are you?

Mayfly Nymph

My name is Mabel. I'm a Mayfly Nymph. Who are you? I've never seen you around here before.

Water Bug

My name is Willy. I'm from the pond over yonder. I was bored living there and thought I'd try this stream. But it moves so fast. How do you keep from being swept away?

Mayfly Nymph

Oh, the rushing water doesn't bother me. I have something special on my feet that helps me hold on to rocks and other things in the stream.

Water Bug

Your feet don't look that special to me.

Mayfly Nymph

Oh, you have to look at them very closely. Here, why don't you look through this magnifying glass? *[holds up large magnifying glass]* I'll hold up my foot so you can get a good look. *[Mayfly goes down and the hooked end of a coat hanger appears behind the glass]*

Water Bug

Wow! You **do** have special feet. Look at those hooks. Why, I bet you can really grab onto rocks and things with those! *[magnifying glass exits; Mayfly reappears]*

Mayfly Nymph

I sure can. Well, so long, Willy. It's back to the stream for me. I'm hooked on rushing water! *[exits]*

Water Bug

Gosh, I don't have any hooks. What's a bug to do?

Walt Water Penny

[from below the stage, whispering] Psst! Listen, buggy, you don't need hooks. I live in the stream, and I don't have any hooks.

Water Bug

Who said that? I heard you, but I don't see you anywhere!

Water Penny

I'm on this rock in the stream. Pull me out if you'd like to talk.

Water Bug

[reaches down and pulls up Water Penny] My, you're really hanging on tightly to that rock. If you don't have any hooks, how do you hold on?

Water Penny

I use something very special to cling to this rock.

Water Bug

Really? I don't see anything.

Water Penny

Well, I just happen to have a magnifying glass right here! *[hold up large magnifying glass again]* I'll show you how I hold on to the rocks. *[Water Penny goes down and plunger comes up behind lens]*

Water Bug

Wow! Now that sure is different from a hook. You were **stuck** to that rock! If I had a suction cup, I could stick anywhere.

Water Penny

That's right! I'd better go now. Take it from me, bug, suction is the way to go! See you around. *[Water Penny exits]*

Water Bug

[pause] I don't really have **anything** special to help me hold on in all this rushing water. I wonder if I should go back to the pond? *[pause]* I'm not even sure how I got here. I wonder if Greg Grasshopper is still around. Gregory? Can you help me call Gregory? *[with audience]* Gregory! Oh Gregory! *[Grasshopper enters]*

Grasshopper

What's up, Willy? How's life in the stream?

Water Bug

Well, life in the stream is fine, if you're equipped for it. It's a bit too exciting for me. In fact, it's downright dangerous.

Grasshopper

You'd better follow me back, Willy. I'll hop you back to your pond.

Water Bug

Thanks, Gregory! There's no place like home.

Ponds

LIFE IN STILL WATERS

A visit to a pond is a sensory delight. Frogs, birds, and a multitude of insects provide a symphony of sounds, while many kinds of plants create visual texture and a fragrant mix of smells. Bare toes relish the warm water and soft, squishy mud. Here is a wonderful outdoor classroom where we can use all of our senses to discover and enjoy a great diversity of creatures.

Most people recognize a pond when they see one, though local names for a body of water are often based more on tradition than particular physical features. The most widely accepted definition identifies a pond as a shallow body of water in which rooted plants grow across it from shore to shore. Yet some bodies of water considered to be ponds have a central region devoid of plants. In the end, nature tends to ignore human definitions, and "pond" creatures have no trouble identifying their preferred habitat.

Although every pond is different, all ponds share some common characteristics. Most important, a pond is a body of standing water. The water contains dissolved oxygen and other gases, as well as nutrients that have flowed into it or have been released through **decomposition**. The amount of oxygen dissolved in the water is a critical factor in the survival of many pond organisms. Aquatic plants add oxygen to the water through **photosynthesis** while animal respiration and the decomposition of dead plants and animals remove oxygen. Dissolved oxygen levels fluctuate dramatically in a pond, reaching a peak during the day, when plants are active, and plummeting at night, when plants stop photosynthesizing.

Temperature also helps to determine the life forms that can survive in a pond. It influences the rate of plant and animal growth and decomposition as well as the amount of dissolved oxygen present (warmer water normally contains less oxygen than cold water). In a shallow pond, the water warms up considerably during the day and loses heat at night. If the water is deep, only the surface layer warms while the lower layers stay cool. Springs that pump cold water into a pond also influence water temperature and can create variable conditions.

The diversity of life in a pond is due to the existence of several distinct habitats within it, including the water's edge, the open water, the surface film, and the bottom. The water's edge is where the land meets the water. Various rooted plants are found growing in this shallow water. They contribute nutrients to the pond, as do the trees and shrubs along the banks that drop their leaves into the water in the fall. Frogs, fish, insects, worms, snails, and an array of microscopic animals find shelter among these shoreline plants.

Open water is the area in the center of the pond where rooted plants do not extend to the surface of the water and, in deep ponds, where plants do not grow at all because sunlight cannot reach the bottom there. Life in open water consists of large, free-swimming animals, such as fish and turtles, and microscopic plants and animals (plankton) that drift suspended in the water.

The surface film, created by the water's **surface tension**, is the habitat of various creatures that get their oxygen directly from the air. Water striders skate across the water prowling for insects and other small animals. Mosquito **larvae** hang suspended from the underside of the surface film where they filter tiny plants, animals, and debris from the water. The aptly named whirligig beetle wheels, spins, and zigzags crazily across the water, sometimes in large groups. They hunt for **prey**, aided by compound eyes that are split so they can see above and below the water at the same time.

The pond bottom is an area of much **decomposition**, where bacteria help to recycle nutrients by attacking dead organisms that sink to the bottom. Many insects, especially those in their larval form, hide from their **predators** here. **Nymphs** of dragonflies and damselflies stalk their prey on the pond bottom. Worms, crayfish, clams, and other animals burrow into the bottom mud.

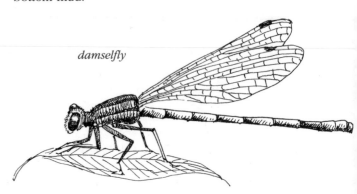
damselfly

A pond's plant life is as varied and fascinating as its animal life, but children can rarely be diverted from their encounters with frogs, newts, and insects long enough to observe it. Most obvious are the **emergent** plants, like the familiar cattails, which grow along the edge of the pond where their stems, leaves, and flowers rise out of the water. Plants with floating leaves, such as water lilies, grow a bit farther out. **Submergent** plants, often referred to as pond weeds, have leaves and flowers that remain below the surface and grow in deeper water, unseen by the shore-bound observer. In addition, free-floating freshwater algae are common in ponds, providing food for snails, polliwogs, water fleas, and mosquito larvae.

water tiger eating a tadpole

A number of predators are found in ponds. Both the immature and adult forms of the dragonfly, water strider, whirligig beetle, water boatman, and predaceous diving beetle (whose larvae are aptly named water tigers) are predatory. Some of their favorite foods include frog eggs, tadpoles, and insects. The red-spotted newt, as well as many species of frogs, turtles, and snakes, are all animals that eat other animals in the pond.

Pond-dwelling creatures have amazing adaptations for breathing in their watery habitat. The whirligig beetle carries a bubble of air under its abdomen when it submerges, using this bubble as an oxygen tank. Some beetle larvae bore into plant stems to get their oxygen. Mosquito larvae and water scorpions breathe through tubes that pierce the water's surface film. Most salamanders use oxygen from the water that diffuses through their skin. A whole variety of animals use gills to obtain dissolved oxygen from the water. Fish gills are familiar, while a damselfly's gills look like three feathers at the end of its body.

Pond-dwelling creatures also have a variety of methods for moving through the water. Turtles and beavers use webbed hind feet for paddling, and fish swim with swift, undulating movements. Some insects, like predaceous diving beetles and water boatmen, have oar-like legs for paddling. One of the most remarkable strategies is that of the dragonfly nymph, which uses a form of jet propulsion. It draws water into its rectal gill chambers and then shoots it forcefully out, causing the nymph to dart forward.

A pond may seem still and static, but it abounds with a multitude of living things. This is demonstrated by the excited squeals of children who come looking for frogs and find so much more.

The water boatman hooks one leg around a plant to anchor itself under water.

Suggested References:

Andrews, William, ed. *A Guide of the Study of Freshwater Ecology*. Englewood Cliffs, NJ: Prentice-Hall, 1972.

Caduto, Michael J. *Pond and Brook: A Guide to Nature Study in Freshwater Environment*. Englewood Cliffs, NJ: Prentice-Hall, 1985.

Coker, Robert E. *Stream, Lakes, Ponds*. New York: Harper and Row, 1968.

Graves, Eric V. *Discover the Invisible: A Naturalist's Guide to Using the Microscope*. Englewood Cliffs, NJ: Prentice-Hall, 1984.

Reid, George K., and Herbert S. Zim. *Pond Life*. New York: Golden Press, 1967.

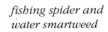

fishing spider and water smartweed

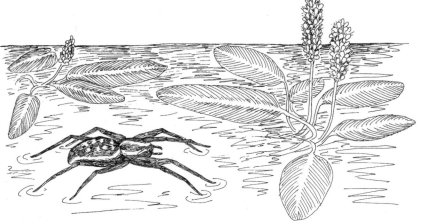

Ponds

FOCUS: A pond is composed of different habitats, each of which hosts a number of creatures specially adapted to live there.

OPENING QUESTION: *What are the challenges of living in a pond habitat?*

PUPPET SHOW

Objective: To identify some adaptations needed to live in a pond.

Perform, or have the children perform, the puppet show. Using the puppet characters as examples, discuss the various adaptations needed for life in the pond.

Materials:
- puppets
- script
- props

PICTURE A POND WITH YOUR EARS

Objective: To become aware of the variety of sounds around a pond.

Upon arrival at the pond, have the children sit quietly in a circle with their eyes closed, listening carefully. Have them hold up one finger for each different sound heard. When most children have heard five or more sounds, ask them to open their eyes and share sounds heard. Which sounds are commonly associated with a pond? Which did they like the most? The least?

POKING AROUND POND HABITATS

Objective: To collect and observe creatures from different pond habitats.

Provide simple collecting equipment and, in small groups, have the children collect pond animals from the different pond habitats. Place the collected creatures in white dish basins filled with pond water. Observe the creatures, watching to see which stay on the surface film, which swim through the water, and which walk on the bottom. Compare the physical characteristics of the different animals. Try using field guides to identify specimens. Follow this activity with Pond Pantomime or end with a special release ceremony so that all animals are returned safely to the water.

Materials:
- white dish basins
- plastic cups
- collecting nets or large plastic kitchen strainers
- hand lenses
- (optional) field guides

POND PANTOMIME

Objective: To observe and recreate the movements of various pond creatures.

In small groups around separate collection basins, each child should choose one specimen to observe, noting which parts of its body are moving, how it propels itself, and/or how it might breathe and eat. Using available materials (reeds, rocks, leaves) for props, each child should imitate the creature's actions for the rest of the group. Others guess the creature's identity by pointing to the matching one in the community dish basin. In which pond habitat does this animal spend most of its time and why? End the activity with a special release ceremony so that all animals are returned safely to the water.

Materials:
- dish basins with collected animals
- hand lenses

RELEASE CEREMONY

Objective: To return animals to their proper home in the pond.

Gather the children together and gently tip the collecting basin into the pond, returning the pond creatures back to their home. Have the children repeat the following poem.

> *Swim away, float away, dive away, leap –*
> *You're free to go, I'm not going to keep*
> *You from living your life.*
> *You deserve to swim free.*
> *Thanks for sharing this time with me.*

POND MURAL

Objective: To review and illustrate the different habitats within a pond and the accompanying pond critters.

On a large sheet of paper, have the children work together to create a pond mural, incorporating the various habitats found within a pond: the surface film, open water, water's edge, submerged plants, and the bottom. Next, have each child draw a picture of one of the pond animals collected. Ask each child to carefully examine the animal's adaptations. Then one at a time, have each child place the animal on the mural in the pond habitat where it might live.

Materials:
- large sheets of paper
- paints, crayons, colored pencils and/or markers
- tape

SHARING CIRCLE

Objective: To share new discoveries about the pond and its inhabitants.

Sit together in a circle and have each child complete the following sentence. "I would like to be a _____ in the pond because _____"

EXTENSIONS

Pond Aquarium: Set up a pond aquarium using water, plants, and animals from a nearby pond. Make sure the water is kept cool and the food is replenished. Observe the animals, noting which are most active and where they hide. Try to identify who's who. Return the animals to the pond within three days.

Summer Day: Have the children pretend they are pond creatures, and ask each to write a story or draw a picture about a summer day in the pond from the perspective of the creature each has chosen.

A Closer Look: Use microscopes to view plankton, the smallest organisms in the pond.

Ponds

Characters: Mother, Polly, Dream Fairy, Wesley Water Boatman, Shirley Whirligig Beetle, Dana Damselfly Nymph

Prop: Pond Polly—girl puppet with swim mask, snorkel, and paddles for extra arms

Mother

Polly, honey, just where are you going at this time of night?

Polly

Out to the pond, of course. I need a quick swim before bed.

Mother

But Polly, you've been swimming in the pond all day. I think that's enough! Now off to bed you go. Come on, hop to it. *[exits]*

Polly

Hmmm. I'd rather hop right to the pond. I wish I were a frog or a fish or a water bug; then I'd spend my whole life in the pond! Well, the sooner I fall asleep, the sooner I'll be able to go to the pond again! *[lies down; Dream Fairy enters]*

Dream Fairy

As an expert on dreams, I think t'would be good
To make sure Polly knows all that she should.
For life in the pond is not all it may seem
As she shall discover in her very next dream.
[Dream Fairy leaves and dream begins]

Polly

Oh, I'm so happy. Mom says I can live in the pond forever. I won't bother packing a suitcase. I'll just wear my bathing suit. *[Water Boatman enters]*

Wesley Water Boatman

Hold on, Polly. You're going to need more than a bathing suit.

Polly

Who are you?

Water Boatman

I'm Wesley Water Boatman, at your service.

Polly

Oh, I've seen you paddling around in the water. You're so quick. It must take a lot of practice to swim like that.

Water Boatman

It takes more than practice, Polly. We water insects are well equipped for life in the pond.

Polly

Well, I'd like to be well equipped, too. What should I pack?

Water Boatman

Well, you'll need more legs, for one thing.

Polly

More legs? Oh, I guess having six of them must really speed you right along. Don't worry – I'll use both my arms and my legs to help me swim.

Water Boatman

But how will you catch food? We use our two front legs for that.

Polly

Well, let's see. I'll have to use my hands to catch food. I know! I'll tie canoe paddles to my waist for pushing. Then we'll both look like little rowboats with oars sticking out to the sides!

Water Boatman

That might work. But maybe you should meet some more water bugs and see what other kinds of equipment you might need.

Polly

But what more could I possibly need? *[Water Boatman exits; Whirligig Beetle enters]* Hey, where's Wes? And who are you?

Shirley Whirligig Beetle

Why, hello. Let me introduce myself. I'm Shirley, a whirligig beetle.

Polly

Oh, you're one of those little bugs that spin about the pond like little bumper cars. It's amazing you never bump into each other.

Whirligig Beetle

Thanks to our antennae, we steer clear of trouble. They feel the tiny waves in the water and tell us where we can and cannot go.

Polly

Well, it's a good thing you have those antennae! Your eyes are split in half, so they must not work very well.

Whirligig Beetle

They are split, but all the better to see with, my dear. The upper half can see above the water, and the lower half can see under the water. We don't miss a trick!

Polly

Hmm. Maybe I should bring along swim goggles, too!

Whirligig Beetle

Good idea! Well, I've got to dive. See you down under! I'll just grab a bubble of air before I go. It lets me breathe under water. *[exits]*

Polly

Breathe under water! Oh no, I hadn't thought about that. I'd better pack a snorkel, too. Gosh! This is getting complicated! *[Damselfly Nymph enters]*

Dana Damselfly Nymph

You could just breathe the way I do – with gills.

Polly

With gills? You don't look like a fish! Who are you?

Damselfly Nymph

I'm Dana the Damselfly Nymph. I live under the water now, but someday I'll have wings. Then I'll be a dainty damselfly hovering over the pond.

Polly

Sounds exciting. Now, did you say you have gills? I don't see any.

Damselfly Nymph

What did you think these tails at the end of my body were for? Wagging?

Polly

Well, yes. I mean, no. I mean, I just thought they were tails.

Damselfly Nymph

Well, these tails have gills. And I couldn't breathe under water without them. Now I'd better dive, or I'll be a damselfly in distress! *[exits]*

Polly

Oh, now I'm in distress! I don't have tails or gills! *[Water Boatman appears]*

Water Boatman

So, Polly, are you all set for living in the pond?

Polly

Well, I'm not so sure anymore. There are so many things I'll need.

Water Boatman

Don't worry. I made a model for you. If we can just get you looking like this, then you'll live happily ever after in the pond! *[Water Boatman disappears, Pond Polly puppet appears]*

Polly

Ahhhh! *[Polly faints, Pond Polly exits, then Polly wakes up]* What a nightmare I just had! I've had bad dreams with monsters in them, but never one where **I** was the monster! Swimming in the pond is one thing, but living there is quite another. Good night, everyone!

Cycles

As we study the natural world, we soon discover that life has few clear-cut beginnings and endings, but rather passes through stages in a continuing cycle. A leaf bud contains the beginnings of new life, but it is also the final product of a summer's growth. A flower dies, but its seeds are left to carry on. Swallows leave their nests and fly to warmer climates for the winter but return each spring to begin nesting again.

The life cycles of organisms often reflect the seasonal cycles of their environment – a resting stage during harsh seasons, an active stage at times when food and warmth are plentiful. Plants bloom and grow in the lengthening days of spring and summer when sunlight and water are available and become dormant as the days shorten and turn colder. Animals, too, change through the seasons. For example, many moths and butterflies survive the winter months in the egg stage from which the tiny, hungry caterpillars emerge just as tender new leaves unfurl in the spring. Likewise, birds nest and raise their young when food supplies, such as plump, juicy caterpillars, are abundant.

Through the activities in this theme, children learn how different organisms change during their lives, and how these changes are linked to the seasons. Children visit each of the topics – insects, trees, birds, and flowers – at different times of the year, witnessing for themselves these cycles of ongoing change.

wild strawberry

Insect Lives

SURVIVING THE SEASONS IN STAGES

Insects are perhaps the most successful forms of animal life on this planet. Like spiders and lobsters, insects are **arthropods,** with external skeletons and jointed legs; however, they belong to a separate class, the Insecta.

There are many reasons why insects have survived and even thrived through 350 million years of climatic and habitat changes. Their small size, the ability of most to fly, the wide variety of foods they can eat, and the rapidity with which new generations are produced have all contributed to their continued existence. But there is one especially effective strategy that insects have developed to a high degree of perfection: **metamorphosis**. Metamorphosis describes the transformation process by which an insect proceeds from the egg stage to that of mature adult.

There are different forms of metamorphosis. **Simple metamorphosis**, sometimes called incomplete or gradual metamorphosis, consists of three stages: egg, **nymph**, and adult. Grasshoppers, true bugs, and cicadas are examples of insects that go through simple metamorphosis. A newly hatched nymph most often resembles the adult insect, but it is smaller and wingless. Dragonflies, damselflies, and stoneflies also undergo simple metamorphosis but the young of these insects look quite different from the adults. The growing nymph molts, usually several times, shedding its **exoskeleton** after having grown a new, larger one underneath it. After its final molt, the insect emerges as a sexually mature adult, generally equipped with wings.

About 87 percent of all the known insect species go through **complete metamorphosis**, including moths, butterflies, bees, wasps, ants, beetles, and flies. This life cycle consists of four stages: egg, **larva**, **pupa**, and adult. The egg hatches into a larva, which does not resemble the adult. Larvae have been given various names; a fly larva is a maggot, a beetle larva is a grub, and butterfly and moth larvae are called caterpillars. Generally, larvae are mobile, and they have chewing mouthparts even though as an adult they may not. Many insects do most of their feeding and growing in their larval stage. As the larva grows, it molts several times until, having completed its growth, it enters a resting stage called the pupa.

To change from larva to pupa, an insect may spin a silken covering over itself, like the cocoon of some moth species, or it may find a sheltered spot where it can shed its outer skin, leaving only a thin membrane of protection, such as a butterfly's **chrysalis**. During the pupal stage, the larval tissues are broken down and slowly rebuilt into organs more suited to adult life. When this transformation occurs, the larval cells begin to die and clusters of adult cells, inactive up to this point, are stimulated by hormones into growth. Essentially the body is reorganized: wings develop, reproductive organs are formed, mouthparts change, and most of the muscular system is transformed. The adult then emerges from the pupal case ready to reproduce.

Metamorphosis contributes greatly to insect survival. Eggs are often encased in hard coats that can withstand extreme cold or dryness, and emergence is sometimes delayed until the proper conditions are available. The larval stage of any given insect is timed and located to coincide with the appropriate food supply. Colorado potato beetle larvae hatch out when the succulent potato leaves are ready to eat. Monarch caterpillars emerge from eggs a few days after they are laid by the mother butterfly on milkweed leaves. For many insects, it is in the egg or the pupal stage that they are best able to survive extreme conditions, so many insects overwinter as eggs or pupae.

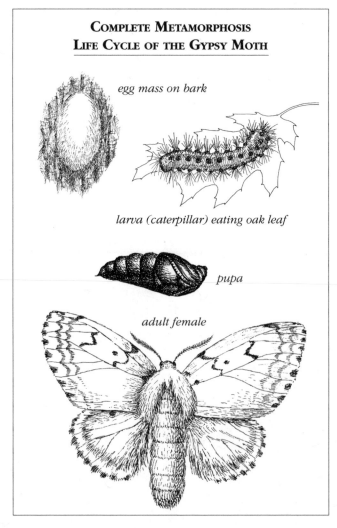

COMPLETE METAMORPHOSIS LIFE CYCLE OF THE GYPSY MOTH

egg mass on bark

larva (caterpillar) eating oak leaf

pupa

adult female

In many species, the changes that occur with metamorphosis allow insects to inhabit very different habitats in different stages of their life cycles. For example, the dragonfly nymph spends its young life under water, feeding beneath the surface and breathing with gills. When ready to emerge as an adult, however, the nymph crawls out of the water, splits its skin for the last time, and is able to breathe air as a winged adult.

The adult stage of insects has but one mission: to make sure that eggs are laid for the next generation. Being short-lived, insects must find their mates quickly. Each species has its own strategy by which potential mates find each other. Different species of fireflies, which are not really flies but beetles, use light signals to attract mates. Females generally perch (some are flight-less) and watch for flying males flashing the right signal. The different species' signals vary by time of day; speed of flight; distance between signals; and number, length, and frequency of flashes. The female firefly then signals the male, and he approaches her. Similarly, the colorful, day-flying dragonflies and butterflies can rely on sight, with males cruising or finding high perches from which to search for females. Night-flying male moths, on the other hand, sense the female's **pheromone** through their **antennae**, in some instances detecting another moth several kilometers away. Sounds are another way many insects communicate. A cricket hears the chirping of others of its kind through the **tympanum** on its foreleg. Grasshoppers, cicadas, and katydids all sing their songs at different times of the day and year.

Having found partners, insects mate and lay eggs. Many die soon afterward. Some adult insects, such as stoneflies and mayflies, live only days or hours; many even lack mouthparts. The lives of butterflies in summer may be only a couple of weeks long; others may over-winter as **dormant** adults and live for many months. In many species of beetles, the adult life is longer than the larval stage, and the adult consumes more food than the larva.

Insects' life cycles enable these small but marvelous creatures to take advantage of widely diverse habitats, while expending a minimum of energy to do so. Metamorphosis is one of the reasons why insects have been successful for millions of years.

Suggested References:

Borror, D. J., C. A. Triplehorn, and N. F. Johnson. *An Introduction to the Study of Insects.* Orlando: Harcourt-Brace, 1989.

Borror, D. J., and R. E. White. *A Field Guide to the Insects.* Peterson Field Guide. Boston: Houghton Mifflin, 1970.

Hubbell, Sue. *Broadcasts from the Other Orders.* New York: Random House, 1993.

Stokes, Donald. *A Guide to Observing Insect Lives.* Boston: Little, Brown, 1983.

Zim, Herbert S., and Clarence Cottam. *Insects: A Guide to Familiar American Insects.* New York: Golden Press, 1987.

SIMPLE METAMORPHOSIS
LIFE CYCLE OF A STINK BUG

egg

nymph stages

adult

Insect Lives

FOCUS: Insects go through different stages as they grow from egg to adult. The stages of their life cycles are timed to fit with seasonal cycles.

OPENING QUESTION: *In what ways do insects change as they grow?*

PUPPET SHOW

Objective: To compare the processes of complete and simple metamorphosis.

Perform, or have the children perform, the puppet show. Ask the children to compare how the grasshopper and the butterfly grew to be adults. Using a diagram to illustrate the various stages, discuss simple and complete metamorphosis. How do insects' life cycles coincide with seasonal changes?

Materials:
• puppets
• script
• insect life cycle chart

METAMORPHOSIS PUZZLES

Objective: To learn the different stages of the two main life cycles of insects.

Give each child an insect puzzle piece that portrays one stage in the life cycle of some insect. Ask the children to look for others whose puzzle pieces fit with theirs. When they've completed the puzzle, groups should discuss whether they think their insect undergoes complete or simple metamorphosis. Have the children share the different insects with the whole group. If they were to find their insect outside today, in what life stage do they think it would be? Have older children look up their insects in a field guide to see in what stage they overwinter.

Materials:
• insect puzzles, including 3-piece simple metamorphosis puzzles (e.g., spittlebug, dragonfly, grasshopper) and 4-piece complete metamorphosis puzzles (e.g., lacewing, housefly, mosquito, honeybee), enough for one puzzle piece per child
• (optional) insect field guides

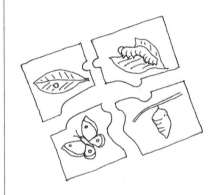

FIND YOUR KIND

Objective: To experience the variety of ways insects signal their mates using different senses to receive the messages.

Explain that because insects are tiny, they need to send and receive messages in special ways, especially when they are seeking mates. The three insects examined in this activity are active at night. What special signals and senses do children think might help insects find each other in the great outdoors at night? Set up three stations: one for fireflies, one for moths, and one for crickets. (If time is short, select 1 or 2 strategies to do and mention the others.) Divide children into groups of 8-10 to rotate through the different stations.

Note: For an added challenge, try these activities at night or in a darkened room, and/or simultaneously, like a real summer night.

STATION I - MOTHS:

Discuss how moths locate each other using their sense of smell. Each child receives a scent container; half of the containers are labeled "female moths" and the others "male moths." Female moths form a large circle around the males. At "go," females hold out open scent canisters, and males move around the circle until all the males have found a female canister with the same scent as their own. Then, switch roles, and have the children who are now the "female moths" swap scent canisters. Repeat the activity with the new roles and swapped canisters.

Materials:
• duplicate sets of scented cotton balls (use natural extracts like anise, lemon, mint) in film canisters labeled "male moth" and "female moth." If desired, place a small color-coded dot on the bottom of the canisters to identify matched scents.

Station II - Fireflies:

Explain that different kinds of fireflies have different signals. In this activity, the leader is stationary like a female firefly, and the children are the searching males. Give each child a firefly code card (use 3 or more different codes for older kids, and just 2 codes for younger kids). Have the group form a line a small distance from the leader. Using a flashlight, the leader flashes one of the codes and all children with the matching code fly forward. Afterward, they return to the line and the leader flashes another, different signal. Repeat until all codes have been flashed. For older children, give flashlights to three or four children to flash different signals simultaneously, and the other children must find their own kind.

Materials:
• flashlights
• firefly code cards
  (some possible codes:
  "long-short-long," "short-short,"
  "long")

Station III - Crickets:

Explain that male crickets sing, and female crickets are attracted to the song of their species. Each child receives a cricket song card instructing them to either "sing" or "listen for" a particular sound. At "go," children designated as singers start singing their particular call, while the others listen for and join the child that sings their call.

Materials:
• matching sets of cricket song cards (possible sounds: bzzz, chirp, tick-tick, pit-pit)

Insect Hunt

Objective: To observe insects in the different stages of their life cycles.

As a whole group, discuss ways to distinguish insects from other tiny animals. Divide the children into small groups or pairs. Give each a pencil and an Insect Stages Search card with the following directions:

Materials:
• Insect Stages Search cards
• clipboards
• pencils
• (optional) insect field guides

Insect Stages Search Card

Look for adult: butterflies/moths, grasshoppers, crickets, bees, ants, flies, beetles.

Look for cocoons or pupa cases.

Look for nymphs: grasshoppers without wings, small earwigs, wingless leaf hoppers, dragonfly nymphs in the water, others.

Look for larval stages: fuzzy caterpillars, inchworms, caterpillars with interesting patterns or colors, larvae with camouflage, leaves eaten by larvae.

Look for eggs (often on the underside of green leaves) or insects laying eggs.

Groups should visit at least three different areas (e.g., playground, field, under trees, hedgerow, building ledges) to conduct their search. Check off findings. Older children should record some specific details, such as where each insect was found, special characteristics, what it was doing. Suggest they use field guides to identify a few.

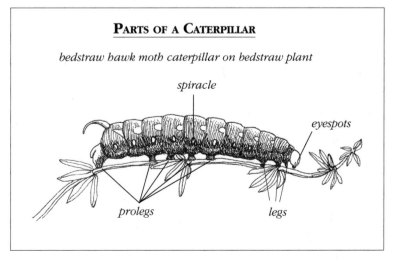

PARTS OF A CATERPILLAR

bedstraw hawk moth caterpillar on bedstraw plant

spiracle

eyespots

prolegs

legs

TRANSFORMATION

Materials:
- insect nametags
- crackers or celery
- gym cones or other place markers
- signs saying "Eggs," "Pupate Here," "Adults," and "Next Generation"
- egg carton
- bowl of marbles

Objective: To experience the stages in complete metamorphosis and illustrate how the cycle repeats.

Ahead of time, set up gym cones in a triangle with 30' between cones. At one cone you'll place a sign saying "Eggs," at the next will be a sign reading "Pupate Here," and the third cone will have a sign reading "Adults." Place an egg carton and a bowl of marbles near the "Adult" cone and mark off a "Next Generation" waiting area near the "Eggs" cone.

Divide the children into groups of 2-3, each group representing a different type of insect that goes through complete metamorphosis. Give the children a nametag identifying their insect: mosquito, lacewing, monarch butterfly, luna moth, house fly, carpenter ant, yellow jacket, honeybee, June beetle, ladybug, or other.

Have one child from each small group assemble at the "Eggs" cone. (The others stand in the "Next Generation" area and watch and cheer for their teammates while they wait for their own turns.) Give "Eggs" a cracker to save for eating once they hatch. Children begin by squatting down and pretending to be eggs — silent, curled up, eyes closed. At the leader's signal, they hatch and become larvae. They must then crawl or walk in baby steps toward the second cone while eating their cracker. When they reach the "Pupate Here" cone, the children stand very still while softly chanting "I'm changing, I'm changing, I'm changing." When the leader calls out "Emerge!" the insects burst out, "flying" around, flapping their wings and moving toward the "Adults" cone. Once there, each child moves one marble from the bowl to the egg carton, and then gets a waiting teammate and takes him or her to the "Eggs" station. At that point, the children in the first group of insects have completed their life cycle and each must shout, "Long live the Luna Moths!" (or Honeybees or Carpenter Ants or …) and "die" dramatically. The leader then gives each new Egg a cracker and, at the signal, the new group repeats the cycle.

With younger children, leaders may wish to act out the stages for the children before asking them to perform. Pictures of insects may be used in place of written words on their nametags.

SHARING CIRCLE

Objective: To personalize the learning and reflect on the various stages in an insect's life cycle.

Sitting in a circle, have the children close their eyes and picture their favorite insect. Where does it live? What is it doing? Then, one at a time, have the children complete the sentence, "My favorite insect to watch is a _____ in its _____ stage, because _____."

EXTENSIONS

Monarch Butterflies: Have the children research the life cycles of monarch butterflies and find out where they spend the winter. Check out the Internet site: Journey North.

Creative Writing: Invite the children to write a story about what it would be like to go through metamorphosis (as an insect or as a human).

Cricket Watch: Research the natural history of crickets. Keep a male cricket in a cage in the classroom for a day. Observe its singing.

Insect Lives

Characters: Grasshopper Egg, Grasshopper Nymph, Grasshopper Adult, Butterfly Egg, Butterfly Caterpillar, Butterfly Adult, Insect Fairy

Props: sign saying "Weeks Later"; pupa case, attached to a plant

Grasshopper Egg

Hey, you up there. Hey, little egg. I see you. Can't hide from me. What's it like being a leaf egg?

Butterfly Egg

I'm no leaf egg. I'm a butterfly egg. I was just placed on this leaf by my mother butterfly. Who are you, anyway?

Grasshopper Egg

I'm a grasshopper egg. Pretty soon I'll be hopping circles around you.

Butterfly Egg

Is that so? While you're hopping in circles, I'll be flying through the air over fields and ponds.

Grasshopper Egg

You? Flying? I can't believe that. No way is there room for wings inside your egg.

Butterfly Egg

You'll see. I'll be flying...

Grasshopper Egg

Oh my, oh my goodness. I think I'm about to be born as a grasshopper. Oh, oh. *[Egg goes down; Grasshopper Nymph appears]* See that? I am a grasshopper. I told you I'd be hopping circles around you.

Butterfly Egg

Not for long. I think my time has come, too. I think I'm about to become a butterfly! Oh, oh. *[Egg goes down; Caterpillar appears]*

Caterpillar

See that, I'm a...Oh, no, I'm **not** a butterfly.

Grasshopper Nymph

Ah ha ha! Why, you're a caterpillar. Where'd you get the idea that you'd be a butterfly? You'd better get used to crawling, instead of dreaming about flying. See you later! *[exits]*

Caterpillar

Oh, *[crying]* I'm so upset. I know I was supposed to be a butterfly. I just know it. *[Insect Fairy enters]*

Insect Fairy

Stop crying, little caterpillar. You're just going to get that leaf all soggy, and then it won't be any good for eating.

Caterpillar

Who are you?

Fairy

I'm the Insect Fairy. I look after all the insects of the earth. Please stop crying, or everyone will think I'm not doing my job.

Caterpillar

Well, you're not. I was supposed to be a butterfly, and look at me! I'm nothing but a caterpillar.

Fairy

Stop worrying. You'll turn into a butterfly and be able to fly, but you need to have patience, and you need to eat a lot. You know, your mother laid your egg on just the right kind of plant for you to eat. So start chewing!

Caterpillar

Patience! Why do I have to wait so long? That grasshopper went straight from an egg to a full-grown grasshopper.

Fairy

First of all, you're different from a grasshopper, and second of all, she's not a full-grown grasshopper.

Caterpillar

What do you mean? I saw her go hopping...

Fairy

I know she **looks** like a full-grown grasshopper, but if you had looked a little more closely instead of wetting your eyes with tears, you'd have seen that she doesn't have wings.

Caterpillar

What! No wings?

Fairy

That's right. Full-grown adult grasshoppers have wings. That little grasshopper is just a nymph. She looks like an adult, but she has not yet grown her wings.

Caterpillar

What did you call her? A nymph? Ha, ha, that's a good one. I'd rather be a caterpillar than a nymph any day.

Fairy

Well, one day you'll be a butterfly. So no more tears. Farewell. *[exits; Grasshopper Nymph enters]*

Grasshopper Nymph

Hi, little caterpillar. Still waiting to turn into a big butterfly?

Caterpillar

As a matter of fact, I am waiting to turn into an adult. If you took time to look at yourself, you'd notice that **you're still waiting, too!**

Grasshopper Nymph

What do you mean I'm still waiting, too? I already **am** an adult grasshopper.

Caterpillar

No you aren't! Adult grasshoppers have wings. You haven't grown your wings yet, so you're just a nymph – nothing but a little grasshopper nymph. But I can't stand around talking, I have to keep eating. Bye. *[Caterpillar exits]*

Grasshopper Nymph

She's right. I don't have any wings yet. I'm not a full-grown grasshopper. I **am** just a nymph. I guess I'd better be on my way, too. It'll be a while before I'm big enough to get my wings. *[exits; hold up sign saying "Weeks Later." Then Grasshopper Adult and pupa case hanging from plant appear]*

Grasshopper Adult

I've been feeding here for almost a month, and I keep looking for that caterpillar, but I can't find her anywhere. Do any of you see her? *[Butterfly enters]*

Butterfly

Here I am, here I am.

Grasshopper

Where did you come from?

Butterfly

I came out of this pupa case that's hanging from the leaf.

Grasshopper

But when I saw you last, you were a caterpillar.

Butterfly

Well, when I grew to be big enough and old enough, I split my caterpillar skin and this pupa case was underneath it. The whole time you were looking for me, I was right there inside the case changing from a caterpillar to a butterfly. Just before you got here this morning, I came out as a full-grown butterfly.

Grasshopper

So you were right. I'm sorry I didn't believe you.

Butterfly

And I'm sorry I called you a silly nymph. I see you already have your wings.

Grasshopper

Oh, yes. I had to shed my skin a few more times, and then I found I had a full set of wings.

Butterfly

You know, Grasshopper, I think it's amazing that we're grown-ups now. We both can fly, but we grew up in such different ways. I changed completely, and you just kept changing a little at a time.

Grasshopper

You're right. And now we'd both better think about finding a good place to lay eggs before winter. Maybe next summer our kids will be having the same argument we had. That would be funny. See you around.

Butterfly

Yeah, I've got to fly, too. I want to find the right kind of plant to lay my eggs on, just like my mother did. Bye, bye. *[both exit]*

Meet a Tree

THE SUM OF MANY PARTS

Suppose you read that someone had invented a machine that is powered by the sun; has an automatic thermostat and humidifier; manufactures its own fuel out of water and carbon dioxide; can split a rock or pump tons of water; produces oxygen, water, food, and fuel; *and,* rather than polluting the air, actually cleans and beautifies its surroundings. Incredible? Yes, but this is an accurate description of a tree. These complex and beautiful organisms grace our backyards and neighborhoods, performing invisible but amazing feats, and changing as they grow through the seasons and the years.

A tree is a sum of many parts, most of which are familiar to us, but their functions are worth mentioning. The trunk supports the branches and twigs, which in turn hold up the crown of leaves to the sunlight. The trunk also contains the cells that transport water and minerals from the roots upward and food (in the sap) from the leaves downward. These water and food channels are called the **xylem** and the **phloem,** respectively. Both are added to each year by a tiny ring of growth cells just inside the bark called the **cambium.**

Xylem cells are the new wood cells formed on the inner layer of the cambium. Nutrient-rich water moves upward through the xylem cells, drawn by **capillary action** as it evaporates from the leaves. New, active xylem cells are referred to as sapwood, while old dead cells nearer the center of the tree help support the tree's bulk and are called the heartwood. The age rings one counts in tree stumps are the xylem growth layers. The (usually) wider part of each ring shows the fast spring growth with its large, thinner-walled cells, and the neighboring ring shows the smaller, thicker-walled cells of summer's slower growth. The annual rings in a stump,

then, tell the tree's history; the wider the ring, the greater the growth that year. Foresters gather information about a tree's past by examining these rings: slower growth may have been caused by drought or insect damage; faster growth may mean ample rains and sunlight; noticeable scars record fire damage or other injury.

The phloem cells, which transport the sugars manufactured during photosynthesis to the rest of the tree, form on the outer edge of the cambium layer. When these cells die, they become bark. The bark of different kinds of trees varies from the thick, shaggy bark of the shagbark hickory to the thin, smooth bark of the American beech. Bark serves as the tree's protective covering, shielding it from disease, infestation, and weather.

The roots of a tree are its most aggressive, persistent part. Roots can split rocks or inch their way along cliffs to find life-giving soil. A growing root three inches wide and four feet long can lift 50 tons. The function of roots is to anchor and support the tree and to absorb water and nutrients from surrounding soil. The actual absorption of water and nutrients is accomplished by miles of tiny root hairs that appear and grow just behind the tip of the constantly elongating roots. Most **conifers** (cone-bearing trees) have a shallow, fibrous root system parallel to the surface of the ground. Many **deciduous** trees (those that drop their leaves in the fall) have a spreading root system plus a long,

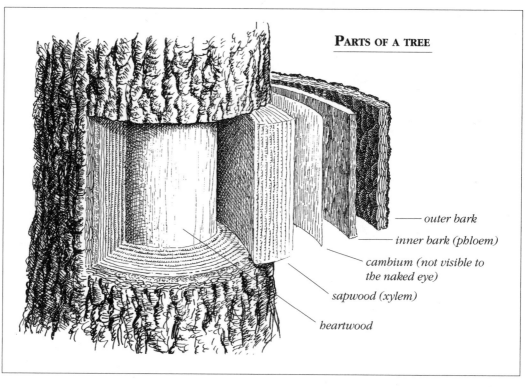

PARTS OF A TREE

— outer bark

— inner bark *(phloem)*

— cambium *(not visible to the naked eye)*

— sapwood *(xylem)*

— heartwood

perpendicular taproot. The trees that topple first in gale winds are the ones with shallow root systems.

Most picturesque, perhaps, are the leaves or needles of a tree. They manufacture the tree's food through a process called **photosynthesis**, converting water and carbon dioxide into sugars and starches. They perform this chemical magic with the aid of **chlorophyll**, produced by the tree, and sunlight, the energy source that drives the reaction. Oxygen is released as a byproduct of photosynthesis, and more than 99 percent of the water absorbed by the tree **transpires** through the leaves. The upward pull of water within a tree has been estimated as 100 times more efficient than the best suction pump ever made. On a hot summer day, a giant oak may release 300 gallons of water into the atmosphere.

As natural organisms, trees must adjust to the seasons of the year. Summer is the growing season, when the tree's leaves manufacture food and its flowers produce seeds. Tree seeds come in many packages, from the winged samaras of the maples and ashes, to the fleshy fruit of apples and cherries, to the woody cones of pines and spruces. No matter the form, however, the function of these seeds is to produce another generation of trees.

During the fall, deciduous trees prepare for winter by shutting down. Winter is a time of drought for trees, since most ground water is frozen and not readily available. If leaves continued to transpire in winter, the tree would quickly dehydrate. Deciduous trees prevent this water loss by dropping their leaves in the fall. In the north then, deciduous trees must accomplish most of their growth and seed production within the few warm months. Evergreen needles, on the other hand, can remain attached year-round as they have less surface area through which to lose moisture and a waxy coating that greatly retards moisture loss. Also, they are filled with resins that resist freezing and the resulting breakdown of cells.

Spring is the tree's wake-up season. Sap that was stored in the roots during the autumn and winter rises through the trunk and travels out into the branches and twigs. There it feeds the buds that formed the previous summer, enabling them to develop into the shoots, leaves, and flowers that begin the next growing season.

Tree leaves that fall to the ground, plus other organic matter in the soil, provide many of the elements needed for the tree to grow. Decomposition by earthworms, bacteria, fungi, and other organisms releases nutrients back into the soil. The breakdown of rocks, water, and carbon dioxide provides other needed elements. In this way, the forest ecosystem recycles its basic building blocks from its physical environment.

In addition to nutrients drawn from the soil, trees also require sunlight to grow, and a tree's shape will adapt to meet this need. Because plants reach for the sunlight that fuels photosynthesis, a tree that is crowded on one side by other trees will have longer branches and more leaves on its open, sunlit side. Plantation pine trees growing in close rows will have live, green needles at the top and dead or dying branches along their trunks because these lower branches receive little sunlight. A tree growing in a field will fill out in all directions.

Trees are a magnificent combination of productive efficiency and quiet beauty. They figure in many stories of the beginning of the world and are frequent characters in legends and myths. People appreciate trees for all they provide: shade, food, fuel, and recreation, and we come to know them as neighbors. We watch them grow and change, a visual reminder of the passing seasons and years.

Suggested References:

Brockman, C. Frank. *Trees of North America*. New York: Golden Press, 1968.

Gray, Robert. *The Tree*. Mechanicsburg, PA: Stackpole, 1993.

Harlow, William M. *Trees of the Eastern and Central U.S. and Canada*. New York: Dover, 1957.

Hutchins, Ross E. *This Is A Tree*. New York: Dodd, Mead, 1964.

Jackson, James. *The Biography of a Tree*. Middle Village, NY: Jonathan David, 1979.

Kricher, John C., and Gordon Morrison. *Ecology of Eastern Forests*. Peterson Field Guide. Boston: Houghton Mifflin, 1988.

a spreading silver maple

Meet a Tree

FOCUS: A tree is the sum of many parts, each designed to perform a necessary function within the tree's life and within its seasonal cycle.

OPENING QUESTION: *In what ways do trees change through the seasons?*

THE SUM OF MANY PARTS

Objective: To learn the different parts of a tree and their functions.

Bring in a sapling hidden inside a cardboard box (decorate with question marks). Tell the children that you've discovered a marvelous machine that runs by sun power, has an automatic humidifier and thermostat, manufactures its own food, cleans the air, is strong enough to split a rock, and is beautiful. What could it be? Present the sapling, then use it and pictures to introduce the characteristics and functions of the different parts of a tree.

Materials:
- sapling tree
- cardboard box
- slides, photos, or magazine pictures of tree parts

A TREE ARE WE

Objective: To review the parts of a tree and experience how a tree changes with the seasons.

Materials:
- slips of paper each noting a different tree part, one per child

Gather the children in a large circle and give each a piece of paper with a tree part written on it: heartwood, sapwood, bark, roots, etc. (Repeat roles as needed in order for each child to have a part.) With younger children, the leader selects children for each role as the tree is being built. Explain that together they are going to act out the role of one mighty tree. Starting with the heartwood, the leader calls out the various tree parts and describes their functions as the children join together to play their parts. After all are in place, ask the children what changes they might observe in their tree over the course of a year.

Heartwood (1 child, standing tall): You are in the very center of the trunk, strong and stiff, holding up the tree. Your cells are not alive anymore.

Sapwood or Xylem (2 children, arms joined around Heartwood): You surround the heartwood. Water and minerals flow up through your cells to the branches. You are not active in winter.

Bark (2-3 children, arms outstretched around Sapwood): You are the strong, protective outer layer of the trunk. Your inner layers grow new wood and transport the sap up and down.

Roots (2 children sitting with backs to Bark and legs outstretched): You grow from the base of the trunk, outward or downward, to anchor the tree. Your toes are the smallest roots that collect water and minerals from the soil.

Branch (2 or more children touching Bark with one hand and other arm outstretched): You come out from the trunk and hold all the twigs.

Twig (2 or more children grasping Branches): Each of you holds up the buds, leaves, flowers, and fruits.

Leaf (2 or more children holding on to Twigs): You grow out of a bud, unfolding in the spring to soak up the sun. Then in the autumn you change color and fall.

Flower (1 or more children holding on to Twigs): You come out in the spring or early summer, and by autumn you become a fruit, seed, or nut.

A LOOK INSIDE

Objective: To observe the parts of wood and see how wood records the life of the tree.

Divide the children into small groups. Give each group hand lenses and tree slices (plus other tree parts, if available). Have the children examine the tree slices, noting these parts: rings, made of larger and smaller cells; heartwood; sapwood (xylem); pith; layers of bark. Display a labeled tree anatomy chart for children to refer to as they examine the samples. Review the way a tree grows, with new layers of wood and bark added by the cambium. Discuss the seasons of growth for a tree and their relationship to the rings. What kinds of stories might the rings tell about the life of the tree? (possibilities include insect damage, drought, crowded on one side, good growing conditions)

Materials:
- tree anatomy chart
- hand lenses
- tree slices, at least two per small group
- (optional) bark samples, root samples, twigs, leaves

REACH OUT AND TOUCH

Objective: To examine a tree using senses other than sight.

Divide the children into pairs and all gather at Point X outside. Explain that partners will take turns keeping eyes closed or being blindfolded for the activity. The sighted partner carefully leads the blindfolded partner to a tree and asks the blindfolded partner some questions to help guide the exploration: Can you reach around the tree? Feel any branches? Any leaves? Describe the texture of the bark.

When the exploration is over, both return, blindfold still on, to Point X. Remove the blindfold and have the partner that was blindfolded find the touched tree. Switch roles – sighted person closes eyes – and repeat the activity with a different tree.

Materials:
- (optional) blindfolds, one per pair

LOOK AND FIND *(grades K-2)*

Objective: To explore the variety of trees in a nearby area.

Divide the children into small groups (each with an adult leader), and give the groups a Look and Find card. Outside, focus on one task at a time to get children looking and talking about the trees that live nearby.

LOOK AND FIND:

~ The largest tree you can. How many children does it take to reach around the largest tree you can find?

~ Two or more different tree seeds

~ The oldest leaf you can

~ A dead tree with a mushroom on it

~ Smooth bark

~ Listen for any sounds made by trees

~ Two or more different tree smells (crush or scratch leaves or twigs)

~ The tallest tree you can see

~ A tree that is smaller than you are

~ Your favorite tree

Materials:
- Look and Find cards

MEET A TREE QUESTIONNAIRE (grades 3-6)

Materials:
- Meet a Tree questionnaires
- clipboards
- pencils

Objective: To examine and record observations about a particular tree.

Divide the children into small groups or pairs. Give each team a questionnaire and clipboard. Within set boundaries, send them off to choose a tree and answer the questions. You may wish to have children visit these trees again in the winter, so encourage them to pick and flag a tree that will remain accessible. When all have finished, compare findings. Have the children share their "interview" questions.

MEET A TREE QUESTIONNAIRE

Do you think your tree is young, grown up, or old?_____

Can you reach the lowest branch?_____   Do your arms reach around the tree? _____

Stand back from the tree. What is its shape? _____

Are the leaves flat and broad____, or narrow and needle-like____, or are they like scales____?

What color are the leaves? _____ Are they all still on the tree ____, falling ____, mostly fallen ____?

What color are the smallest twigs?_____

Is your tree standing alone ____, near other trees ____, or as part of a forest ____?

Are the trees nearby taller ___, shorter ___, or about the same height ___?

Are there seeds or fruit on your tree?____ Flowers?_____

Look for plants growing under or on your tree. Record what kinds (mosses, lichens, mushrooms, grasses): _____

Look for insects, birds, and other animals on your tree. Record your observations: _____

What is special about your tree? _____

If you could interview this tree, what question would you like to ask it?

_____

TREE POEM

Materials:
- clipboards
- paper
- pencils

Objective: To describe a tree from different perspectives.

Divide the children into small groups. Designate one person in each group to be the recorder with paper and pencil. Each group should select a tree, and then each child in the group should be placed at a different distance from the tree (20 feet away, 5 feet away, under it looking up, nose to the bark, arms around it). After a moment's study, each person should give three words to describe the tree from his or her perspective, and the recorder should write them down. The group then creates a poem about its tree, using all the adjectives. Afterward, groups share poems, either introducing the tree or challenging the other groups to find it from the description.

SHARING CIRCLE

Sit together in a circle. Each child should complete the sentence: "I'd like to thank a tree for_____"

EXTENSIONS

Tree Measurements: Have children experiment with various ways to measure the height of trees.

1) The artist's quick method – one child stands next to a tree. Another child stands well away from the tree, holds a pencil out at arm's length with the eraser end even with the top of the head of the child by the tree and thumb on the pencil even with child's feet. Note the spot on the tree that corresponds to the tip of the eraser and then scoot the pencil up to measure again. In this way, count how many of that person would fit into the height of the tree. Multiply this number by the height of the child to get an estimate.

2) Triangulation method – you'll need a sunny day and a tree whose shadow falls across open, level ground. Hold a yardstick exactly upright and measure its shadow. Now measure the length of the tree's shadow. The relationship of the tree's height to its shadow will be the same as that of the stick to its shadow.

Tree Girth: Find the diameter of trees as foresters do, at "breast height" (4.5'). First measure the circumference of the tree at 4.5' above the ground, then divide that number by Pi (3.14) to get the tree's diameter.

Tree Questions: Help the children search for answers to their Interview questions from the Meet a Tree activity. Answers could uncover some interesting local history or lead to identification skills.

Stay in Touch: Encourage the children to visit their tree in different seasons and to notice how it changes. A class or group could mark their tree for the year, and make regularly scheduled visits to note leaves, buds, flowers, fruit, animal inhabitants, and so on.

More Trees: Initiate a project to plant some trees where they might give food or shelter to people or animals. Care for them. The Arbor Day Foundation, the local garden club, or state forestry and extension services are good resources.

hemlock roots growing over rocks

Seed Dispersal

INGENIOUS WAYS TO GET AROUND

Anyone who has blown the fluffy seeds from a ripe dandelion or tossed an apple core onto the ground has unwittingly contributed to one of the most important missions in the plant world – seed dispersal. Without the dispersal of seeds to new locations, young seedlings would be competing with their parent plants, often unsuccessfully, for sunlight, soil, water, and nutrients. Plants have evolved a wide variety of strategies, including dispersion by wind, water, and animals, to assist in the dispersal of their seeds.

Seed production and dispersal may not seem especially significant to those of us who are distracted by a plant's larger, showier structures, but for a plant, these processes are of primary importance. The two large plant groups that produce seeds are the flowering plants (**angiosperms**) and the cone-bearing **conifers** (or **gymnosperms**).

What are seeds? Biologically, a seed is a fertilized, ripened **ovule**. A seed contains an embryo that will develop into leaves, stem, and roots. Surrounding the embryo is a layer of tissue that stores enough food to nourish the tiny plant from the time it sprouts until its roots can take nutrients from the soil and its leaves can produce the plant's food. Outside this food-storage layer is an exterior coating that shelters the vulnerable embryo.

The outer seed coat protects the embryo from desiccation, freezing, and other dangers. A tomato seed is apt to be swallowed, but its seed coat is relatively smooth and hard, so the seed passes through an animal's digestive system intact. Each kind of seed, no matter how tiny, has its own distinctive seed coat. A hand lens will reveal the ridges, indentations, and sometimes tiny hairs that give a seed its characteristic markings.

The formation of viable seeds is a plant's primary goal; seed dispersal to a favorable location is the next assignment. Plants don't move about, so how can seeds

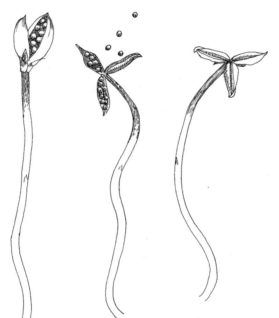

The blue violet's seeds pop out as the pod dries.

travel? Among flowering plants, it is at this stage that the seed container plays a vital role, whether it be an apple, an acorn, a peanut, or a cranberry. Plants package their seeds in ways that best guarantee dispersal.

Technically, a fruit is simply the ripened **ovary** that surrounds one or more seeds. Many fruits are designed to appeal to animals that might act as dispersal agents for their seeds. Fruits may be colored, scented, or contain nutrients that appeal to a certain kind of animal: mammals, birds, fish, turtles, even worms or insects. In eating the fruit, an animal may swallow the seeds, discard the seeds, or store them for future use. Squirrels hide acorns and forget to retrieve them all. Cherry seeds pass unharmed through the birds that eat them and get deposited in the birds' droppings. In some dry areas of the world, ants gather seeds for the oils they contain but then discard the rest of the seed in "trash" chambers of their ant hill, where conditions are ideal for germination. In all of these cases, the seed has been successfully dispersed.

Other seed containers rely on ingenious design rather than sensory appeal in order to disperse the seeds they contain. Some have wings or blades to propel them through the air whichever way the wind takes them. Maple, ash, elm, pine, and basswood trees have such seeds. Some seeds are borne along by the wind on parachutes or fluffy hairs. Airborne dandelion and milkweed seeds can travel long distances. Other seeds have sharp hooks or barbs that attach to passersby. Burdocks and beggar ticks are well-known hitchhikers. Many of us have learned through experience that the outer covering of the burdock seed head has hooks, and any attempt to remove it makes it fall apart, scattering the seeds.

There are even seed containers with seams that burst open with such force the seeds seem to explode from them. Jewelweed gets its other name, touch-me-not, from the sudden expulsion of seeds that follows touching a ripened pod. A few seed containers are buoyant and carry their seeds on the water to new destinations. Cranberries disperse their seeds in this way, as do other plants that grow in or near water. Peanuts are unusual in that they actually plant themselves – with the flower's ovary tip growing downward into the soil after fertilization – so no elaborate dispersal mechanism is needed.

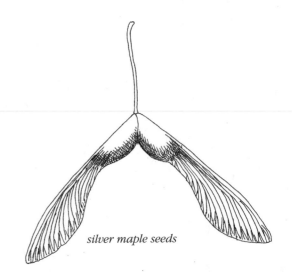

Since the time when humans began tilling the soil and traveling to all corners of the earth, we have become the primary dispensers of many seeds. As settlers sailed from Europe to the New World, they brought seeds for agriculture, thus dispersing them to a new continent. But many European seeds emigrated as accidental stowaways on clothing or in hay for livestock. When the pioneers cut down trees and plowed the land, they created ideal openings for many of these tag-alongs. Red and white clovers and timothy grass were deliberately brought to North America from Europe; the Canada thistles and bull thistles were probably accidental introductions. Some plants, like dandelions, purple loosestrife, Japanese knotweed, and lamb's quarters, were brought in because they were pretty or useful, but they quickly became weeds in their new homes.

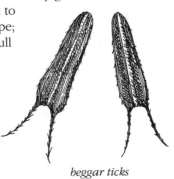

pin oak acorns

beggar ticks

Once a seed reaches a suitable habitat, germination is the next step. But sometimes special conditions must be met before growth can occur. Apple seeds and those of many other temperate-zone plants must undergo a winter chill before they will sprout. The seeds of the jack pine tree are locked inside their cones, sometimes for years, until the heat from periodic fires opens the cones and releases the seeds onto the fire-prepared seedbed. The dust-like seeds of the lady's slipper, like most orchids, store very little nutrition. The seeds germinate but cannot continue to grow until joined with a special **fungus** that helps them absorb nutrients from the soil. Willow seeds are viable for about a day after they blow off their tree or shrub; if they don't land in wet mud right away, they can't grow at all. Other seeds, including many weed seeds, germinate only in light. They can be buried in soil or shaded by other plants for years, waiting until a disturbance of the soil brings them up to the surface. This is why gardeners and

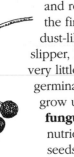

viburnum berries

farmers get a new crop of weeds every time they cultivate, even after weeding and mulching meticulously for several seasons.

Just as farmers and gardeners become aggravated by the persistence of weed seeds in their cropland, so too are consumers sometimes annoyed by seeds in the foods they eat. So horticulturists have redesigned some types of fruits. Did you ever look for seeds in a banana? There aren't any – nor are there seeds in seedless grapes, navel oranges, and some other fruits that plant breeders have developed for our convenience. These plants have to be propagated by taking cuttings, grafting, or other cloning methods. Of course, all these plants do have ancestors with seeds, ancestors that produced new generations through the usual methods.

The story of seed dispersal is a story of variation in design. But no matter how this process is carried out, the objective is always the same – to distribute the ripened seeds to suitable habitats far enough away from the parent plant to reduce competition. Each of the ingenious strategies for dispersal increases the likelihood that the seed will germinate and a new plant will grow.

Suggested References:

Attenborough, David. *The Private Life of Plants*. Princeton: Princeton University Press, 1995.

Capon, Brian. *Botany for Gardeners: An Introduction and Guide*. Portland, OR: Timber, 1990.

Harrington, H. D. *How to Identify Plants*. Athens, OH: Ohio University Press, 1985.

Stokes, Donald, and Lillian Q. Stokes. *A Guide to Enjoying Wildflowers*. Boston: Little, Brown, 1985.

silver maple seeds

Seed Dispersal

FOCUS: An important part of the life cycle of many plants is the dispersal of ripe seeds to new locations, and this is accomplished in a variety of ways.

OPENING QUESTION: *What role do seeds play in the life cycle of a plant and how are these seeds dispersed?*

SEE INSIDE A SEED

Objective: To examine the parts of one seed.

Give each child an unshelled peanut. Have them open the shell (the ripened ovary) and look at the peanuts inside (the seeds). Discuss the different parts, including the brown papery seed coat, the halves of the peanut (which is food stored for the plant when it is first growing), and the tiny leaflets between the two halves of the peanut. Then let the children eat these seeds!

Materials:
- unshelled peanuts (or soaked lima or other big beans in case of allergies)
- hand lenses

PEANUT PARTS

- stem to parent plant
- cotyledon or seed leaf (contains stored food for the seedling)
- plumule (becomes first leaves)
- hypocotyl (becomes stem)
- radicle (becomes root)
- ovary wall (peanut shell)
- peanut seed covered with seed coat (papery covering)

PUPPET SHOW

Objective: To learn some of the different strategies plants have for the dispersal of seeds.

Materials:
- script
- puppets
- props

Perform, or have the children perform, the puppet show. What dispersal method did the children like best?

MIX AND MATCH

Objective: To examine the seeds and seed-bearing structures of some familiar plants.

Give each child either a seed (taped on to a card) or a fruit, seed head, or pod. Allow the children to move from person to person examining the seed or seed-bearing structure each holds. Have the children match the seeds with the appropriate seed-bearing structures. Then, briefly discuss how each type of seed might be dispersed.

For younger children (K-2): Give each small group a variety of fruits, berries, pods, and seed heads to look at and take apart. Have the children determine how the seeds are dispersed.

Materials:
- seeds, taped to 3" x 5" cards (e.g., aster, peach pit, milkweed, burdock, poppy, squash, apple, avocado, cucumber)
- matching seed-bearing structures (e.g., aster seed head, peach, milkweed pod, burdock bur, poppy seed head, squash, apple, avocado, cucumber)
- hand lenses (optional)

MILKWEED RACE

Objective: To test how far a seed can travel with a little help.

Give each child a milkweed seed. Have the children see how far they can make the seed go without letting it drop to the ground or using their hands. How is seed dispersal important to a plant's survival?

Materials:
• milkweed seeds

SEED SCAVENGER HUNT

Objective: To discover what seeds can be found outside and how they are dispersed.

Divide the children into small groups (or pairs) and give each group a task from the following list. When teams complete one task, they return to the leader for another. Children should collect seeds only if there is a large area with many seed-bearing plants. After each group has done a search or two (or more, depending on interest), they should gather to report their findings. If groups have collected seeds, have them introduce a seed they found interesting. Discuss its means of dispersal.

Materials:
• Seed Hunt list of tasks
• collecting bags
• old wool sock tops or sweater sleeves
• hand lenses

SEED HUNT

~ Find two different seed containers that look good enough to eat. (Don't eat them yourself!)
~ Find two different seeds that travel at least a meter when you blow on them.
~ Find two different seeds with hooks to stick to fur. (Place an old sweater sleeve or sock top over your arm or leg and see what is collected.)
~ Find a seed head with more than twenty seeds in it. How will they travel?
~ Find a plant whose seeds have already dispersed. How can you tell?
~ If there are trees nearby, look for two different tree seeds that are carried by wind, and two that animals might eat.
~ Look at some seeds through a hand lens. Are they smooth? Rough? Hairy? Describe one.

FRUIT AND NUT SNACK

Objective: To enjoy a fruit snack while learning about seeds.

Bring in a collection of common and exotic fruits and nuts and dissect them with the children. Examine and discuss where the seeds are (or aren't, in the case of bananas, seedless grapes, seedless oranges). How do people contribute to seed dispersal? Then eat!

Materials:
• a variety of fruits and nuts such as apples, oranges, kiwis, figs, pomegranates, grapes, bananas, avocados, almonds, hazelnuts, coconuts

EXTENSIONS

Seeds' Needs: Have the children collect some seeds during the scavenger hunt or snack. Take some of the collected seeds and plant them in containers in the classroom. Plant some of the seeds on top of the dirt (keep moist), and plant others half an inch down. Put others in a plastic bag in the freezer for about two weeks, then plant them in the same ways as the first batch. Are there any differences in the germination rates? If so, can the children speculate why this might be?

Seeds as Feed: Use various types of seeds (thistle, sunflower) to fill a birdfeeder. On a regular basis, have the children observe the birds that use the feeder. Record what types of seeds the various birds prefer. Maintain the feeder throughout the winter months.

Seed Journey: Ask the children to pretend they are seeds. In story, cartoon, or picture form, have them describe their travels from the mother plant to the place where they might land and start to grow.

Seed Dispersal

Characters: Mary Maple Seed (on a dowel so she spins), Mama Maple Tree, Milkweed Seed, Carol Cranberry, Charlie Chipmunk, Maple Seedling

Prop: sign saying "NEXT SPRING"

Mary Maple Seed

Mama, what am I going to be when I grow up?

Mama Maple Tree

A maple tree, just like me, dear.

Maple Seed

Well, where am I going to live, Mama?

Maple Tree

I don't know, my little maple sugar, but someplace far enough away from me so that your roots will have space to grow and your leaves can get all the sunshine they need.

Maple Seed

How am I going to get there?

Maple Tree

I'm not sure. You might fly or float in water or be carried by an animal.

Maple Seed

Boy, that sounds exciting! I wonder which I'll do. Do all seeds fly and float and get rides to get away from their parents?

Maple Tree

No, dear. Some seeds have such special designs that they can only travel one way. I'm sure you'll meet many different kinds of seeds on your journey. *[big wind noises]*

Maple Seed

Yikes! I guess I'm going by air. Bye, Mama. *[Maple Tree exits]* Wow, I'm getting dizzy spinning around like this. *[Milkweed Seed enters]*

Milkweed Seed

You poor thing. I'm glad I don't spin that way. I'm so well rounded, I just float along like a parachute.

Maple Seed

You sure do, Milkweed Seed. Me? I'm like a helicopter propeller spinning out of control! I just know I'm going to crash-land somewhere.

Milkweed Seed

Well, I'll be drifting along. Good luck and Goodbyyyee! *[exits]*

Maple Seed

Hey, where am I? I've stopped spinning, and I'm all wet – this must be a river. It's a good thing I can float. Wow, I'm really going fast – Ouch! That rock hurt. *[Cranberry enters]*

Carol Cranberry

It doesn't hurt if you're built like me. I'm so bouncy and light, I can hardly feel those things. But I sure wish I could get back to the bog…I wanted to travel, but not **this** far!

Maple Seed

Gosh, you are a **really** good floater. Do cranberries always travel by water? How did you get here?

Cranberry

Whoa! Too many questions. We do usually travel by water, if we don't end up in sauce…but I got here because that silly bird that picked me didn't eat me in the bog. It dropped me in a brook instead, and I've been going downstream ever since.

Maple Seed

Well, at least you didn't get eaten.

Cranberry

Oh, I don't know – I'm actually designed for that – the little seeds inside my red coat would've gone right through that bird and done just fine. Now I don't know where I'm headed…but life's an adventure, isn't it? Whee – here I go – goodbye, little Maple Seed! *[exits]*

Maple Seed

Goodbye, Cranberry! Gee, she sure is a better water floater than I am. I'd like to get **out** of this river. It's rough! *[bobs up and down, then stops]* Whew! Lucky thing that big wave just washed me up on the shore. But now what? *[sniffs a little]* I can't live here. It's too cold and muddy for a maple. How can I get to a good place, like my Mama said? *[Charlie Chipmunk enters, bustling back and forth]*

Charlie Chipmunk

Gotta hurry. Gotta hurry. Have to collect all my seeds before winter. Gotta hurry.

Maple Seed

Goodness, who is that? I wish he'd slow down. He makes me nervous.

Chipmunk

Hello, Maple Seed. I'm Charlie Chipmunk, out collecting seeds.

Maple Seed

Well, that's funny. I thought I just saw you getting rid of seeds. Weren't you rubbing against that rock to scrape some seeds off your back?

Chipmunk

Yes, I was, but I didn't want those prickly seeds. They were just trying to hitchhike to a new spot. They stick in my fur and get it all tangled. They're awful! But **you** look good. Want to come home with me? I'll give you a ride over to my secret hiding place. *[secretly to the audience]* I won't tell her that she's next winter's dinner!

Maple Seed

Sure. My mama said an animal might carry me to a good place to grow up. Where are we going?

Chipmunk

Oh, you'll see soon. *[Charlie picks up Mary, runs across stage, puts her down]* Here we are. A nice, cozy hole in the ground. I'll just leave you with these other seeds. Catch you later. *[Chipmunk exits]*

Maple Seed

My goodness, there are a lot of seeds here. I wonder why. I seem to be way over on the side. I wonder if that matters. *[Chipmunk reappears]*

Chipmunk

Here I am back again, in time for a little supper. A couple of maple seeds will taste good, then maybe some sunflower seeds for dessert. *[nibble, nibble]* Lovely – now I'm off again, got to get a lot more seeds. Winter's coming. Gotta hurry. *[Chipmunk exits]*

Maple Seed

Whew, that was close! Now I know why the chipmunk brought me here. I'm sure glad he dropped me way over here at the edge of this hole. If I'm lucky, he'll forget about me.
[Maple Seed exits; sign saying NEXT SPRING appears briefly; Maple Seedling enters]

Maple Seedling

Hurray! I made it through the winter, and here I am starting to grow, with leaves and roots and everything! It'll probably take me a long time before I'm as tall as my mother, but I'm off to a good start. Goodbye, everyone. *[exits]*

Fly Away or Stay?

WHERE DO THE BIRDS GO AND WHY?

Konk-laree! Konk-laree! A red-winged blackbird announces his return in early spring. In August hundreds of swallows line up on telephone wires, the signal that soon they will all be gone. Since people first began watching birds, they have wondered about the birds' disappearance in the fall and their return in the spring. This seasonal movement from one area to another for the purposes of feeding and reproducing is called migration. Some birds migrate only short distances, but many fly thousands of miles each year. In fact, more than one third of all the world's bird species migrate with the seasons.

How and why did bird migration originate? Most bird experts agree that the glacial periods affected migratory habits and routes, but the combination of climatic or environmental factors that initially motivated migration remain a mystery.

One widely held theory proposes that the ancestors of migratory birds lived year-round and nested on what is now their southern wintering range. When the climate and environment became suitable farther north, birds gradually moved northward during the nesting period to take advantage of abundant space and food, and then returned to their southern homes for the winter. Some bird families, like the tanagers, are very species-rich in the Tropics, suggesting that's where they first evolved. Another theory suggests that migratory birds were formerly permanent residents of northern lands and were driven periodically southward by wintry climates.

Apart from the original reasons for migration, why do many species of birds migrate today? The answer lies in the climate of northern regions. The northern winter is an extremely challenging season for several reasons: 1) the food supply is reduced or unavailable, 2) the length of day and thus time in which to forage is shortened, and 3) the amount of energy needed for a bird to keep warm is increased.

Some birds, such as the black-capped chickadee and tufted titmouse, can survive these pressures and remain in the same environment all year. The adaptations of these permanent resident birds to winter conditions are really remarkable. For instance, chickadees manage to store twice as much fat on a day in February as in spring or fall, with almost all of it being used each night as the bird sleeps. Also, chickadee feathers provide excellent insulation, and when fluffed are even better.

Bird migration occurs in a variety of forms. Complete migration takes place when all members of a species migrate every spring and fall. In some species, some members may migrate while others stay put. In other species, such as the common redpoll and the snowy owl, migrations occur during some years but not in others, perhaps depending on food availability. This type of movement is known as irruptive migration. Some birds, like the blackpoll warbler and the upland sandpiper, are long-distance migrators, traveling thousands of miles between breeding and wintering sites. Other species, such as the American robin, yellow-rumped warbler, and eastern bluebird, are short-distance migrants, often stopping along the Gulf coast or southern U.S. The Clark's nutcracker migrates from breeding sites high in the Rocky Mountains or the Sierra Madre to wintering sites at lower elevations.

Among the many mysteries of migration is the question of how birds know when to set forth. Birds undergo hormonal changes just before and just after the nesting season, and most birds migrate during these periods of change. Experiments show that these changes are triggered by an innate clock. Another factor affecting take-off time is light – the lengthening days of spring and the shortening days of autumn. Weather also has an influence; birds are sensitive to barometric pressure and other meteorological conditions.

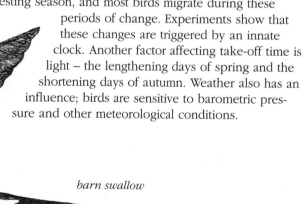

barn swallow

Most birds that cover long distances, especially smaller birds, travel at night. While they usually stop to rest and feed around sunrise, they will also fly without a break when crossing water or unsuitable habitats. Some birds migrate during daylight hours; these include the soaring birds (such as hawks) that ride thermal updrafts, and the swallows, which feed on the wing.

How can birds traveling long distances, often in the dark, keep on a regular course? How do some find their way back to exactly the same breeding or wintering grounds used in previous years? It is thought that day-flying land birds tend to use eyesight to guide them, following river valleys, coastlines, and ridges. Some birds may also orient themselves to the position of the sun. This may be especially important for birds crossing large bodies of water that lack landmarks. As the sun's angle changes, the birds adjust their inner compass in order to stay on course. Similarly, experiments with night migrants show that they navigate by the stars. Some birds can also use the earth's magnetic field to find their way. It seems clear that migrating birds use a combination of tactics to get where they are going.

While geographical and astrological clues help orient birds during migration, meteorological conditions can help birds conserve energy during their travel. Small land birds, flying mostly at night, generally travel with the airflow in order to ease their journey. They go northward in the spring on warm air masses from the south, and southward in the fall on cool winds from the north.

Birds prepare for migration by accumulating fat to fuel the arduous flight. They may change their diets in the fall, looking for the highest-calorie sources around. The ability of a hummingbird to cross the Gulf of Mexico is a metabolic miracle. Some of the journeys documented by bird banding are extraordinary, such as the voyage of one Arctic tern, banded on the Arctic coast of Russia, that flew more than 14,000 miles to spend the winter off Australia.

Certain routes are so commonly used by so many birds that they have been called flyways. There are four such routes on the North American continent that follow prominent north-south geographical features. They are the Atlantic, Mississippi, Central (along the Rocky Mountains), and Pacific flyways. These routes are more heavily used in the fall. Springtime's northward migration is much more diffuse, with the birds spreading out and taking advantage of the food available on the way.

Migration poses great risks, and hundreds of millions of migrating birds never reach their destination. It is estimated that over half of all first-year birds perish before they complete their first round-trip of migration. Bad weather claims many birds. Strong winds frequently carry them so far off course that they are unable or too weak to find their way. Fog seems to confuse their sense of direction. On misty nights light attracts them, causing many to crash into tall, lighted buildings. Light beams over airports are distracting. At Robbins Air Force Base in Georgia, 50,000 birds were killed in a single night. Birds that congregate at certain places along the route can become **prey** for migrating **raptors** and other **predators**. They can also fall victim to altered conditions in their stopover habitats; coastal development and deforestation can deprive them of essential resources en route.

Birds' remarkable adaptations, including migration, allow them to survive the hardships of winter. For those attuned to watching birds through the seasons, the tenacity of the chickadee enduring a long northern winter strengthens the spirit while the sight of geese on their journey south touches a responsive inner chord.

Suggested References:

Elphick, Jonathan, ed. *The Atlas of Bird Migration: Tracing the Great Journeys of the World's Birds*. New York: Random House, 1995.

Griffin, Donald R. *Bird Migration*. New York: Doubleday, 1964.

Harrison, Kit, and George Harrison. *The Birds of Winter*. New York: Random House, 1990.

Kerlinger, Paul. *How Birds Migrate*. Mechanicsburg, PA: Stackpole, 1995.

Stokes, Donald W., and Lillian Q. Stokes. *A Guide to Bird Behavior*. Vols. I-III. Boston: Little, Brown, 1979-1989.

golden-crowned kinglet

Fly Away or Stay?

FOCUS: Migration is one of a variety of ways that birds meet the challenges of winter.

OPENING QUESTION: *In what ways is winter a difficult time for birds?*

PUPPET SHOW

Objective: To understand the concept of migration and its related benefits and risks.

Perform, or have the children perform, the puppet show. Afterward, review the reasons why Willy and Thelma chose to migrate and the difficulties they encountered on their trip. What hardships might Cappy the Chickadee face during his winter stay in the north?

Materials:
- puppets
- props
- script

WHERE'S DINNER?

Objective: To illustrate that a bird's food requirements may determine whether or not it migrates.

Have each child pretend to be a bird. Pass out "You eat" cards. Some possible foods are mosquitoes, worms, berries, fish in small ponds, squirrels, caterpillars, frogs, nectar, ants, fish in streams, seeds, mice, snakes, rabbits, grasshoppers, insects under bark.

Each child says what it eats, whether or not that food is available in your area in winter, and whether or not it will migrate.

Materials:
- cards with "You eat..." written on them and the name and/or picture of a food

MIGRATION DIARY *(grades K-2)*

Objective: To connect, through a story, with the experiences of a migrating songbird.

Ahead of time, read the Migration Diary to yourself and refer to a map to determine the path a barn swallow from your area might follow during post-breeding migration. Barn swallows' summer range includes most of the United States and southern Canada; their winter range is from the Lesser Antilles and Puerto Rico south to the southern tip of South America.

On a large map that shows the Western Hemisphere, stick a picture of a barn swallow atop your state. Ask the children where they think birds from your area go when they migrate. Read the Migration Diary aloud to the children, and as the bird travels, move the picture of the swallow on the map. Afterward, use the map's scale to estimate how far the swallow traveled. Ask the children what abilities might help birds find their way when migrating such long distances, and discuss these.

Materials:
- large map showing the Western Hemisphere
- a picture of a barn swallow
- Migration Diary story
- double-sided or masking tape

WHO GOES THERE? *(grades 3-6)*

Objective: To compare the migration patterns of a selection of birds and the distances they travel during migration.

Ahead of time, hang up a large map of the Western Hemisphere and tape on pictures of different species of birds within the northern part of their range. Include birds that have a variety of migration patterns plus at least one species that does not migrate.

Divide the children into pairs and give each pair an 11" x 17" geopolitical map of the Western Hemisphere and a passport card for one species of bird. Passports should include a picture of the bird, plus information detailing its food preferences and wintering destination *(refer to Passport Cards – Information)*. On the 11" x 17" maps, have the pairs locate the winter destination of their bird and then, using the map's scale, estimate the distance their bird must migrate. When all are finished, have the pairs, one at a time, share one fact about their bird and tape their bird's passport on the large map in the area where it overwinters. Compare the distances the different birds travel. Ask the children what abilities birds might use to find their way when traveling such long distances, and discuss these.

Materials:
- a large laminated map showing the Western Hemisphere
- photocopied 11" x 17" maps of the Western Hemisphere
- 2 sets of bird pictures (one for the map, one for passports)
- bird passport cards
- (optional) bird field guides and other information resources

MIGRATION OBSTACLE COURSE

Objective: To experience the many hazards of migration and thus understand why many birds do not survive the trip.

Outdoors, mark a rectangular area as a migration flyway, noting one end as north and the opposite end as south; adjust the length and width as appropriate for your group. In a large circle, ask the children to name some obstacles that birds might encounter during migration, and create or have neck signs ready for these. Designate three-quarters of the children as migrating birds while the other one-quarter will be obstacles. Obstacles receive a neck sign and place themselves within the boundaries; they must keep their left foot planted while they try to tag the migrants. At "go," migrating birds attempt to fly south without being tagged by an obstacle. If tagged, the bird "dies," stopping in place or falling dramatically to the ground, until all birds have either safely migrated or been waylaid by obstacles. Ask the children what obstacles they think could reasonably be removed or made less harmful as part of human efforts at conservation. Switch roles and try the game again, minus the obstacles the children thought could be removed. Compare migration success rates.

Materials:
- cones or flags for boundary markers
- obstacle neck signs (with string for hanging around neck) depicting fog, high wind, towers, lack of food, cold rain, hunter, new building, lots of cats, trees cut down at resting spot, and others

WHO'S HERE NOW?

Objective: To observe birds in their winter habitat.

Note: Two weeks or more ahead of time, put up a bird feeder and keep it filled. Also, sprinkle seeds on the ground in a protected place with nearby perches.

Divide into small groups and give each group a Who's Here Now? sheet, a pencil, and a clipboard. Tour the outdoors nearby as early in the day as possible to listen and look for birds. Have the children write down their answers to the questions. Regroup and share observations. Periodically tour the same area and compare findings.

Materials:
- Who's Here Now? sheets
- clipboards
- pencils

Who's Here Now?

Record – number, describe, name, or draw – each bird you see and answer the following questions about it:

Where is it?

What is it doing?

Is it feeding? Can you tell what it's eating?

Is it making any sound? Singing? Or calling?

How many of this kind of bird do you see? Are they doing something together?

Do you think this bird lives here all year round?

What have you observed that makes you think this?

Winter Survival Strategies

Objective: To examine some ways other than migration that birds meet the challenges of winter.

Divide the children into small groups, outdoors if possible. Give each group a card (or whisper to them) describing a winter survival strategy used by birds that stay in the north. Let all of the groups practice for a few minutes. Then introduce the groups one at a time, naming their winter bird and having them act out their bird's strategy. The rest of the groups watch the pantomimes and guess what strategy is being acted out. (Props and sound effects are allowed.)

Winter Strategies

Chickadees: In a small flock, you visit a bird feeder for sunflower seeds, which you take away in your beaks to a branch and then hold with your feet and peck open.

Ruffed Grouse: On a very cold evening, you dive headfirst into a soft snow bank and spend the night there. In the morning, you come flying out suddenly. You head to the top of a spruce tree to nibble buds.

Golden-crowned Kinglets: As night falls, you and several others are feeding on insects or seeds in the trees. You call to each other, then gather and huddle together for the night in thick evergreens with your feathers fluffed up.

Nuthatches: After spending the day looking for insects under the bark of trees, you spend winter nights in hollow trees.

Blue jays: You fill your mouths with many seeds at a time and then stash them in cracks in trees or small holes. You eat the seeds later.

Sharing Circle

Objective: To summarize and personalize an understanding of the challenges birds face in winter.

Gather children in a circle and have each child complete the sentence, "If I were a bird in winter, I would _____."

Materials:
- Winter Strategy cards
- (optional) props: cardboard box, white sheet, large branch or two
- (optional) seeds or jelly beans for children to snack on

EXTENSIONS

Feeder Survey: Keep a daily journal of the birds visiting your bird feeding station. Include information on the day's temperature and weather. Encourage the children to record other observations they make or questions they have about bird behavior or habits.

Hawk Watch: If there are hawk watches nearby, have the children join one. Study the silhouettes of different species of hawks first.

Gardening for the Birds: Invite an extension service agent or Fish and Wildlife officer to talk about what shrubs, bushes, and other plants provide good food for migratory or overwintering birds. Encourage the children to take on the purchasing, planting, and maintenance of such a garden.

My Favorite Bird: Have each child select a favorite bird, draw and color it, and write a story about its adventures, either as a migrating bird or a permanent resident.

FLY AWAY OR STAY? — MIGRATION DIARY

As you listen, pretend that you're a young barn swallow, and this is the story of your adventures.

Dear Journal: I was hatched early this summer, in a nest on a beam inside a big barn. I have four brothers and sisters. It's midsummer now, and I can fly. Every day I fly low over fields and along the brook, catching insects in my open mouth. When I'm thirsty, I dip down to the pond for a drink.

Dear Journal: Today the late-summer sun warmed my back as I flew around catching flies and grasshoppers. For weeks now I've been hungry, eating and eating to get strong for a journey. The days are shorter, and the nights cooler. The old swallows say soon we'll go south, but I don't know where that is. All they say is, "You'll see."

Dear Journal: When I woke up this morning, all of my family and cousins and neighbors were gathered on the wires. I felt the wind coming out of the north, and the sky was blue, with no clouds. This was the day to travel! All together, my flock took off, heading south.

Dear Journal: Now I've traveled many miles south. I guess we'll stay here awhile, sleeping in the reeds by the river at night, searching for food during the day. Though there are plenty of insects here, the chilly evening air tells me that soon we will move farther south.

Dear Journal: Late last night, it began to rain. I was snuggled down in the tall reeds where we were sleeping, but the rain turned cold and soaked my feathers. I felt awfully cold and hungry. I hopped down a little deeper and hoped the morning sun would come soon to warm me up.

Dear Journal: In the past few days I've crossed mountains, flown over valleys, followed rivers, and passed big towns. As long as there are insects to eat and some fresh water to drink, I'm not in any hurry. Some of my own flock aren't with us anymore. They just weren't strong enough. I'm one of the lucky ones, hatched early in the summer, so I was fat and strong when I left home.

Dear Journal: I've been traveling such a long time. Today I came to what seems like the end of the earth, where all you can see to the south is open, salty water. All the birds stop here before trying to fly across the sea.

Dear Journal: When I got up this morning, I just knew that it was time to move on. The wind was behind us as we headed across the water. There was no place to land and rest. The wind changed, blowing me sideways. My wings felt heavy. I was tired and hungry. The wind and fog made it hard to fly the way I wanted to. Finally, I spied land! I dropped to the beach. Hundreds of other birds crowded the ground and the trees along the shoreline.

Dear Journal: The air felt warm and wet as I crossed the land today. I ate new kinds of insects. I even tried some berries. We don't eat those at home, but here they taste good. There seems to be enough food here. It was a long, hard journey, but I made it. Now I feel welcome in my South American winter home.

Dear Journal: I can't believe it's been so long since I wrote. It is now almost spring, and the old swallows say we will go north soon, back to the barn and the fields and the ponds. Maybe I will help my mother with her new nest and babies. Or maybe I'll find a mate and make a nest of my own. I can't wait!

Passport Cards — Information

Eastern Bluebird

Winter Range: Winters across much of North America, depending on food availability. May go as far south as northern Mexico and Florida.

Food: Berries, insects.

Other info: Bluebirds may remain near their nesting area during winter if food is available.

American Robin

Winter Range: Winters across much of mid to southern North America, south to Bermuda and Guatemala.

Food: Earthworms, insects, fruit.

Other info: The Latin species name is *migratorius*, which means wanderer.

Cedar Waxwing

Winter Range: From Canadian border to Mexico and occasionally the Caribbean. Winter movements are very erratic.

Food: Berries.

Other info: A longer intestinal tract allows waxwings to digest berries very rapidly and still get greater nutritional value compared with robins or bluebirds.

Golden-crowned Kinglet

Winter Range: Winters across much of North America, depending on food availability. Stays farther north more consistently than robin or bluebird.

Food: Insects, spiders, fruit, seeds. Drinks tree sap.

Other info: Kinglet is the tiniest bird to reside in the north in the winter. It weighs about as much as one nickel.

Brown Thrasher

Winter Range: Southern third of eastern U.S., as far south as Florida.

Food: Insects, berries.

Other info: A few successfully winter in northern states. Thrashers have in excess of 3,000 song types.

Eastern Towhee

Winter Range: Southern half of the United States.

Food: Insects, seeds, berries.

Other info: Only migratory in northern parts of its range. Those that live in the southern U.S. stay there for the winter.

Brown-headed Cowbird

Winter Range: Widespread – agricultural areas from southern Ontario to southern Florida.

Food: Insects, seeds, berries.

Other info: Will search cow-pies for undigested seeds in winter.

Tree Swallow

Winter Range: Gulf coast to Mexico, Central America and Cuba.

Food: Insects, rarely berries and seeds.

Other info: Migrates in large flocks during the day.

Ruby-throated Hummingbird

Winter Range: Southern Texas to Costa Rica.

Food: Nectar and insects.

Other info: Crosses Gulf of Mexico in nonstop flight during migration.

Red-winged Blackbird

Winter Range: Southern U.S., coastal areas, Mexico.

Food: Insects, seeds, some berries.

Other info: Forms large flocks, in the many thousands, in the fall.

Gray Catbird

Winter Range: Gulf coast to central Panama, Bermuda, Greater Antilles.

Food: Insects, spiders, berries, fruit. Feeds in trees and bushes.

Other info: Both sexes give a "meow" call when alarmed.

Mourning Dove

Winter Range: Through most of U.S., south to central Panama.

Food: Seeds, including waste grain from cultivated fields. Ground feeder.

Other info: Common at bird feeders. Courting male coos, puffs out throat, and bobs tail.

Eastern Phoebe

Winter Range: From southeastern U.S. to southern Mexico.

Food: Insects, occasionally small fish and frogs; few seeds and berries in the winter.

Other info: Flies off perch to catch flying insects.

Downy Woodpecker

Winter Range: resident.

Food: Insects, some fruit, sap from sapsucker holes.

Other info: Drums on resonant surfaces like signs or mailboxes.

Fly Away or Stay?

Characters: Thelma Thrush, Willy Thrush, Cappy Chickadee

Props: sign saying "Monday at 6:00 p.m."; sign saying "Days Later"; black strip of cardboard with yellow star on top

Thelma Thrush

Boy-oh-boy, am I excited!

Willy Thrush

Hi, Thelma! What are you so excited about?

Thelma

Oh, hello, Willy! I'm about ready to migrate...not just a hop and a jump to the nearest bug, but a trip of many hundreds of miles! And fall's a great time of year to travel. Want to come along? All the other thrushes are going, too, you know.

Willy

Gee, Thelma, that does sound exciting!

Thelma

Think of the views – the meals out – the friends we'll meet – and when we arrive, ahh, the warmth, the sun, and plenty of bugs and berries.

Willy

Wow! It sounds too good to be true. Plus your timing couldn't be better, Thelma. It's starting to get quite cold here in the north. I have to eat more and more bugs just to keep warm.

Thelma

Yes, and on top of that, have you noticed how early it gets dark now? There's hardly enough time to find our food in the daylight.

Willy

Before you know it, Jack Frost will have frozen all the little caterpillars and covered the ground with snow, and we won't be able to find enough to eat.

Thelma

That's right! So the sooner we head south, the better! You'd better stuff yourself the next few days, Willy. Flying takes heaps of energy.

Willy

You bet! Now, when shall we leave?

Thelma

How about next Monday at 6 o'clock?

Willy

Sounds good. I like to get up early.

Thelma

Get up? I mean 6 o'clock at night!

Willy

At night?

Thelma

Of course at night. We'll use the setting sun to get us started in the right direction. And then we can stop and rest during the day, when we can see to find our food.

Willy

Oh, okay. I guess that makes sense. Bye for now. *[both leave; sign pops up saying "Monday at 6:00 p.m."]*

Cappy Chickadee

Chickadee dee dee dee. Say, Thelma, you've seemed so busy lately. What's up?

Thelma

Hi, Cappy. "What's up?" you ask. I hope I will be, tonight. Willy and I are heading south for a warmer winter. Why don't you come with us?

Chickadee

Leave? Now? No way! I get plenty of exercise just traveling from the woods to the bird feeders. And the cold isn't so bad, really. I can always find some snug hole where I can keep warm.

Thelma

Well, I'll think about you, Cappy, while I'm enjoying balmy breezes and some juicy southern worms. *[Willy enters]* Uh oh, there's Willy. I gotta go. See you next spring, Cappy.

Chickadee

Bye, Thelma. *[louder]* Bye, Willy! Send me a postcard when you get there! *[Chickadee exits]*

Thelma

Well, Willy, are you all set?

Willy

Oh, yes, Thelma, but it may take a little work for me to get off the ground. I've been eating ever since I saw you. I'm fat, and I'm ready to go.

Thelma

Then, let's go! *[birds fly for a while]*

Willy

Hey, this is fun, Thelma. But slow down. If I lose sight of you, I'll get lost for good!

Thelma

Don't worry, Willy, just follow the stars. They're like a lit-up road map.

Willy

Uh oh. Here comes one and we're mighty close. *[put up and take down the black strip of cardboard with yellow star on top]*

Thelma

Yikes! That's what I call a narrow escape.

Willy

What happened, Thelma?

Thelma

I thought that was a star, but it turned out to be a tall tower with a light on top of it. Glad we saw it in time.

Willy

Me, too! Boy, this takes a lot of work. Are you having to beat your wings faster, Thelma?

Thelma

Yes, I sure am. I think we've headed into a storm. *[wind noises]* These winds are terribly strong, and I'm afraid we've been blown a little off course. I'm glad daylight's coming so we can stop and rest a while. That'll give us a chance to eat and get our bearings.

Willy

Let's head down to the bushes at the edge of that big field. Maybe we can find a few bugs.

Thelma

Yes, and there might be some berries on those bushes, too. *[land on ground]* Phew! It's nice to finally have my feet on the ground.

Willy

And it sure feels good to stop and rest. *[birds exit; hold up sign saying:"Days Later." Birds come up, sitting on the ground.]*

Willy

This trip isn't exactly all you cracked it up to be, Thelma.

Thelma

What do you mean, Willy?

Willy

Well, to begin with, you said the view would be terrific. The only view I got was of that huge tower from about two inches away. How can you have a view when you only fly at night?

Thelma

Well, uhh...

Willy

And meals out sound like a real treat, but when you have to find your own meals, and you're someplace that you've never been before, it's not that much fun.

Thelma

That's true.

Willy

Besides, we've been flying for days and days, and we haven't had time to meet **any** friends. If you ask me, migration is for the birds.

Thelma

Well, you have to admit it's warm and sunny here, and it looks as if we won't have any trouble finding breakfast.

Willy

Yes, you're right about that. It's so nice that I think I might spend the rest of the winter here. Want to join me?

Thelma

You know, that sounds like a good idea. This is enough traveling till next spring when we fly north again. Guess we can go write that postcard to Cappy. Bye, bye everyone. Have a good winter. *[both exit]*

Galls

SMALL HOMES FOR TINY CREATURES

What are those odd swellings on some plant stems and the lumps and bumps on certain leaves? Look closely; perhaps they are galls. A gall is an abnormal growth of plant tissue produced by a stimulus outside the plant itself. Stated more simply, some irritation or chemical causes that part of the plant to swell or grow in a particular way. Gall growth can be induced by a number of organisms, including fungi, viruses, mites, and, most commonly, insects. Once you start looking, you may find galls on grasses, vines, trees, and flowers – they are surprisingly common.

Galls come in a fantastic variety of colors, shapes, and textures. They may be round or conical, or shaped like a kidney, disc, or spindle. They may be nearly woody or almost fluffy, smooth or bumpy, papery or scaly. They are sometimes known by creative and confusing common names, such as the spruce pineapple gall on spruce trees, the rose moss gall on rose stems, the willow pine cone gall on willows, and the oak apple gall on oak trees.

A gall may be found on any part of a plant, but it can develop only while that part of the plant is growing. This explains why most galls form in the spring or early summer. However, although they develop during the growing season, many galls are easiest to spot in the

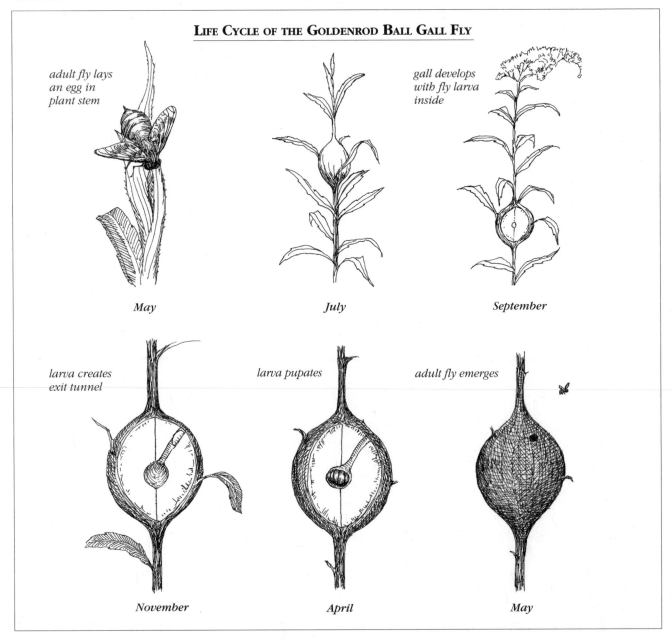

LIFE CYCLE OF THE GOLDENROD BALL GALL FLY

adult fly lays an egg in plant stem

May

gall develops with fly larva inside

July

September

larva creates exit tunnel

November

larva pupates

April

adult fly emerges

May

autumn, after the leaves fall. Look for an unusual swelling, especially with an exit hole. If a gall has no exit hole, the inhabitant is still inside. Some insect galls are used only briefly, during warm seasons. Others are inhabited through the winter. In every case, they are an indispensable part of the life cycle of the insect.

The purpose of an insect or mite gall is to provide a home and food for the developing young. And what a good home it is, with its solid outer walls and constant food supply. Besides producing food and providing shelter, the gall also keeps its occupants comparatively safe from parasites and **predators** and protects them from drying out.

Exactly what initiates gall growth on a plant is not clear. It appears to be caused by a chemical substance either injected by the mother insect as she lays her eggs, or secreted by the larvae as they bore into the plant for food, or produced by the plant itself in response to physical irritation by the insect. Sometimes the adult female insect seems to start the process and her off-spring, carrying precisely the same chemical, continues to stimulate the gall's growth. The initial stimulant changes starch in the plant to sugar, resulting in an excess of food, which causes the plant cells to enlarge and/or multiply and the gall to grow. Usually the presence of a gall does not harm the host plant.

Although galls grow on a great variety of plants, oaks are the most popular, claiming 800 of the 2,000 kinds of American galls. Small wasps belonging to one family, the Cynipids, produce almost all of the oak galls. Many kinds of galls are also found on willows, poplars, plants of the rose family, such as blackberries and raspberries, and **composites** such as goldenrods and asters. Fossils of galls on ferns have been found, formed as many as 300 million years ago.

Each gall-making insect produces a specific kind of gall that usually grows on just one species of plant. Often, especially in winter, you can identify a plant by the gall that grows on it. The goldenrod ball gall offers a good example. It is an easily recognizable ball-shaped swelling found on the stems of either tall or late golden-rod. A small, brown-winged fly, *Eurosta solidaginis*, causes this particular gall.

The life cycle of the goldenrod ball gall fly demon-strates why the formation of a gall is vital to this insect's survival. In the springtime, a female fly lays an egg in the bud of a growing goldenrod stem. She may lay up to 100 eggs, one egg per plant. She "tastes" the goldenrods with her feet and her **ovipositor** to identify the right kind before depositing her egg. Within a week, the egg hatches, and the young **larva** burrows into the plant stem causing a ball-shaped gall to form around it. If all goes well, the soft, white larva eats and grows through the summer and fall. It stays sheltered within the gall through the winter until it transforms into a **pupa** in the spring.

Prior to pupating, the larva chews an exit tunnel through to the outermost layer of plant tissue. Then it retreats back to the center of the gall where it pupates. After pupating, it emerges as an adult, equipped with a tiny balloon on the front of its head that it can inflate with fluid. The young adult fly then crawls along the tunnel and batters through the final barrier of plant tissue by inflating and deflating the balloon. Once it has emerged from the gall, the fly absorbs the balloon into its head. Emergence, of course, must be timed to coincide with the growing season of the goldenrods upon which it will lay its eggs. After spending nearly 50 weeks of its life within the goldenrod plant, this gall fly lives only a couple of weeks as an adult. In this brief time, the fly mates, lays eggs, and dies, and the cycle continues.

Many galls are convenient winter shelters for other visiting insects. Investigators studying willow pine cone galls found that each contained dozens of other insects, besides the gnat that made the gall. Many gall insects are themselves hosts to parasites, the young of other species that feed on the gall insect and sometimes on the gall itself. Gall insects are part of the food chain in other ways as well. For example, goldenrod ball gall fly larvae are a favorite food of many winter birds, such as downy woodpeckers and chickadees, which peck a hole through the gall to get the plump insect larva inside.

Whether seen as beautiful or bizarre, galls serve a critical function in the life cycle of many insects. Although the life cycles of the different species of gall insects vary greatly, and some are very complex, each of these insects depends on the gall to some degree to provide food, shelter, and protection for their young. Galls vividly demonstrate the dependency of animals on plants, a relationship evident throughout the natural world.

pine cone gall on willow made by a midge

Suggested References:

Borror, Daniel J., and Richard E. White. *A Field Guide to Insects: America North of Mexico.* Peterson's Field Guide. Boston: Houghton Mifflin, 1970.

Felt, E. P. *Plant Galls and Gall Makers.* New York: Hafner, 1965.

Hutchins, Ross. *Galls and Gall Insects.* New York: Dodd, Mead, 1969.

Marchand, Peter J. *Life in the Cold: An Introduction to Winter Ecology.* 3rd edition. Hanover, NH: University Press of New England, 1996.

Stokes, Donald W. *A Guide to Nature in Winter.* Stokes Nature Guides. Boston: Little, Brown, 1976.

Weis, Arthur E., and Warren G. Abrahamson. "Just Lookin' for a Home." *Natural History,* Sept. 98, pp. 60-63.

Galls

ACTIVITIES

FOCUS: Some insect species spend a part of their life cycle inside a gall, a growth on plants that provides food and shelter for the young insect.

OPENING QUESTION: *What is a gall?*

FIND YOUR OWN

Objective: To examine and notice differences among individuals of one kind of gall.

Give each child an example of the same kind of gall (goldenrod ball galls work well). Ask the children to examine their gall carefully so they can distinguish their own from other galls. Put all the galls together in a pile. Have the children find their own galls and then, one at a time, share their gall's distinguishing features. Have them guess what's inside.

Materials:
- galls of one species, one for each child

GALL FANTASY

Objective: To understand the life cycle of one gall insect.

Explain to the children that you will read them a story in which they are the characters. They should listen carefully and follow the directions silently. Give them each a cracker, and remind them that they should eat it only when directed to do so. Read the Gall Fantasy story aloud. Afterward, ask the children what it was like to be a gall insect. In what ways does the gall benefit the insect? Use a diagram to review the life cycle of the goldenrod ball gall fly.

Materials:
- Gall Fantasy story
- crackers or carrots
- diagram illustrating the life cycle of a goldenrod ball gall fly

PUPPET SHOW

Objective: To discover the variety of galls and gall insects.

Perform, or have the children perform, the puppet show. Afterward, review the different kinds of galls and gall insects introduced in the show.

Materials:
- script
- puppets

VIEW A VARIETY OF GALLS

Objective: To observe and describe similarities and differences among a variety of galls.

Divide the children into two or three small groups (with an adult in each). Give each child a gall to examine, noticing color, shape, size, texture, where it grows on the plant, and whether there are holes in it. Now have each child make a drawing of his or her gall. One at a time, have the children hold up their gall (and their illustration, if desired) and introduce it, telling one special thing they noticed or asking a question they have about it. Why are there different kinds of galls? Try to include these points in the discussion:

Materials:
- a variety of galls, labeled if possible, one for each child or pair
- hand lenses

gall on willow stem

~ Each kind of gall insect causes its own type of gall to form.
~ Gall-making insects must choose the correct species of plant or the gall will not form.
~ Certain types of plants (oaks, rose family, willows, goldenrods) are host to many kinds of galls.
~ Most galls are formed by insects, but some are caused by worms, mites, fungi, or viruses.

152 / Cycles

GALL HUNT

Materials:
• hand lenses

Objective: To notice as many different kinds of galls as possible outside.

In small groups, search outdoors for galls growing on flowers, bushes, or trees. Look on leaves, stems, and ends of twigs. If there are many specimens of any particular kind, samples could be picked for display or further study. Check for exit holes, or invasion holes by other insects; sometimes it's hard to tell the difference.

WHAT'S INSIDE?

Materials:
• galls, one for each pair of children (goldenrod ball galls work well)
• hand lenses
• knives
• cutting boards
• (optional) plastic bag(s) to take larvae home (if no bird feeder is handy)

Objective: To examine the inside of a gall and its insect inhabitant.

Divide the children into pairs or small groups and give each team a gall. Adults with knives and cutting boards should cut open the galls. Have children examine the halves and look for the following:

Insect – in its white larval stage or – in spring – pupating in a case

Exit route – hold each half of the gall up to the light for easier inspection

Interior – how much has been eaten by the larva or by other insect invaders?

Share the findings.

Note: Children may express legitimate concern about destroying the insects and their homes. Discuss this. Possible answers include: 1) we have cut open a kind that is very, very common – most remain growing where they belong; 2) scientific knowledge is often gained by observing live specimens, and our knowledge helps us understand and respect each creature's place in the natural world.

Larvae can be returned to the food chain by placing in a bird feeder, where they will be eaten by wild birds.

EXTENSIONS

Keeping Galls: Put goldenrod ball galls or other inhabited galls into glass jars. Make sure that the lid is ventilated (plastic wrap with holes poked in it, held on by a rubber band, works well) and place damp soil at the bottom for humidity. Watch for the emergence of adults. Observe, draw, and then release them when they do appear. Check similar galls in the wild and compare dates of emergence. Try to identify the galls and insects using field guides.

I'm Living in a Gall: Have the children draw or write about themselves spending the winter inside a gall. What would their gall home be like? What would they look like when they came out?

Papier-mâché Galls: Using an inflated balloon as a form, make papier-mâché or glue and water and tissue paper galls. When dry, cut away one side, mount on sticks, and use paper and craft sticks to make the interior and the creature(s) inside.

spindle galls on maple leaf made by mites

Please, very quietly, get your jackets – but don't put them on. Find a spot on the floor where you aren't touching another child, but where you can easily hear my voice. I'm going to give each of you a cracker that you may nibble on each time you hear the word "eat" while I tell you the story of your life.

I am a mother gall fly, and you all are my tiny eggs. I have placed each of you in the bud of a young goldenrod plant. *Crouch down just as small as you can. Pretend your jacket is the outside of the plant, and pull it over your head. Close your eyes and be very silent. When I count to three, you will hatch as a small, squirmy, hungry little larva inside a goldenrod. One, two, three!*

Now, little larva, you start to **eat,** taking tiny bites of the goldenrod and burrowing farther down into the stem. As you do this, the plant's stem grows around you, enclosing you in a round ball.

It's now the middle of summer, and you're hungry all the time. You **eat** and **eat** some more. You nibble the soft, rich-tasting walls of your goldenrod ball gall home. It is dark inside your gall and you are all alone. You **eat** as your gall sways gently in the summer breeze. Nibble by nibble you **eat** the soft, moist pith of the goldenrod stem.

It is fall now, and you're nearly grown. The days are getting shorter and the nights are cold. You are snug inside your gall. All you have to do is reach out to the nearest wall for food. Still, something tells you to make a tunnel. So you **eat** one! Up and outwards, until you get to the very tough, brown, outside layer of your gall. There you stop. You turn around and go back to the center of the gall to rest, away from the cool autumn air.

Autumn turns to winter. Snow blankets the ground; the ponds are iced over. Winter buries food for many creatures. But you are safe, even though the cold wind blows and shakes your home.

The sun is higher now, warming the air and melting the snow. Streams are thawing; owls are nesting; the sap is running. You **eat** your final meal. The time has come for you to change. You become a pupa, quiet and motionless deep inside your private gall.

Finally the days grow longer. Buds open and the grass is green, though your gall home is brown and dry. You awaken from your deep sleep feeling very stiff and cramped! You stretch and move. Slowly you make your way toward the end of your tunnel where the light from the spring sun beckons. Putting your head against the thin outside layer of your gall, you stretch and push, and stretch and push. Suddenly you are out of your gall, out of the small, dark world you have known. Soaking in the sunlight, you dry your wings. Your new wings! You are now an adult, a beautiful, brown-winged fly. Soon you'll be ready to fly off to find a mate, and perhaps lay a tiny egg in the new green bud of a young goldenrod plant.

Galls

PUPPET SHOW

Characters: Woody Woodchuck, Goldie Goldenrod (with ball gall), Oak Apple Gall, Will Willow (with pine cone gall), Raquel Raspberry (with knot gall)

Woody Woodchuck

Mmm Mm. Look at all this lovely goldenrod. Just waiting to be eaten.

Goldie Goldenrod

Well, don't eat me. I have a baby.

Woodchuck

Huh. You're just a plant. You can't have a baby!

Goldenrod

Well, I do have one, but it's not mine. I'm just babysitting. I've made a cradle for it – here on my stem.

Woodchuck

You mean that round ball? I always wondered what those were. I see them on lots of you goldenrod plants. So there's a baby plant inside?

Goldenrod

No, no, there's a baby fly inside – a larva. Even after I get brown and dry in the fall, the larva will stay snug inside this gall in my stem. Then in the spring, out comes a brand-new fly.

Woodchuck

Well, gosh, I guess I'd better find a different plant to eat. I wouldn't want to eat…a fly maggot. I'll just leave you to your babysitting and go somewhere else.

Goldenrod

Thanks, Woody Woodchuck. You're a life saver. *[sings and waves]* Rock-a-bye larva, in a wee gall…*[exits]*

Woodchuck

[Oak Apple Gall appears] Hey, look. Here's an apple! I bet a nice, juicy apple would taste good. I'll try a nibble. *[bites Oak Apple Gall]*

Oak Apple Gall

Who's that nibbling on my wall!

Woodchuck

Huh? A talking apple?

Oak Apple

I'm not an apple. I'm a gall.

Woodchuck

You're a gall? But there isn't a goldenrod plant near here.

Oak Apple

Not **that** kind! I'm an oak apple gall. Quite different from a goldenrod ball gall. I grew on an oak leaf, not a goldenrod stem. And I housed a wasp larva, not a fly.

Woodchuck

So there might be a baby wasp inside? I guess I'll have to be careful from now on when I take a bite of an apple.

Oak Apple

You'll be okay so long as you don't try gathering apples under an oak tree again! Now, if you don't mind, please put this gall back where you found it!

Woodchuck

Oh sure. Sorry. Here you go. *[Oak Apple Gall exits]* Boy! So far, I've had lots of food for thought, but none for my belly. *[Raspberry enters]* But here's a nice raspberry bush – I'm sure I can fill up my belly here.

Raquel Raspberry

I have zee fresh leaves for you to eat today.

Woodchuck

I like your leaves, but your prickly stems make it hard to reach them. Hey, this stem has got a gall on it.

Raspberry

Eet eez knot gall.

Woodchuck

If it's not a gall, what is it?

Raspberry

Eet eez a Knot Gall. A Raspberry Knot Gall. Shaped like a knot.

Woodchuck

Oh I get it! It is kind of bumpy like a knot. I suppose you must have a fly or wasp baby in there.

Raspberry

There are lots of baby wasps in here.

Woodchuck

Oh yeah? Kind of a gall apartment house, huh? Well, I don't want to stir up a wasp's nest! I think I'll just go over to the meadow and find some nice plain clover to eat. Goodbye now. *[Raspberry exits]* I don't see any galls on these clover plants. *[Willow appears]* And I can hide under this pine tree if anyone comes.

Will Willow

You've got a lot of gall, calling me a pine tree. I'm a willow.

Woodchuck

Come on. You can't fool me. You've got pine cones. Of course you're a pine tree.

Willow

Willow!

Woodchuck

Pine Tree!

Willow

Willow!

Woodchuck

Pine Tree!

Willow

Willow! Look, I've got **leaves.** Pine trees have needles.

Woodchuck

Well, what do you know! You've got pine **cones,** but no pine **needles.**

Willow

That's what I've been trying to tell you. I'm a willow! And those aren't pine cones. They're insect galls.

Woodchuck

Oh, I know all about galls. I'll bet there was a fly, or maybe a wasp, that laid an egg on your branch.

Willow

No, it was a tiny gnat, and it laid its eggs on my new buds.

Woodchuck

Then those pine cones, I mean willow cones, are gall homes for gnats.

Willow

Gnaturally! Isn't it amazing that a tiny insect can make one of my buds **swell** up to be home sweet home?

Woodchuck

That just goes to show, galls really are swell! And I've seen enough of them for one day. I think I'll head back to my own little home sweet home. Nighty night, Will!

Willow

Sleep tight, Woody. *[both exit]*

Winter Twigs

SIGNS OF FOUR SEASONS

When a tree loses its leaves in the autumn, many people feel it also loses its identity, but a bare tree's branching and twig pattern creates an individual profile for the observant eye. And careful examination of its winter twigs reveals many distinguishing features about each kind of tree. The various patterns, textures, shapes, and colors teach us many things about a twig's role in the life of a tree.

Twigs give a miniature account of the tree's past, present, and future. At the tip of the twig, and along its length, there are numerous buds. Formed the previous summer, these buds contain next spring's leaves, stems, and flowers. The buds are protected by bud scales, the arrangement of which is characteristic for each tree species. Bud scales are really modified leaves designed to protect the delicate growing point within from drying out or from becoming injured. Willows have only one bud scale, which unzips and comes off like a hood in spring. Maples have several overlapping scales, while oaks have many scales arranged in five rows. And the buds on some trees are "naked," with no scales at all. Butternut and witch hazel buds are examples. Evergreens also make buds, of course. These trees hold their needles for an entire year and look quite similar in winter and summer, but their buds and twig branching patterns are also worthy of study. There are a few **conifers**, like the larches and baldcypresses, that lose their needles and that are distinctive in the winter, having both cones and bare twigs.

The bud at the tip of the twig is called the **terminal bud**. It marks the end of one season's growth and contains the embryonic stages of the next season's growth. A twig elongates from its terminal bud, shedding the bud scales as it grows. Farther down the twig, the remnants of last year's bud scales are also evident; look for a narrow band of fine rings around the twig. The distance between the new terminal bud and last year's bud scale rings shows how much the twig grew in one year. A long section indicates the previous growing season was productive; a short one may indicate a tough year for the tree, most often because of drought. Different twigs on the same tree record various rates of growth because of their placement. Twigs on the sunny side of a tree usually grow more than those in the shade, and twigs near the top of the tree grow more than those near the bottom.

When studying terminal buds, you may notice one, three, or even many buds at the tip of the twig. Aspens have a single terminal bud; red and sugar maples have one terminal bud flanked by two smaller **lateral buds**;

the oaks have a cluster of terminal buds. Some trees – like basswood, staghorn sumac, and elms – have false terminal buds that are not centered but instead point slightly to one side. These are actually the final side buds formed on the end of the twig during the growing season.

Buds along the sides of the twig are called lateral buds and may contain stems and leaves or flowers or both. When two sizes of lateral buds occur on one twig, the larger one usually contains flowers and the smaller one leaves. The location of buds in an **opposite**, **alternate**, or **whorled** pattern is useful in identifying certain kinds of trees. Only a few trees have an opposite arrangement, where pairs of buds, twigs, and even whole branches, grow directly opposite each other throughout the tree. An easy way to remember these opposite trees is to think of the acronym MADCAP HORSE. The letters stand for different families or genera: M - maple; A - ash; D - dogwood; CAP - Caprifoliaceae (the family of plants including honeysuckle, elderberries, and viburnums); and HORSE - horse chestnut tree. Trees with an alternate arrangement show a zigzag pattern along twigs and branches, especially noticeable in the elm, which has all the buds on a branch in one plane. Most alternate buds are arranged on all sides of the twig. Whorled buds and twigs are arranged in circles; these can be seen particularly on evergreens like pine and fir, but also on catalpa.

Bud patterns and branching arrangement are not the only guides one has when trying to identify tree species in winter. The leaves of a tree, now generally absent, leave behind clues for the astute observer. Each autumn, a corky layer of cells develops across the leaf stem where it joins the twig. Called the **abscission layer**, it gradually cuts off the supply of water and food and separates the leaf from the tree. When the leaf drops off, a scar is left behind on the twig. The shape of the leaf scar reflects the shape of the end of the leaf **petiole** or stem. In most

red oak buds opening

trees the scar is an oval, a crescent, or a triangle, but in a few trees, such as sycamore and staghorn sumac, it is almost circular, enclosing a lateral bud. The veins that serve to conduct food and water between the leaf and the twig also leave scars (within the leaf scars). Referred to as bundle scars, they vary in number and pattern specific to different types of trees. In examining twigs carefully, one can find some very interesting leaf scars; the butternut scar resembles a monkey face, with the bundle scars forming the monkey's eyes and mouth.

Careful examination of a winter twig will reveal yet another identification tool. **Lenticels** are the corky vents through which carbon dioxide and oxygen are exchanged between the tree's tissues and the outside air. The size, color, and density of these marks vary. On white birches, lenticels appear as dark horizontal lines; on cherry trees, the horizontal lenticels are light colored and smaller than those of the birches; and on maple and alder twigs, they are pale dots.

Color is another characteristic of twigs that helps to differentiate them. Some twigs are red (dogwoods and striped maple), others are golden yellow (weeping willow), and others vary from soft gray and brown to deep purple and bronze green. Sometimes two distinct colors are observed on the same branch of the same tree. The end of the twig (the most recent growth) is often a different color than the rest of the branch. For example, the sugar maple's newest twigs are a deep russet brown while older growth is a lighter tan. Look for the faint bracelet of scars left by the bud scales between the old and new growth.

Each twig has a distinguishing feature that can only be seen by cutting through the twig itself. The very center of the twig, called the **pith**, is soft food-storage tissue. It varies in color, shape, and structure from tree to tree. For example, red-berried elder has orange pith while its close cousin, the common elder, has white.

Elderberry and staghorn sumac have large central piths that, years ago, people pushed out so they could use the twigs to make whistles or spouts for collecting maple sap for syrup.

Sometimes the smell of a bruised twig is a noseworthy feature. Spicebush, sassafras, and tulip tree twigs have an extremely spicy odor. Both black birch and yellow birch smell and taste like wintergreen. Cherry twigs have a strong, bitter, almond-like odor and taste. Silver maple and basswood smell rank and "green."

Many wildlife species also recognize different tree species and value their winter twigs as an important component of their winter diet. White-tailed deer, snowshoe hare, and ruffed grouse, among others, nibble buds and twigs when the more lush vegetation of summer is unavailable. They tend to be quite selective and seek out the most palatable kinds to browse. Although less nutritious than summer fare, twigs provide critical calories during the harsh winter season.

Winter trees may appear as lifeless skeletons against a somber landscape, but careful examination of their twigs reveals the prophecy of spring and the history of seasons gone by.

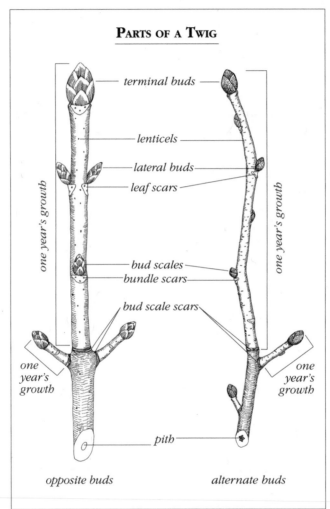

PARTS OF A TWIG

terminal buds
lenticels
lateral buds
leaf scars
bud scales
bundle scars
bud scale scars
one year's growth
one year's growth
one year's growth
one year's growth
pith
opposite buds
alternate buds

Suggested References:
Campbell, Christopher, and Fay Hyland. *Winter Keys to Woody Plants of Maine.* Orono, ME: University of Maine Press, 1975.
Eastman, John. *The Book of Forest and Thicket: Trees, Shrubs, and Wildflowers of Eastern North America.* Harrisburg, PA: Stackpole, 1992.
Harlow, William M. *Fruit Key and Twig Key to Trees and Shrubs.* New York: Dover, 1946.
Petrides, George. *A Field Guide to Trees and Shrubs.* Peterson Field Guide. Boston: Houghton Mifflin, 1986.
Stokes, Donald W. *A Guide to Nature in Winter.* Boston: Little, Brown, 1976.

Winter Twigs

FOCUS: The winter twigs of each kind of tree are unique, but some of the features common to all show evidence of seasons past and hold promise for the coming spring.

OPENING QUESTION: *What features can be found on a tree's twigs in winter, and how are these features important to the tree?*

PUPPET SHOW

Objective: To discover some features of winter trees and their place in the winter food chain.

Perform, or have the children perform, the puppet show. Afterward, discuss how twigs are important to the tree and to the animals visiting it.

Materials:
- puppets
- script

TOTAL TWIG

Objective: To examine the parts of a twig and understand their functions.

Hand out twigs, all from the same kind of tree, one for each child. Have the children study their twig and tell what they see. As they point out the various parts, dress up one child as a twig, explaining the function of each part as it is added. Use a paper cone hat for the terminal bud, smaller paper cones on the hands for the lateral buds (one with a rolled-up felt leaf inside), sticky dots for the lenticels, yarn wound around the waist several times for the bud scale rings, felt pieces for the leaf scars, and dots for bundle scars. Talk about alternate or opposite buds and branches (have child hold arms each way), and point out how to see a year's growth by looking at the bud scale scars. Remove the cone that covers the rolled-up "leaf" and unfold it. If time permits, have the children draw their twig, including each detail they've observed.

Materials:
- twigs, all the same kind
- hand lenses
- construction paper
- tape
- scissors
- felt
- yarn
- dot stickers
- (optional) drawing paper and pencils

TWIN TWIGS

Objective: To observe and describe twig characteristics.

Divide into groups of four or five and give each a set of twigs, then put one duplicate set in the center of the room. Taking turns, one child from each group should select a twig from the duplicate pile and, keeping it hidden, face away from his or her group and describe the twig, noting the shape, color, and texture of the various parts. Are the buds alternate or opposite? Does the twig have a smell? (scratch bark at lower end with a thumbnail) The rest of the group, with their set of the twigs spread before them, chooses the correct twig from the description. If stumped by the description, allow the group to ask questions. Check to make sure the two twigs match. Repeat to match other twigs.

Materials:
- sets of five or six kinds of twigs, one for each group plus a duplicate set
- (optional) twig diagram for vocabulary

Be a Tree Fantasy *(grades K-2)*

Materials:
• Tree Fantasy script

Objective: To experience the life cycle of a tree through the seasons.

Tell the children that they will now pretend to be trees. They should stand a little more than arm's length apart, listen to the story, and pantomime it. (Leader demonstrates while reading the script.) At the start of the story, their arms should be spread like branches, their fingers should be closed in fists like unopened buds, and their toes, like roots, should press downward. After the story, review the seasonal cycles of the tree. Ask the children what time of year they think is most difficult for the tree.

Tree Hunt

Materials:
• sets of fresh twigs collected from nearby trees, including an evergreen if any are present; one set for each group
• colored yarn, a different color to mark each species of tree
• sets of 3" colored yarn pieces that match the yarn on the trees, one set per group
• yarn to mark a special tree

Objective: To locate different kinds of trees by matching them with their twigs, and observe other differences among the various tree species.

Ahead of time, mark five or six different kinds of nearby trees, each with a different color of yarn. Divide the children into small groups, and give each group a set of twigs from the kinds of trees marked, and a set of yarn pieces. Have the children look carefully at each twig, noting special features to look for outdoors. Point out that those with opposite buds will have opposite twigs on the trees; alternate buds will have alternate twigs. Outdoors, groups match twigs to trees, mark each twig with the corresponding color of yarn, and, if possible, collect a seed or a leaf from those trees. Observe the branch patterns, bark, and other winter clues to tree identification. Afterward, gather the groups together to compare results, observations, and questions. Have the whole group mark a branch of a nearby tree to watch as spring comes.

Twig Key *(grades 3-6)*

Materials:
• sets of twigs, labeled (adjust for age levels)
• masking tape to cover labels
• large paper
• markers

Objective: To learn the use of a key by making one.

A key is a tool used to identify unknown things. With the children, generate a sample key for a group of hats. Get six of the kids' hats, put them on a desk in the middle, and ask: What is one characteristic that we could use to divide these hats into two groups? What could divide each subgroup into two? (Example: hats with brim, without brim; hats with pompom, without; hats blue, not blue; hats plain color, with pattern.) When any group contains only one hat, end with the name, e.g., Joe's hat.

Here is an example:

A. Hat has a brim.. go to B
A. Hat has no brim... go to C

B. Hat is yellow, says "Woodbury Bears"......................... Jack's hat
B. Hat is green, says "Ide's Blue Seal Feed Store".............. Ricky's hat

C. Hat has a pompom or tassel..go to D
C. Hat has no pompom or tassel.......................................go to E

D. Hat is over ten inches long, purple and green............Monica's hat
D. Hat is less than ten inches long, yellow and blue.........Emma's hat

E. Hat is plain blue ....Michael's hat
E. Hat is red and white, with snowflake pattern................Jody's hat

Once children understand how to use a key, give small groups the labeled twigs of three to five common trees, some paper, and markers, and ask them to write a key to those twigs. (Select more or less obviously different twigs according to the children's ages.) When groups have finished, tape over the labels and have them switch places and try another group's key. What did they learn while designing their key? While trying to use one?

TWIG NIBBLE

Objective: To construct edible twigs while reviewing the parts.

Materials:
- large and small pretzel sticks;
- peanut butter and/or marshmallow spread
- other edible construction materials like raisins, licorice whips, candy corn, etc.
- paper plates or plain paper

Working in small groups, have the children use the edible materials provided to create a twig. Give each child a paper plate, a large pretzel stick for a branch, and access to other materials. Each child designs a twig and then describes it to the others in their small group using the appropriate vocabulary: lenticels, terminal and lateral buds, opposite, alternate, and so on. Nuts, catkins, or other parts could be included, and children may want to name their tree or shrub. Then eat.

EXTENSIONS

Opening Up: Anytime from late winter on (after February or so), you can force twigs by bringing them inside and placing them in water for several days to a week or more. Observe the buds opening, the bud scales dropping, and the leaves and possibly flowers unfurling. Have the children check the buds every 2 days and record/draw any changes they note.

Spring's Arrival: In winter, allow small groups to visit a tree they've studied before or choose a favorite tree outside and mark it with a ribbon on a lower branch. Encourage them to revisit often as spring comes to watch the buds swell and open.

Twig Display: Have the children make a twig board with common twigs taped and labeled on poster board. Add a tree picture beside each twig.

Dissect a Bud: Give small groups of children a bud to dissect. They can either carefully cut down the center and look at the parts, or remove the scales and notice the layers. When was the bud formed and what will it become?

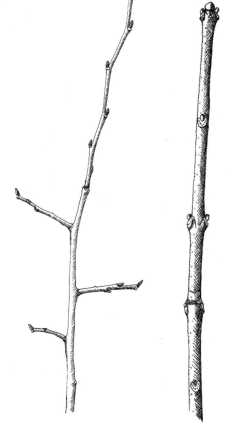

butternut   *American elm*   *white ash*

Winter Twigs

Characters: Tree (maple in winter), Henry, Dosie Deer, Gracie Grouse, Porcupine

Tree

Ah, what a good, long nap I've been having. I needed this winter rest. Last summer just wore me out, growing all those leaves and seeds. *[Boy enters]*

Henry

Hello, Tree. *[big sigh]* Boy, am I sick of winter. Everything is so cold and gray and dull. Nothing is happening out here. How can you stand it?

Tree

Gosh, Henry, are you sleepwalking? Open your eyes. There's lots going on in these woods.

Henry

Are you kidding me? All I see is a bunch of dead trees and some crusty old snow.

Tree

Dead trees?! Oh my, where?!...Oh, you mean me. I can see how I might look dead to you, but I've got a secret. Why don't you go hide over there at the edge of the woods and just watch and listen for a while? I think you'll find there's more going on than meets the eye!

Henry

Okay, I'll do that. And I hope you're right. Then I might even like winter better! *[moves to side; Deer enters]*

Tree

Hi, Dosie Deer, how are you today?

Dosie Deer

I'm starving! It's **so** hard getting around in winter, and deer like me can't always find enough food. So I've come to nibble on some of your twigs.

Tree

My twigs are a very important part of me, so please don't eat too many.

Deer

[munch, munch] They sure taste good – tender and juicy.

Tree

They should be tender! Those twigs are my newest parts. I grew them just last summer.

Deer

Is that right? Mmm Mm.

Tree

Hey! Try not to rip, please! Those buds on my twigs hold the leaves and flowers I'll need this coming spring. Without them I couldn't make my own food or produce seeds.

Deer

Okay, I'll try to be careful. You know, your twigs are important to me, too. *[chomp, chomp]* But you're getting too tall – I can't reach any more. I think I'll go look for some apple twigs next. Oh look, Gracie Grouse is coming in for a landing! *[Grouse lands in top of tree]*

Gracie Grouse

You should fly, like me, Dosie – then you could reach even the highest branches! But I don't eat the whole twig. I have a beak instead of teeth, and I use it to nip off the sweet little buds. Thanks for sharing, Tree.

Tree

Humph. I don't seem to have any choice about it. **I** should learn to fly. Did you know those buds have next year's leaves rolled up inside them?

Grouse

Are you pulling my feathers? You couldn't fit a leaf inside these puny buds.

Tree

Well, it's **true.** That's next summer's shade you're gobbling up!

Grouse

Oh, there are plenty. Stop worrying. You'll never miss what I eat. Still, I like the idea of apple buds. Dosie, I'm coming with you! See you later, Tree. *[Deer and Grouse leave]*

Henry

[talking from the side of the stage] The tree is too far away for me to look at the twigs, and I can hardly see the buds. I want to go look at them close up...Wow! Who's that?! *[Porcupine enters]*

Tree

Uh oh! You're not coming after me are you, Porcupine?

Porcupine

Well, I'm pretty hungry, and that tender inner bark of yours sure tastes good.

Tree

But if you eat the bark all around my trunk, then the water can't get up and the sap can't get down! So watch it!

Porcupine

Oh, I'll just chew through this little bit here. *[chomping on tree]* I won't eat too much. I'm on my way to the hemlocks, anyway – they taste even better to us porcupines. *[Porcupine exits]*

Henry

Boy, porcupines are strange. Imagine wanting to eat tree bark for lunch! *[approaches Tree]* Gee, Tree, you were right. There's a lot going on in these woods – everybody wants to eat you!

Tree

Yes, I feel like a fast food restaurant. But I don't mind as long as they don't eat every twig. My buds are so dear to me. Did you know that you can tell each kind of tree by its twigs and buds? If you study us, you'll see we're all different.

Henry

What do you mean? You all look the same to me.

Tree

You have to look more closely, Henry! See how my buds grow straight across from each other? And look at their shape and color. Why, no other kind of tree looks like this.

Henry

You mean, I can tell what kind of tree I'm seeing from the buds?

Tree

The buds, and the twigs, and the bark. You can fit all the pieces together to figure out who I am!

Henry

Gee, this is like being a private eye, searching for clues to a secret identity. Guess I'll get started.

Tree

Well then, goodbye for now, Henry. Have a good time. *[both exit]*

WINTER TWIGS — BE A TREE FANTASY

You are small trees standing in a big field. It is the end of winter, and your roots are stiff and frozen in the ground. Your branches are bare. Last year's leaves fell off months ago, and next year's leaves and flowers are inside tight buds, hiding from the cold. Animals like grouse and snowshoe hares nibble some of your buds for winter food. The sky is gray. The last snowstorm of the year starts to swirl around you. Your branches move stiffly in the wind. Your trunk sways back and forth.

Finally, spring comes and the ground thaws, allowing water to reach your roots. Feel the cold water trickle between your roots. Your branches and twigs are thirsty. They need water so that the tight buds can grow into new twigs, leaves, and flowers. Pull the water up through your roots. Stretch tall so that the water can work its way up into your trunk and out to your branches. As more and more water rises and finds its way to the tips of your twigs, the once tight buds now begin to swell. Suddenly they break open, and the leaves and flowers emerge. *[fingers uncurl and hands spread open]* One by one, the leaves unfold and reach toward the sky, and the flowers bloom.

It is the growing time for trees. You are hungry and your leaves are working hard to make food for you – spread them wide to catch the sun and feel the warmth soaking in. Your flowers are making seeds for the next generation of trees – feel the small seeds growing. For all this work, you need water from your roots – reach for more water; and you need sunlight from the sky – reach for more sunlight.

Summer passes, and you have grown taller and wider. Stretch your roots, your trunk, your branches. As the weather grows colder, your leaves stop making food. Suddenly, one fall day, the wind blows hard and the rain pelts down. Feel the cold and the wind. Sway and move your branches. Now that your leaves have weakened, they fall off and blow away, once again leaving tight buds behind. *[fists clenched again]* Those buds hold next spring's leaves and flowers.

Bird Songs

MUSICAL MESSAGES

Bird songs have inspired people for centuries. We love to hear the first returning bluebird or robin. The sound of birds singing is a joyful affirmation that winter is past and spring has come. Rachel Carson's choice of title for her book, *Silent Spring*, was extremely effective as it raised the dreaded prospect of an eventual spring without birds, without song.

Why do birds sing? And why especially in the spring? Song plays an important role in the annual cycle of many birds as it initiates courtship proceedings and defines territorial holdings vital to producing and successfully rearing offspring. When one considers the small size of birds, the distances that may separate them, and the many obstacles that act as visual barriers (like leaves and trees), it is no wonder that songs and calls are birds' most effective ways to communicate.

Songbirds are also known as the passerines (from the Latin for "sparrow"), though plenty of nonpasserines have songs as well. Male birds are usually the songsters of a species; they most often establish a territory, defend it, and attract a female (or females) to share it. There are a few common exceptions in which females occasionally sing as well, such as the bluebird, the Baltimore oriole, the cardinal, and the white-throated sparrow.

Most songbirds are small and usually undergo grueling annual migrations, so their life span is relatively short. Since both partners in a given breeding cycle rarely survive many years, most songbirds find a new mate each season rather than mating for life. It could happen that a female and male instinctively return to the same area and coincidentally mate again. But in general, it is thought that the pair bond does not last beyond caring for the young of that breeding season. The song is thus critical to reproductive success as it quickly and effectively announces the presence of a particular species of bird and its desire to mate, and advertises the territory that individual bird has staked out. Males sing more rapidly and for a longer period of time if they are unattached or fending off a rival than if they are mated and simply delineating their territory.

Each species of bird has its own specific song, although frequently there are regional dialects. This song is what enables potential mates to recognize, respond to, and find each other. It also serves to keep intruders of the same species out of established territories so that the established breeding pair has the pick of nest sites and food sources. Within the song pattern of a given species, individual characteristics enable females to distinguish their mate's songs.

Researchers have used sonograms to make sophisticated recordings and visual renditions of bird songs, which have allowed them to document songs and study subtle differences inaudible to the human ear. They have learned that 75 percent of all songbirds possess at least two different songs. These "song repertoires" include various singing patterns, or song types, each with a repeated set of syllables and phrases. The male brown thrasher is a most versatile singer – one male that was studied had more than 3,000 song types.

How do birds produce these varied songs? Birds do not have vocal cords within a larynx, as mammals do, but rather a bony structure called a **syrinx** surrounded by an air sac. Air, when released from the lungs under pressure, causes membranes within the syrinx to vibrate. Birds produce their songs by controlling the frequency of vibrations made by these membranes. The number of syringeal muscles in a given species determines its ability to vary the vibrations. Crows, thrushes, and mockingbirds, which produce a great variety of sounds, have eight pairs of muscles, while pigeons, with their simple cooing, have only one pair.

Young birds of some species, usually those with simple songs, know their song by instinct, but most songbirds have to hear it and practice it. Young birds begin singing anywhere from eight days to thirteen

house wren

weeks after hatching. Their early songs, called subsongs, consist of notes or phrases from what will become the primary song. If reared among members of their own species, young birds usually perfect their song by the spring after the year of hatching.

The time of year when birds sing is related to their reproductive cycle, which in turn is timed to coincide with the maximum food supply for the offspring – usually spring and summer. It is thought that increasing daylight in late winter and early spring stimulates the production of hormones that result in breeding behavior and readiness. The ability and desire to sing is part of that process. The time when they cease to sing varies; some stop as soon as mating has occurred, others not until all care of the young is completed.

The daily song cycle also varies. Most birds sing early in the morning, triggered by the amount of available light. Members of the thrush family are often the first heard at dawn, followed by insect eaters, which can find their flying food outlined against the early morning sky. Many seed eaters and hole nesters seem to wait for better light. By 11 a.m., most birds are quiet, although one account described a red-eyed vireo that sang its song 22,000 times in the course of 10 hours. By 3 p.m., many birds resume their singing and continue until sunset, roughly in reverse order from the morning.

Songs serve well to attract mates and to mark territories, but other messages are communicated vocally as well: aggression, alarm, distress, location of food. Most of these are conveyed by brief, relatively simple call notes. While songs are specific to individual species, calls are more general and often communicate across species. The blue jay's warning scream is a good example of a call that is understood by many birds. Calls are also used during migration and are thought to keep a flock together and help night-flying migrants avoid collisions with each other. Members of a foraging flock of chickadees keep in touch with their constant call notes. Young fledglings make characteristic begging sounds, reminding the parents to keep bringing more food. Birdwatchers learn to make a variety of sounds that attract songbirds in to investigate. One – the commonly used, loud *pishh, pishh* sound – is thought to mimic a general distress call.

The study of bird songs can last a lifetime. First, one becomes aware of differences among the sounds. Then gradually certain songs are associated with specific birds until eventually it is ears more than eyes that reveal bird identities to the experienced birder. To learn the language of bird songs is to add a new dimension of auditory pleasure to your experience of the natural world.

Suggested References:

Ehrlich, Paul R., David S. Dobkin, and Darryl Wheye. *The Birder's Handbook: A Field Guide to the Natural History of North American Birds.* New York: Simon & Schuster, 1988.

Jellis, R. *Bird Sounds and Their Meaning.* Ithaca, NY: Cornell University Press, 1984.

National Geographic Society. *Field Guide to the Birds of North America.* 3rd edition. Washington, DC: National Geographic Society, 1999.

Peterson, Roger Tory. *The Birds.* New York: Time-Life Books, 1963.

Stokes, Donald W., and Lillian Q. Stokes. *A Guide to Bird Behavior.* Vols. I-III. Boston: Little, Brown, 1979-1989.

Bird Songs

FOCUS: Most songbirds sing special songs in the spring to attract mates and to establish and defend territories; other calls may communicate messages such as danger or food sources.

OPENING QUESTION: *Why do birds sing?*

PUPPET SHOW

Objective: To learn some reasons why birds sing.

Perform, or have the children perform, the puppet show. Discuss the advantages of each kind of bird having its own distinct song. Review Mr. Bird's reasons for singing.

Materials:
- puppets
- props
- script

WHICH BIRD SINGS WHICH SONG?

Objective: To hear some common bird songs and notice the differences among them.

Play one bird song at a time for the children. Have them repeat the song in words (see Clock Chorus for common word representations of some familiar bird songs), try to imitate the cadence and rhythm of the bird songs, then listen again to the tape. Show the children a picture of each bird. Do this for six to twelve bird songs, depending on the ages and interest of the children. Play through all the songs again and see if they can remember which bird goes with which song.

Note: The same list of birds can be used for this activity, Clock Chorus, Bingo, and Singing My Song. Pick favorite and backyard birds. For Clock Chorus, they should be ones whose songs can be imitated with words.

Materials:
- audio tape of six to twelve bird songs (birds should be common to your area)
- color pictures of the birds
- tape player

CLOCK CHORUS

Objective: To understand that each species of bird has its own song and favorite time to sing.

Lead a brief discussion on the general times of day when birds sing. Give each child a bird tag with one of the following possibilities written on it:

- American Robin: "Cheerio cheery me cheery me" 4-9 a.m.
- Ovenbird: "teacher-Teacher-TEACHER-**TEACHER**" 4-9 a.m.
- White-throated Sparrow: "Poor Sam Peabody-Peabody-Peabody" 4-9 a.m.
- Mourning Dove: "Coo-ooo, coo coo coo" 5-10 a.m.
- Eastern Wood-Pewee: " Pee-a-wee" 5-10 a.m.
- Red-winged Blackbird: "Konk-la-ree" 5-10 a.m.
- Common Yellowthroat: "Witchity-witchity-witchity" 6-11 a.m.
- Black-capped Chickadee: "Chickadee-dee-dee" or "Fee-bee" 6-11 a.m.
- Red-eyed Vireo: "Going up – Coming down" 6-11 a.m.
- Yellow Warbler: "Sweet sweet sweet, I'm so sweet" 6-11 a.m.
- Northern Cardinal: "Cheer, cheer, cheer, birdy-birdy-birdy" 6-11 a.m.
- American Goldfinch: "Potato chip, Potato chip" 7-12 a.m.
- Eastern Phoebe: "Fee-bre, fee-bre" 7-12 a.m.
- White-breasted Nuthatch: "Yank-yank" 7-12 a.m.

Materials:
- large clock with movable hands
- bird tags – cut in bird shape, 2 of each kind, with names and songs, on strings to hang around necks or with loops of masking tape

Have children put on the tags. Ask all to read their nametag to learn the name of their bird, what their song sounds like, and the hours at which they sing. Have everyone practice the songs. Leader moves the clock hands from midnight to noon and, as the appropriate time arrives, the respective birds should begin singing. By 7 a.m. all birds should be singing, and by noon all should quiet down. If desired, the clock can progress till evening when songs are often sung again.

BIRD SONG BINGO

Objective: To practice bird song recognition.

Tell the children they'll be playing a variation of Bingo that lasts until all the squares are covered and the birds are identified. Hand out Bird Bingo cards and chips to the children. Play the taped songs one at a time and help the children reach agreement on the identity of the singing bird. The children cover the correct bird on their card. Continue until the whole card is covered. Then eat the Bingo chips! Play several times if desired.

Materials:
- bird Bingo cards with names (and pictures) of the birds that are on your tape
- audio tape (you may wish to tape several different song sequences ahead of time, to make the game more fun) and tape player
- animal crackers or other edible Bingo chips

SINGING MY SONG

Objective: To experience how male birds attract mates with their songs.

Materials:
- tags from Clock Chorus, divided into two sets, both identical

Divide the children into two equal groups (with an odd number of children, have one adult play). Hand out the bird tags used for Clock Chorus – identical sets to each group, and have the children review their tag information. One group forms a very large circle, and the other group gathers in the center. Explain that the group forming the outer circle will play male birds who, at a signal, should sing their assigned songs. The other group represents female birds who find mates of their species by listening for their song. At "go," the singing starts, and the search is on. When all have found partners, discuss some of the ways birds pick mates: by song, display, nesting sites, or food supply. Repeat if desired, swapping roles, and have the new male birds swap their tags around so each must find a new bird.

Note: If kids are reluctant to be labeled "males" and "females," just call them "singers" and "searchers," and discuss the roles afterward.

SINGING BIRD SURVEY

Objective: To see and hear how individual birds stake out territories with their singing, and to notice habitat preferences of different birds.

Note: best done as early in the day as possible since most birds are very quiet in the middle of the day.

Materials:
- maps of school grounds
- pencils
- clipboards
- binoculars, if possible

Have the children draw, or the leader may prepare in advance, a sketched map of the area to be surveyed, showing main features such as buildings, roads, trees, playground, brook, and shrubs. In small groups of four to six children, take maps and walk around the area until a bird song is heard. Stop and mark the spot on the map with a letter. At the edge of the map, write each letter and describe the song. Continue on your survey, using a new letter for each new song. Try to spot the birds that are singing. Can you observe an individual bird singing from different spots? What kind of places do the birds choose to sing from? Are any birds listening nearby? (In early spring, look for birds in flocks that have not yet split up to find territories and mates.)

Back in the classroom, groups can share their observations and combine all on a master map. How many different birds did your group hear? Were the same kinds of birds found in similar habitats in your study area?

EXTENSIONS

Still Singing: Encourage the children to return with their song maps to the same area on subsequent days and see if the same songs can be heard from the same places. Try to go out at different times of day and note the differences in singing levels. Several classrooms, or even the whole school, could cooperate to make a big map of the schoolyard's bird songs.

More about Me: Have each child look up the bird he or she portrayed in the Clock Chorus, or another favorite songbird. Ask each child to draw the bird, color it, and place it in its favorite habitat, either in a group mural or by adding habitat details to the drawings.

Bird Walks: Using field guides and binoculars (if available; better for older kids) take the children on bird walks and learn to identify some common birds and their songs. Observe their habitat preferences. Back indoors, check your identifications with a bird song tape or CD.

Schoolyard Habitat: Places that are good for birds are good for people, too. Make it a class, grade, or school project to improve the bird habitat around your school. Some projects could include feeders; nest boxes; nest platforms; and plantings for food, cover, and nesting places. Local birdwatchers and gardeners, state conservation agencies, Audubon chapters, and extension services can help with suggestions.

Bird Songs

Characters: Rocky Raccoon, Mr. Bird, Ms. Bird
Prop: sign saying "Next Day, 5 a.m."

Mr. Bird

Twee Tweedle Dee, Titter Tatter Teer!
Twee Tweedle Dee, Titter Tatter Teer!

Rocky Raccoon

[waking up] Uh, Mr. Bird? Mr. Bird! You have been singing that same song with those same words over and over and over again since five o'clock this morning. No offense, but it's **driving me crazy!** If you insist on singing for so long, can't you at least change the song?

Mr. Bird

Change the song?! I can't just change my song. It'd be like asking you to start barking like a dog.

Raccoon

What do you mean? I hear lots of different bird songs.

Mr. Bird

Yes, but they're coming from lots of different birds. Each different kind of bird has a different special song of its own, and mine is Twee Tweedle Dee, Titter Tatter Teer, Twee Tweedle Dee…

Rocky

Yes, yes, I know what your song is. Okay, I accept the fact that you only sing one song. But **why** do you have to keep singing it?

Mr. Bird

Because I'm looking for a mate, a partner, a Mrs. Bird…

Raccoon

Oh, brother, I should have known, the same old story.

Mr. Bird

That's what my song means:
Twee Tweedle Dee
Come see me
Titter Tatter Teer
I'd like you here.
[flies off singing]

Raccoon

Twee Tweedle Dee, Come see me. I think this bird is going to drive me crazy. I think it **has** driven me crazy. I'm starting to talk to myself. Well, I think the only way I'll get that bird to stop singing is to find him a mate. Here goes: Twee Tweedle Dee, Come see me. Titter Tatter Teer, I'd like you here. Come on audience; I could use some help. Twee Tweedle Dee, Come see me. *[Ms. Bird appears]*

Ms. Bird

Why, I could have sworn I heard a Twee Tweedle Dee, Titter Tatter Teer coming from here. But I don't see any bird like me around. *[starts to leave]*

Raccoon

Don't leave yet, Ms. Bird. If you go right over there to that tree, I'm sure you'll find yourself a handsome mate.

Ms. Bird

Oh! Thanks for the advice, Rocky. *[exits]*

Raccoon

Oh, I think this is going to work. Maybe I'll be able to sleep late now. *[Mr. and Ms. Bird fly around together then exit]* It looks good. It looks very good! Thank you so much for the help, audience. No more five o'clock alarms for me. *[lies down, snores a little; sign saying "Next Day 5 a.m." appears and Mr. Bird enters]*

Mr. Bird

Twee Tweedle Dee, Titter Tatter Teer!
Twee Tweedle Dee, Titter Tatter Teer!

Raccoon

[waking up] I don't believe it. He's still singing at five o'clock in the morning. Mr. Bird, what is going on? I found you a mate.

Mr. Bird

Yes, thank you! But my song is not only to find me a mate.

Raccoon

You told me it meant Twee Tweedle Dee, Come see me. Titter Tatter Teer, I'd like you here. Isn't that right, audience?

Mr. Bird

Yes, but it also means:
Twee Tweedle Dee
Stay away from me
Titter Tatter Teer
Don't come near.

Raccoon

One song means two different things?

Mr. Bird

You said it. To a Ms. Bird it means one thing, and to a Mr. Bird it means something else.

Raccoon

But why are you trying to keep birds away?

Mr. Bird

Because this is **my** territory, **my** home. Mrs. Bird and I are going to build a nest and raise our young here. My song will keep other birds like me away.

Raccoon

Have you ever considered No Trespassing signs?

Mr. Bird

I'd rather sing.

Raccoon

So when spring is over and Mrs. Bird and you have built your nest and raised your young, **then** will you be quiet?

Mr. Bird

Well, not completely quiet. I won't be singing my Twee Tweedle Dee so much, but I'll still be making my short calls to warn others of danger and tell them where there's food.

Rocky

As long as there won't be quite so much Twee Tweedle Deeing.

Mr. Bird

Speaking of which, I'd better get singing. Twee Tweedle Dee, Titter Tatter Teer! Goodbye, everyone. Twee Tweedle Dee, Titter Tatter Teer! *[flies away]*

Rocky

Oh brother, I guess I'd better be going, too – going to go buy myself some earplugs, that is. *[exits]*

Inside a Flower

MAKING A NEW GENERATION

From the fistful of dandelions and violets collected by a child to the cheery blue and white roadside gardens of chicory and Queen Anne's lace that brighten an August morning's commute, flowers are gifts of pure beauty from nature. It's hard to believe, when looking at the soft pink blossoms of an apple tree or smelling the heavy sweetness of lilies, that these beautiful flowers exist for one purpose only: to produce seeds. Their shape, color, size, and smell all contribute to success in this vital mission.

The structure of a flower includes all the parts necessary for producing seeds, although there are many variations. In the center is the **pistil**, the seed-producing organ. (The old song title "Pistol Packin' Mama" helps some remember its gender.) At its outer tip is the **stigma**, which is sticky or feathery to trap the male **pollen** that lands there. A stalk called the **style** holds the stigma aloft. At the base of the pistil is the **ovary**, where the **ovules** with their egg cells await fertilization.

The male parts of the flower that produce the pollen are called the **stamens**. Flowers may have two or more stamens, often in a ring around the pistil. Pollen is formed in the **anther**, which sits atop the **filament**. Under a hand lens, many anthers look like canoes or hot dog rolls. When the pollen is ready for dispersal, the anthers split open and curl back to release it. The details of pollen grains can be seen under a microscope – each kind of flower's pollen looks different.

Showy flowers, like black-eyed Susans or columbines, easily catch our attention, but how many people notice the wind-pollinated flowers on a grass plant or an elm tree? Flowers that rely on wind pollination have unobtrusive petals or none at all. Typically, their pollen is tiny and abundant, the stamens dangling, and the pistils exposed.

Flowers that must attract animal pollinators advertise with big, colorful petals and the reward of extra pollen and a sweet liquid called nectar. Many of these petals have landing platforms and vivid colors or lines to guide pollinators to the nectar. Some flowers smell sweet,

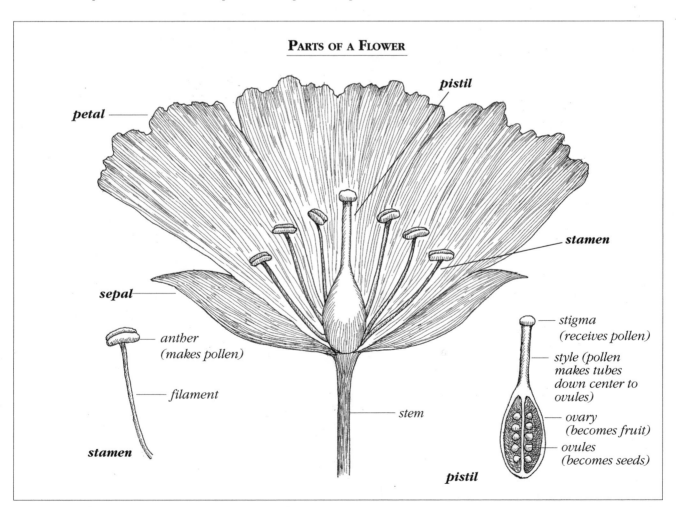

PARTS OF A FLOWER

petal

pistil

stamen

sepal

anther (makes pollen)

filament

stamen

stem

pistil

stigma (receives pollen)

style (pollen makes tubes down center to ovules)

ovary (becomes fruit)

ovules (becomes seeds)

attracting butterflies, bees, and moths; others, like the red trillium, have a color and odor resembling a dead animal to attract the flies that lay eggs in carrion.

Surrounding the petals are leaf-like, usually green, structures called **sepals** that once enclosed and protected the flower bud. Some sepals are very showy and look like petals, like those of tulips and daylilies, which can only be distinguished from the petals because they are on the outside of the buds.

There really is no such thing as a "typical" flower. With over 250,000 species of flowering plants, the structural arrangement of the flower parts can vary greatly among plant species. Some kinds have ovaries above the circle of sepals, petals, and stamens, like the tulip and the tomato; others have ovaries below the circle, like apples and daffodils. Even inside the pistil, the growing seeds are arranged in various ways. When the ovary has ripened to a fruit, the pattern of the seeds within reflects the original design of the pistil. Examine a lily or a daffodil and notice the three-chambered pistil. Cut an apple horizontally in half to see the five-pointed, star-shaped seed container echoing the five compartments in the pistil of an apple blossom.

Before the seeds can develop and before the fruit can form, however, the flower must be pollinated. Most plants cross-pollinate with other plants of the same species, ensuring a mix of genetic material within their seeds. Strategies to insure cross-pollination are varied. More than half of all flowering plants have some guard against their own pollen, whether in the design of the flower, the timing of pollen and stigma readiness, or a chemical way to recognize and reject their own pollen.

While some flowers must guard against self-pollination, all flowers must be designed to receive the pollen that will fertilize the ovules. For some plants that pollen is borne on the wind; for other plants it is carried by pollinators. As they visit numerous flowers, pollinators brush against the pistils and stamens and transfer pollen from one flower to another. Many flowers are designed so that pollinators must brush the female part first, dropping off previously gathered pollen, before they reach the male part where they pick up more pollen for the next flower. Occasionally, as in the beech tree, male and female flowers grow separately but on the same plant. Wind carries pollen from the male flowers on one beech to the female flowers on another. In some other species, such as staghorn sumac, individual plants are either male or female, and grow only male flowers or female flowers accordingly. And in rare cases, the gender of a plant depends on growing conditions. The jack-in-the-pulpit plant produces a female flower when its underground stem has stored lots of food from the year before, and a male flower when the plant is less well nourished.

In North America, the vast majority of pollinators are insects, but the hummingbird does its share of pollen transfer as well. Bats are important pollinators in many tropical areas. Some flowers have evolved unusual ways to get animals to transport their pollen. The common milkweed encloses its pollen in little pairs of attached sacs that look like saddlebags. When an insect lands on the flower to gather nectar, its legs pick up the saddlebags of pollen that it then carries to a new milkweed flower. As Darwin discovered and described, the pollination strategies of many orchids are elaborate, often involving the trapping and tricking of the pollinating insects.

Some plants, such as wheat and garden peas, have excelled through self-pollination, where the pollen in a flower goes no farther than the female parts of that same flower. In such cases, no outside pollinators are needed. The offspring of this union are genetically identical to the parent plant. Dandelions don't even need self-pollination, but can grow seeds from cells in the ovary that have full sets of chromosomes, rather than the half-sets of normal ovules and pollen grains.

Whether wind-borne or animal-ferried, each pollen grain, once deposited on the stigma of a flower of the same species, sends forth a microscopic pollen tube to penetrate the ovary wall. Pollen tubes connect with ovules inside the ovary and pass along male cells, which then fertilize the eggs. Once that has occurred, the ovules grow into seeds and the ovary wall becomes the encasing fruit around the seeds. Picture a tomato or a milkweed pod; both are ripened ovaries containing seeds.

A plant's life cycle varies with the species. Plants that grow, flower, produce seeds, and die within one year, such as ragweed, wild mustard, or wild cucumber vine, are called annuals. Those that require two years to complete the cycle are called biennials. Mullein, foxglove, hollyhocks, and evening primroses are common examples. And plants that live and flower for several years are perennials. Trees and shrubs are obviously perennials, but so are buttercups, bluets, and goldenrods.

The timing of flowering also follows a predetermined yearly schedule to maximize fertilization and germination of seeds. Spring wildflowers and many wind-pollinated trees bloom before the trees leaf out in the spring. Others bloom during the summer or fall season, or continuously over several months. The flowers of the witch-hazel tree appear after leaves have fallen in late fall and remain well into winter.

To look closely at a flower is to find perfection in miniature. Observing that flower's transformation into seeds and fruit is to see a miracle.

Suggested References:
Hunken, Jorie. *Botany for All Ages: Discovering Nature Through Activities Using Plants*. Chester, CT: Globe Pequot, 1989.
Newcomb, Lawrence. *Newcomb's Wildflower Guide*. Boston: Little, Brown, 1977.
Stokes, Donald W., and Lillian Q. Stokes. *A Guide to Enjoying Wildflowers*. Boston: Little, Brown, 1985.

Inside a Flower

FOCUS: Whether they bloom in spring, summer, or fall, flowers have the all-important task of producing seeds. Their structure, smell, color, and blossom time all contribute to this purpose.

OPENING QUESTION: *Why do plants have flowers?*

PUPPET SHOW

Objective: To learn the function of flowers and the basic process of pollination.

Perform, or have the children perform, the puppet show. Discuss the roles played by the flowers, the bee, and the pollen. What if the bee had not come along? How did the birch pollen get around? What would be some advantages of each method?

Materials:
- puppets
- props
- script

FLOWER PARTS

Objective: To learn the different parts of a flower.

Using a diagram or the flower puppets, introduce the parts of a flower: petals, sepals, stamens, anthers, pollen, pistils, stigma, ovary, ovules. Introduce vocabulary appropriate for the ages of the children. Divide the children into small groups and give each group a set of flower-part pieces and a felt board. Ask them to put the pieces together in a way that makes sense to them. After all the groups are finished, join everyone together to discuss the different ways the puzzles were arranged, noting that there are many variations in design among real flowers. Designate one child to be a bee who then acts out delivering pollen so that each felt-board flower is cross-pollinated. Briefly summarize the functions of the various flower parts as well as the importance of cross-pollination to most flowers.

Materials:
- felt boards or other large backgrounds
- flower parts made of felt or paper; removable tape if parts are paper
- flower puppets or large flower diagram

DISSECT A FLOWER

Objective: To examine the parts of different kinds of flowers.

Give each small group one or two kinds of flowers to open up and examine using hand lenses and white paper as a background. Ask them to try to identify the different parts: sepals, petals, stamens, pollen, pistils, stigma, ovaries. Cut some pistils in half lengthwise, others across, to look for ovules in the ovaries. If possible, bring in some flowers whose seeds are further along in development, and cut those open to see the changes. Also, cut open an apple or green pepper to observe that the ovary turns into the fruit, containing the ripening seeds. On the apple, observe the stem at one end and the remains of the sepals and stamens at the other; on the pepper, the stem and sepals are at one end and the tiny pistil at the other.

Materials:
- flowers (wild, garden, or florist – daffodils, tulips, gladioli, and lilies are large and easy to see)
- apples, green peppers, or other fruits with seeds
- hand lenses
- white paper
- knives and cutting boards

FIELD OF FLOWERS

Objective: To construct models of flowers that include the parts necessary for seed production.

Each child or pair makes a flower picture or sculpture, which should include the parts a flower uses to attract pollinators or to allow pollination by wind, and the parts to make seeds. These can be imaginary or real flowers. Share the flowers; end the activity by making a bouquet of these special flowers on a wall or bulletin board.

Materials:
for 2-D flowers:
- white and colored paper
- white paper plates
- crayons/markers, tape, glue, scissors
for 3-D flowers, all of the above, plus:
- straws
- pipe cleaners
- modeling clay

FLOWER HUNT

Materials:
- Flower Hunt questionnaires
- drawing paper
- clipboards
- pencils
- hand lenses

Objective: To observe the variety of flowers in the area and notice differences among them.

Divide the children into small groups. For younger children, have an adult leader in each small group to assist the children in making and recording observations. Older children can work together in pairs or threes. Review the rules for collecting: only with the leader's permission, and only if there are so many that one won't be missed. Give each group Flower Hunt questionnaires, hand lenses, drawing paper, and pencils. Within set boundaries and a given time limit, send the groups out to investigate flowers, answering the questions on the form for each flower observed. (Remind them to look for trees in flower as well as other types of flowering plants.) If there is time, encourage each child in the group to draw one of the flowers.

Afterward, gather the small groups together to share their discoveries and questions.

Note: In bad weather or if no flowers are growing nearby, bring in a variety of flowers and set up a flower hunt inside.

FLOWER HUNT QUESTIONNAIRE

Closely examine a flower and note the following:

What color are the petals? (Or are there no petals?)
Is there a different color in the center?
Are there lines ("bee guides") leading to the center?

Does the flower smell, sweet or otherwise?
Are there any insects on the flowers?

Are the petals separate, or united in a tube?
How many petals are there?
How many sepals, if any?

How many stamens are there?
Can you find the pistil? Is there one, or many?

Is the flower greenish and inconspicuous? Or showy and attractive?

Do you think the flower is insect-pollinated or wind-pollinated? What makes you think so?

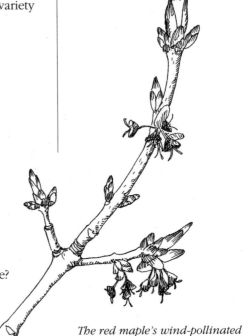

The red maple's wind-pollinated flowers come out before the leaves.

EXTENSIONS

Flower Design: Using magazines and seed catalogs, have the children collect as many photographs of flowers as possible. Sort by color, shape, or other characteristics, and make into a collage on poster board.

Learn the Names and Stories: Together with the children, select ten or more common flowers to identify and learn about (a local gardener, naturalist, or botanist might be able to help). If the flowers are abundant, they could be picked for display and labeling. Try to learn both the common and Latin names, where the flowers first came from, and the uses and folklore associated with the plants.

Flower Calendar: As a classroom or school project, keep track of the blooming dates of wildflowers, shrubs, and trees around the school or in a natural area nearby. Place colored pictures of the blooming flowers on a large calendar. Observe favorites for a longer time. See if you can keep track of one or two until seeds are formed.

Inside a Flower

Characters: Peter Pollen Grain, Paul Pollen Grain, Tiny Pollen Grain, Esther Egg
Props: Flower #1 and Flower #2 attached to puppet stage; Honeybee

[At Flower #1- Hold Peter and Paul on top of the stamens]

Peter Pollen

You know, Paul, it's pretty boring being a pollen grain. I mean, all we do is sit here on the end of this stem coming out of this flower.

Paul

Peter, I've told you a million times – we're not sitting on the end of a stem. The stem is what the flower's sitting on. We're on top of the **stamen.**

Peter

Okay, the **stamen.** What's the difference? It's still boring. And I feel awfully small and unimportant next to these big, colorful petals.

Paul

I know what you mean. These petals **are** very attractive, Peter.

[Tiny Pollen Grain floats by; repeat several times using same puppet in same direction]

Tiny Pollen Grain

[each time it goes by] Whee! Whee! Whoopeee!

Peter

Hey, Paul, how come those little pollen grains can just fly like that when we're still sitting here?

Paul

Oh, aren't they silly? I heard they're from the birch tree. They look as if they're having fun – but they don't know where they're going, and they could land **anywhere.**

Peter

Oh, dear. Where do they want to go?

Paul

To the flowers on another birch tree, of course. A few of them will make it. I've heard we're going on a trip, too, Peter. But we're too big to float like the birch pollen – we have to get a ride.

Peter

You mean I won't be sitting on top of this **stamen** all my life?

Paul

No, no! We pollen grains are really very important. We help make seeds for new plants. I'm not sure how, but I know it means traveling to new places.

Peter

Wow, I can help make a seed?! I can't wait. There's sure not much to do around here.

[At Flower #2 - with Esther inside the ovary]

Esther Egg

Oh gosh, I'm so lonely. Here I sit, day in and day out, at the bottom of this dark flower pistil. Of all the wonderful parts of a flower, I had to be the little egg that sits by itself at the bottom of the pistil. I have a feeling that some day, someone will come down here and visit me. But meanwhile, I'm so lonely. I wonder if anyone even knows I'm here.

[At Flower #1 Honeybee enters and buzzes around]

Peter

Oh no, Paul, what's that coming over here?

Paul

It's a bee. I told you those petals were attractive – they've attracted the attention of that bee! She's coming to get the sweet nectar from this flower.

[Bee lifts up Peter and flies toward Flower #2]

Peter

Oooh no, what's to become of me now? I'm getting airsick. *[Bee lands on Flower #2]* Ooof! That was a rough landing! Oh my, I'm caught on this sticky thing and there's a tube growing out of me! I'm going down it! Yikes! *[Peter slowly goes down the pistil until he's next to Esther]*

Esther

Who are you?

Peter

I'm Peter Pollen grain. Who are **you?**

Esther

I'm Esther Egg. I think maybe I've been waiting for you a long time.

[From Flower #1]

Paul

[to audience] So Peter Pollen joins Esther Egg, and together they become a very important part of the plant – the seed! That's what flowers are for, to make seeds for more plants to grow. It's all part of our life cycle. And speaking of cycling, we'd better get going. I mean, get growing! Goodbye!

Dandelions

SURVIVORS IN A CHALLENGING WORLD

Dandelion!
You'd make a dandy lion
With your fuzzy yellow ruff.
But when you're old
You're not so bold.
You're gone with just one puff.

Dandelions may well be the first flowers that many of us learned to call by name. Their abundance, their early blooming, and their use/nuisance reputations bring them to the attention of children and adults alike. The name dandelion is also easy to remember, especially if you know its derivation from the French *dents de lion* or teeth of a lion, describing the deeply toothed leaves.

Dandelions are members of the **Composite** family (also called the Daisy or Aster family), which gives them a number of famous relations, such as daisies, artichokes, sunflowers, chicory, goldenrod, burdock, and aster, to name a few. Among flowering plants, the composites are second only to the Orchid family in number of species, with something over 13,000 species. "Composite" describes the flower heads, each of which is composed of many tiny individual flowers. When you pick one dandelion flower, you actually pick approximately 300 dandelion flowers or florets.

People familiar with dandelions probably recognize the long, narrow, deeply indented or toothed leaves; the round, compact, fuzzy-textured yellow head; and the ball-shaped, white, symmetrical seed head. How do these three familiar parts connect with one another to create a successful plant? During the first year of a dandelion plant's growth, the leaves, which grow in a circular rosette close to the ground, are the only visible part of the plant. These first-year leaves make food that is stored in the long carrot-shaped root, thus enabling the

mature dandelion seed

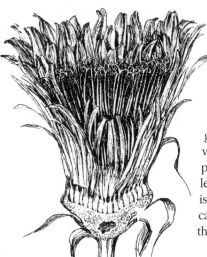

dandelion flower, closed

plant to get a head start making flowers in the plant's second and succeeding years.

One dandelion plant can produce many flower heads at a time and in rapid succession. The unopened buds are found nestled in the center of the leaf rosette, each flower head protected by green **bracts** folded up around it. These bracts curl down when the flowers emerge, but are ready to shutter in the flowers on rainy, cloudy days and at night. The stem lengthens as the flowers develop, holding the flower up above the leaves (though in a closely mown lawn, the flowers will bloom low to the ground). When a dandelion flower head blossoms, the individual florets, each a complete miniature flower, mature in circular rows starting at the outside rim. First they produce **pollen** from the **stamens**, whose heads, or **anthers**, are fused together in a tube around the female part, or **pistil**. The growing pistil then pushes the pollen grains up out of the tube as the pistil emerges. At the pistil's tip, the **stigma** divides into a Y, which curls up as it ages.

Contrary to what one might think, dandelions are rarely pollinated by insects. The many insects that visit dandelions gather pollen, but they only cross-pollinate one in 10,000 dandelions. Rather than containing genes that come half from the **ovule** and half from the pollen of another plant, dandelion seeds form from cells in the ovule that contain the full number of genes. Thus, most dandelions are clones of their parent plant. One can see variations in dandelions, but those growing close together look very much alike.

If a flowering dandelion head were bisected, one could see rows of tiny seeds-to-be nestled on a concave, saucer-like receptacle, each with a minuscule stem topped with a fringe of white hairs that lead up to the yellow tube-like floret. When all florets have matured, the bracts once again close, temporarily, around the flower head. At this time it is easy to mistake the closed green head for an unopened bud. One clue as to its age is the yellowish fluff at the top – the no longer needed flower parts. Gradually the seeds enlarge, the seed stems elongate, the receptacle swells into a convex platform that separates the seeds, and the white fluffy

bracts closed around developing seeds

hairs called pappus become parachutes ready to carry the seeds away. While the seeds are thus preparing for flight, the stalk of the flower head quickly grows taller, providing a raised launching platform for the wind-borne seeds.

After studying dandelions, it's hard not to admire them. Humans, however, have very mixed feelings about the plant. On the positive side of the ledger, new spring dandelion greens are delicious and nutritious, the blossoms make tasty wine, and, to some, green fields blanketed with yellow dandelions are one of May's most beautiful sights. Those who prefer only grass in their lawns and only planted vegetables in their gardens may well have a legitimate complaint against dandelions; they are hard to battle and nearly impossible to conquer. Why? One, the long central taproot, which can reach scarce water and nutrients, can't be pulled up; it must be dug up. This is easier said than done because of its length and because of the tiny secondary roots which grow out from it and help anchor it to the soil. If any fragment of the root is left behind, it can give rise to another entire plant. Two, the leaves, being both bitter and flattened to the ground, are avoided by grazing animals and passed over by lawn mowers. Three, dandelions produce flowers and viable seeds from April into November (few actually blossom in mid-summer). And four, once formed, the seeds are both great travelers and not too particular about where they land because they can germinate in a great variety of habitats.

With adaptations like these in every stage of their life cycle, it's no wonder dandelions are so successful.

dandelion pollen x 250

Suggested References:

Hunken, Jorie. *Botany for All Ages: Learning about Nature through Activities Using Plants.* Chester, CT: Globe Pequot, 1989.

Stokes, Donald W., and Lillian Q. Stokes. *Guide to Enjoying Wild Flowers.* Boston: Little, Brown, 1985.

Uva, Richard H., Joseph C. Neal, and Joseph M. DiTomaso. *Weeds of the Northeast.* Ithaca, NY: Cornell University Press, 1997.

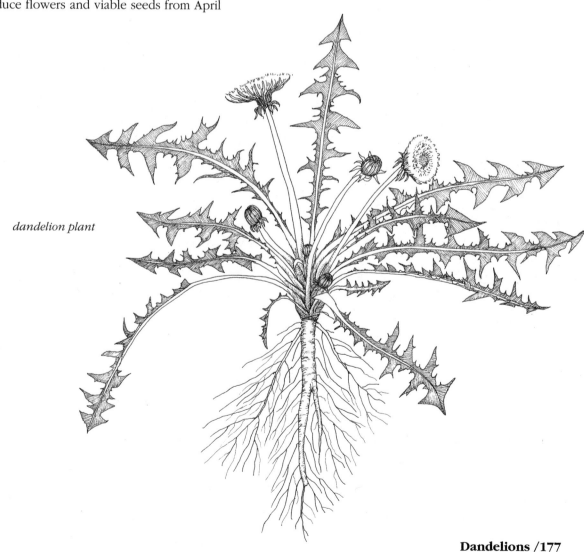

dandelion plant

Dandelions /177

Dandelions

Focus: An efficient and rugged design allows dandelions to survive and reproduce in often inhospitable situations. Their flowering and seed cycle can be observed in every lawn.

Opening Question: *What special characteristics allow dandelions to survive and multiply, even though people often try to get rid of them?*

Dandelion Details

Objective: To inspect the whole plant and learn how each part relates to the rest.

Note: Although dandelions aren't usually allergenic, a few children may be bothered by them. Check to make sure children with asthma or allergies have their medication.

Divide the children into small groups of four to six. Give each group a whole dandelion plant. Pass the plant around the circle. As each child receives it, he or she should examine it, and either make an observation or ask a question about some part of the plant. Record the questions for later review.

Materials:
- whole, washed dandelion plants, one for each small group

Puppet Show

Objective: To learn some special characteristics of dandelions.

Perform, or have the children perform, the puppet show. Afterward, review the various problems encountered by the dandelion, and discuss its defense strategies.

Materials:
- puppets
- props
- script

A Closer Look

Objective: To examine a flower head and its individual parts.

Give each child a dandelion head and a hand lens. Ask the children to first study the dandelion head carefully and then gently take it apart.

Here are some things to look for (introduce vocabulary as appropriate for children):

- The green bracts that surround the flower head. How many rows are there?
- The florets – each a complete flower – with:
  - The yellow petals – called straps;
  - The white hairs, called the pappus, that become the fuzzy part of the seed parachute;
  - The female part – with an ovary at its base and the stem-like style ending at the stigma that divides into two curlicues;
  - The small white ovary that later will cover the seed;
  - The male parts – the anthers, that are joined together around the style.
- Powdery yellow pollen.
- Florets in the center that aren't open yet.

Ask the children to work in small groups to determine how many florets are in one head. Each small group of children could take one head apart and share the counting.

Materials:
- dandelions
- hand lenses
- diagram of individual floret

Note: The flowers may be closed on a rainy or cold day – they will open up in a plastic bag indoors.

INDIVIDUAL FLORET

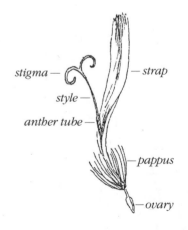

stigma —
style —
anther tube —
— strap
— pappus
— ovary

BEAUTY BEFORE AGE

Objective: To recognize different stages of flowering and seed development.

With the help of pictures or live dandelions, explain the different stages of flowering and seed development.

- Closed bud – the bracts tightly surrounding the bud, then curl down as the flower opens.

- Open flower – note how far the blooming has progressed inward on the flower head.

- Closed-up old flower – the bracts up around the flower again, the brownish-yellow petals pushed up, the white fluff visible. (Cut flower in half to show enlarged seeds and stems.)

- Mature seed head – observe how long the stem on each seed is, how the platform for the seeds has swollen to help separate the seeds, and how the white fluff has expanded.

Then in pairs, have the children arrange some dandelion flower heads in order of age, and display on poster board.

Materials:
- examples (pictures or live) of dandelion flowers in all stages of development
- dandelion flower heads in different stages of development, enough for all small groups to sort
- tape
- poster board

DANDELION INVESTIGATIONS

Objective: To observe dandelions as they grow outdoors.

With older children in pairs and younger children in small groups with an adult leader, give each group a clipboard, a pencil, and a card with one of the following dandelion investigation tasks on it. (Adapt tasks for older and younger children. More than one team can be assigned the same task.)
If some groups complete their assignment early, give them another task card. Once all have completed their tasks, gather to share small group observations and findings.

Afterward, have children each find a dandelion with ready-to-go seeds. Blow the seeds off and follow them. How far do they travel? How is this adaptation important to the flower's life cycle?

Materials:
- Dandelion Investigations task cards, with room to record data
- rulers
- tape measures or measured 10' strings
- thermometer
- clipboards with white paper
- wooden spoons
- pencils
- trowels or spading fork

DANDELION INVESTIGATIONS

- Look for the largest and the smallest dandelion plants you can find. What do you think could account for the size differences? Count the buds and flowers on each. Which one has more buds and flowers? Record your observations.

- Dig up a dandelion plant with the whole root intact. (Fill up the hole neatly.) How long is the root? Write down three words a piece you'd use to describe it.

- Observe and record the weather and dandelion flowering conditions today. What is the temperature? Is it windy? Cloudy? Are the dandelion flowers open today? Can you see any insects on the flowers? What are they doing? What do you think will happen to the flowers in different weather conditions?

- Check the leaf shapes on dandelion plants in different areas. Collect several different shapes and sizes of leaves. Make rubbings of several dandelion leaves: place the leaves under a sheet of white paper, and rub the paper with the back of a spoon. You'll find a good print of the leaf on the back of the paper.

- In an area with lots of dandelions, observe the height of the different flower stages: buds, open flowers, closed flowers, and seed heads. Measure and record the heights of 1) 3 buds; 2) 3 flowers; 3) 3 closed-up old flowers (measure the height from the ground to the flower, not the stem length); 4) 3 seed heads. Older children can take more measurements and average the heights at each stage. What do you think is the reason for the differing heights at different stages?

DANDY LION COLLAGES *(grades K-2)*

Materials:
- paper plates or poster board
- markers
- white glue
- scissors
- lots of dandelion flowers and leaves

Objective: To enjoy the shapes and colors of dandelion flowers and leaves, and illustrate the dandelion's name.

Tell the children that the dandelion gets its name from the French *dents de lion* or "teeth of a lion," referring to the toothed leaves. Using the leaves for teeth and the flowers for fur, have the children create their own lions, or other toothy monsters, by gluing the flower parts onto paper plates or poster board. Allow to dry; then create a wall display with all the creatures.

SHARING CIRCLE

Materials:
- dandelion flowers with stems

Objective: To review what has been learned about dandelions.

Sitting in a circle, give the children some dandelion flowers with stems and show them how to make a dandelion chain by making a slit in the stem of one dandelion one inch or so below the flower head and inserting a second flower stem through this slit. Add to the chain by creating a slit in the second dandelion's stem and inserting another flower. (You may want to secure the ends by tying the stems together.)

Now create a group-sized dandelion chain by having each child add a flower to the chain while completing the sentence, "One interesting thing about dandelions is _____ "

What other ways to make dandelion chains can the children invent?

EXTENSIONS

New Plants: Cut a dandelion taproot into various-sized pieces. Plant these in small pots. At the same time, plant dandelion seeds in other pots. Keep moist. What size root piece will grow a new plant? How long does it take the seeds to grow? Which way makes a new plant faster?

Dandelion Diary: Have the kids mark several dandelion plants for observation, some that get mowed and some that don't. Make a sketch of the plant and note the number of buds and flowers. Keep track of the blooming and going-to-seed process.

Dyeing with Dandelions: Use dandelion plants to make a vegetable dye by boiling them in a large pot of water, removing the plant material, adding white wool yarn, and simmering. Use a guide to vegetable dyeing for suggestions on mordants and other additions if desired. Yarn can be dyed to different shades by leaving it in the dyebath longer. Use the dyed yarn to make pompoms or tassels, or other simple yarn crafts like "God's Eyes."

Dandelion Delectables: Have the children find recipes for dandelion greens, dandelion jelly, sautéed dandelion flower buds, and other dandelion treats. Work together to make and eat them.

Dandelions

Characters: Father, Mother, Polly, Dandy Dandelion (use live plant), Dosie Deer

Props: trowel, lawn mower (paper props to tape to Polly)

Father
> Polly, do you want to earn some extra allowance this week? Our lawn could use some work.

Polly
> Sure, Dad, but I think the grass looks great right now, all bright yellow and green.

Father
> All yellow and green is right. Dandelions are taking over our backyard!

Polly
> I know, isn't it great?! They're one of my favorite flowers. They look so cheerful.

Father
> Well, **I'd** be cheerful if I never saw another dandelion. In fact, I'll pay you a nickel for every dandelion you yank out.

Polly
> You've got a deal, Dad! If you need me, I'll be out back, getting rich. *[Father exits; Dandelion appears]* Gee, it's a beautiful day today. And what an easy way to make money, weeding dandelions from our yard.

Dandelion
> You won't get rid of me that easily!

Polly
> What? Who said that?

Dandelion
> Me, Dandy Dandelion. I've been growing here for two years now, and I like this spot. I'm not going to let you just pull me out.

Polly
> Sorry, Dandy, but my dad will pay me a nickel for every dandelion I get rid of. And you are the first to go.

Dandelion
> Not so fast! I'm sure we can work out something.

Polly
> What did you have in mind?

Dandelion
> How about this? You can have three tries to get rid of me. If I survive all those, you'll leave me here to bloom and grow.

Polly
> That sounds fair. Listen, I'm going to go get my trowel. You watch out for deer while I'm gone. A nice juicy flower like you must look pretty yummy to a hungry animal. And I'd hate to have you eaten before I can collect my nickel from Dad. *[Polly exits; Deer enters]*

Dosie Deer
> Wow, am I hungry! This yard is full of tender new leaves. I'll just grab a few bites while the people are gone. *[approaches Dandelion]*

Dandelion
> Hey, look where you bite! You don't want to eat me.

Deer
> No? Actually, I do. I'm awfully hungry, and you look so green and sweet. Just a little bite. Chomp, chomp. *[nibbles Dandelion]* Phew! Yuck, you taste terrible, and I didn't even get a big mouthful. Oh dear, here comes that girl again. Gotta run. *[Deer exits; Polly enters with trowel]*

Polly
> You're still here, Dandy? I thought I saw that deer eat you for breakfast.

Dandelion
> She tried to! I'm just lucky that I taste so bitter. And my leaves grow flat against the ground so the deer couldn't get much of a mouthful.

Polly
> Well, I'll count that deer as one try, because I'm glad you hung around for me to dig you up.

Dandelion
> Dig me up? I thought you were trying to get **rid** of me. Don't you know that if you leave any of my root behind, a whole new plant will grow from it?

Polly
> **Really?** Dad wouldn't like **that!**

Dandelion
> Well, unless you are a master with that trowel, you'll never dig up all of my long root.

Polly
> Gee, I guess that plan won't work. That's two strikes against me, but you won't survive the next thing I try! *[Polly exits, returns with mower; mower sounds. Polly mows, Dandelion bends, pops back up]*

Dandelion
> **Ha!** Did you think that would kill me? You can mow all you want, I'll just grow flatter and make shorter flower stems. That's three tries, isn't it?

Polly
> Oh, phooey! I give up; I'm telling Dad. *[Polly crosses stage; Dandelion exits; Mother enters]*

Mother
> Hi, Polly, where have you been?

Polly
> **Trying** to get rid of a dandelion because Dad said he'd give me a nickel for each one. But I can't do it. They are smart plants!

Mother
> Well, we can outsmart **them.** Let's put those dandelions to work. Why don't you come out with me and pick a bunch of flowers for a dandelion chain? I'll get some of the new leaves to cook for dinner.

Polly
> That sounds good, Mom. I'll even pick a dandelion bouquet for the table. Maybe I can still get a few nickels from Dad after all!

Designs of Nature

Designs are everywhere in nature, from the spiraling arrangement of scales on a cone to the lacy symmetry of a spider's web to the pattern of cells in a honeycomb. Nature's designs are beautiful to behold, and many of them also have a specific purpose. The closer we look, the more we learn to appreciate the relationship between form and function.

milkweed

While viewing some of the intricate designs found in nature, children will discover how these forms contribute to an organism's chances of survival. The overlapping scales on a pine cone, like the shingles on a roof, help to shed water and protect the developing embryos inside. The beautiful web of an orb weaver is also a deadly trap created by the spider to ensnare unwary insects. The cells in a honeycomb, hexagon-shaped so they share walls and fit together tightly, make efficient use of the space in a hive.

The activities in this theme invite children to look for designs in nature, both to enjoy their beauty and to marvel at their functional efficiency. Children will discover how nature's creations effectively address the many challenges of survival in the natural world.

Spiders and Webs

WEBS AND THEIR WEAVERS

Among nature's beautiful designs, few can surpass the intricate beauty of a spider web glistening with dewdrops in the early morning sun. And few are so immediately and obviously functional. To watch a fly's unsuccessful twisting and turning in the sticky strands of a web is to observe an effective food-trapping device in action.

To many, spiders are far less worthy of admiration than their webs, but in fact, spiders are marvelous creatures. Some people think spiders are insects, but this is not so. They are related to insects in that both are **arthropods**, having jointed legs and external skeletons, but much of the similarity ends there. Spiders have two body parts (**cephalothorax** and abdomen), and insects have three (head, **thorax**, and abdomen). Spiders have eight legs, but insects have only six. Most insects possess both **antennae** and wings, whereas spiders lack both. Spiders have **pedipalps**, appendages located between the jaws and the front legs, but insects do not. These pedipalps are sense organs that also function as sex organs in males.

Spiders belong to the class Arachnida, as do scorpions, mites, and daddy longlegs (harvestmen). The scientific name is derived from the Greek word for spider, Arachne, which commemorates the name of a legendary Greek maiden who challenged the Goddess Athena's spinning ability and was turned into a spider for her audacity.

The English word spider is a corruption of "spinder," one who spins. Almost all spiders can spin silk and are able to do so from birth. The spinning organs are finger-like projections called **spinnerets** that can be extended, withdrawn, compressed, and to some extent, aimed. They are located near the end of the abdomen on the undersurface. These spinnerets are tipped with many "spigots" from which the silk is released. The silk is produced from glands within the abdomen; as the fluid leaves the spider's body, it hardens quickly to form the familiar silken thread. Scientists have identified at least seven different kinds of spider silk, each used for a specific purpose.

Spider silk has considerable strength and elasticity. A rope of spider's silk one inch thick would be stronger than a one-inch steel cable. Some of the threads will stretch nearly one-half their length before they break. The thinnest lines are only one-millionth of an inch wide, and thus invisible to humans, but other lines are much heavier.

Not all spiders spin webs, but those that do, use them to catch insects. When an insect is caught in a web, the spider (often hiding off to the side) feels its struggles to escape. A spider can determine from the pattern and strength of the vibration whether **prey** has been caught, a mate is signaling his arrival, or a **predator** is approaching. Generally, if an insect is caught, the spider rushes toward the prey and injects it with venom or throws a strand of silk over it to disable it. Many spiders wrap their prey in silk to trap and store them before eventually ingesting them. Spiders have small mouths and cannot eat solid food. They must either inject digestive fluids into the insect's body or secrete these fluids over it to dissolve the tissues that they then suck in. If an unpalatable insect is caught, the spider will cut the threads around it until the insect drops out of the web.

Webs vary greatly in complexity and structure, but there are a few fairly common and distinctive types. Sheet webs are easily recognized. The principal part of the web consists of a more or less closely woven sheet in a single, usually horizontal, plane. A funnel web is similar to a sheet web, the difference being that a funnel descends from the web to form the spider's hiding place. A large family of common spiders weaves cobwebs. These tangled, irregular webs are sometimes made under or in objects like leaves or stonewalls. The black widow spider is a cobweb weaver.

The large conspicuous webs often seen on tall grass or suspended between dead tree branches during the summer are orb webs. These resemble large wheels and sometimes have a zigzag band of silk running through the middle that is thought to serve as a lure for flying insects because it reflects ultraviolet light just like a flower. The characteristic design of this kind of web includes a number of supporting spokes made with dry and inelastic silk on which has been spun a spiral of sticky elastic thread. The spider does not become entangled in its web because it steps only on the dry spokes and not on the sticky lines. Also, many web-spinning spiders secrete an oil that prevents them from sticking to their webs.

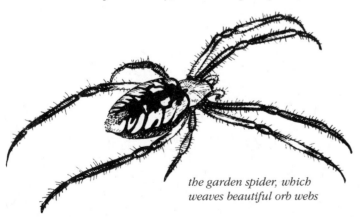

the garden spider, which weaves beautiful orb webs

One of the most familiar orb weavers is the black and yellow garden spider. Some species of orb spiders such as this one remain at the center of the web. Others hide in a nearby retreat where they can feel the vibrations of struggling prey along a so-called trap line that is stretched tightly from the center of the web to the den. The sensitivity of the spider to these vibrations, and its ability to interpret them, is remarkable.

Despite the effectiveness of silken webs for catching food, some spiders do not build webs, but instead stalk or ambush their prey. Wolf spiders, jumping spiders, and fishing spiders all go out and hunt their prey, whereas crab spiders wait in ambush for unsuspecting insects.

During the late summer and early fall, web-spinning spiders are apt to make or repair their web every day, as large insects, once entangled, quickly destroy the webs. Most of this activity takes place around sundown. Often spiders will eat the old web before spinning a new one.

Besides being used for webs, silk is also used for draglines. Wherever the spider goes, it always plays out a silken line that acts as a securing thread, preventing falls and helping spiders to escape predators. Young spiders of most species (and adults of very small ones) spin unattached draglines in conditions of warm, fairly still air. Rising air currents lift the dragline and carry it away with the spider in tow. This is called ballooning, and it helps spiders reach new habitats. Additionally, some spiders use silk as a coating around their egg masses. Others line their burrows with silk. Some water spiders even use silk to trap air under water.

It would seem logical that spider silk, being so abundant and strong, might be used commercially by humans. It is usable as fabric material in the same way as the silk of the silkworm. Although using spider silk has been tried, it was found to be impractical. One of the main problems is that spiders are cannibalistic, making it difficult to rear and feed large numbers in a small space. Therefore, you will have to observe these creatures and their wonderful silk creations in the natural world, or perhaps in your very own kitchen corner.

Suggested References:

Foelix, Rainer F. *Biology of Spiders*. New York: Oxford University Press, 1996.

Graves, Eleanor, ed. *Wild, Wild World of Animals: Insects & Spiders*. Time-Life Films, 1977.

Headstrom, Richard. *Spiders of the United States*. New York: A.S. Barnes, 1973.

Kaston, B. J. *How to Know the Spiders*. 3rd edition. The Pictured Key Nature Series. Dubuque, IA: William C. Brown, 1978.

Levi, Herbert W., and Lorna R. Levi. *A Guide to Spiders and Their Kin*. New York: Golden Press, 1968.

Preston-Mafham, Rod. *The Book of Spiders and Scorpions*. New York: Crescent Books, 1991.

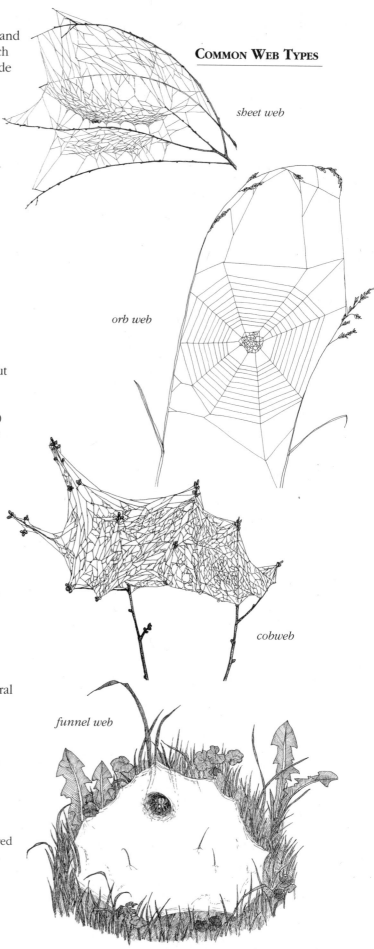

COMMON WEB TYPES

sheet web

orb web

cobweb

funnel web

Spiders and Webs

FOCUS: Spider webs are beautifully designed to carry out their food-trapping function, and so are the spiders that make them.

OPENING QUESTION: *What are some of the special characteristics of spiders?*

PUPPET SHOW

Objective: To learn some of the differences between spiders and insects.

Perform, or have the children perform, the puppet show. Afterward, discuss the differences between spiders and insects.

Materials:
- script
- puppets
- props

SPIDER SPYING

Objective: To observe a variety of live spiders and examine the design of their bodies.

Bring in a variety of live spiders (only one per jar) for the children to look at closely. Have the children use hand lenses to examine the various parts of the spider's body. Notice the number of legs, eyes, and body parts, the coloration, and the hairiness. Have them look for the spinnerets (on the underside of the abdomen) and the pedipalps (on either side of the mouth parts). Is the spider male or female? (mature male spiders' pedipalps have bulbous tips) Make sure the children examine a number of different spiders to note similarities and differences. Look for silk in the jar, draglines, or silk threads coming out of the spinnerets. Remember to release the spiders back where you found them.

Materials:
- live spiders in clear jars with perforated lids, one spider and a moist cotton ball in each jar
- hand lenses

Note: Some people are afraid of spiders. Encourage but do not force the children to hold a spider jar; perhaps you could hold it. If appropriate, discuss some common fears about spiders and the facts or fictions behind these fears. Explain that learning about something often helps people to overcome their fears. If poisonous spiders live in your area, give the children information they should know about these species.

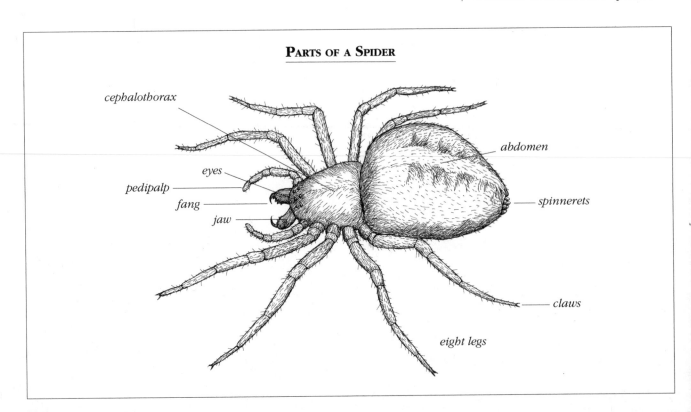

PARTS OF A SPIDER

cephalothorax

abdomen

eyes

pedipalp

fang

jaw

spinnerets

claws

eight legs

FELT BOARD SPIDER

Objective: To review the special features of a spider's body.

After the children have had an opportunity to view live spiders, have them work in small groups to build a model of a spider using a felt board and cutouts. Groups may want to re-examine some live spiders as they work. When all are finished, use a spider anatomy diagram to review the structure of spiders and the functions of the various body parts.

Materials:
- felt board
- spider anatomy diagram *(see Parts of a Spider)*
- felt cutouts of spider body parts: cephalothorax, abdomen, eight legs, eight eyes, pedipalps, spinnerets

SPIDER SENSATIONS

Objective: To experience how web-spinning spiders use their sense of touch to detect motion made by prey captured in their webs.

Ahead of time, make a "web block" for each group by hammering a staple into a block of wood and tying five six-foot lengths of yarn to the staple. Divide the children into groups of six and give each group a web block to put on the floor in the center of their circle. Designate one child to be the spider; the remaining five will be insects. Have the spider crouch next to the web block with eyes closed while each insect takes one strand of the yarn. Insects should spread out so the strands radiate in all directions from the web block; the strands are held taut, next to but not touching the ground. The spider's hands rest lightly on top of the five strands in order to feel any vibrations. The leader points to one insect who plucks its strand once. The spider crawls to the end of the strand that vibrated and, if correct, changes places with the insect. If incorrect, the spider gets another try.

Materials:
- blocks of wood (about 2" x 4" x 6") with five six-foot lengths of yarn tied to a large staple in the center

SPIDER SUPPER

Objective: To learn the process by which many web-spinning spiders capture and eat their food.

In small groups, gather around a spider web wheel. Explain that children are going to act out the process by which spiders capture and consume their food. The leader provides props and assigns a stage in the process to each child. Have the children act out their part (sound effects encouraged!) as the group addresses the following questions.

Materials:
- bicycle wheel with yarn woven in concentric circles through the spokes
- 1 spider (made of pipe cleaners)
- flies (decorated clothespins work well)
- pieces of yarn

For our spider, the story begins when:		Question:
Child #1	Places spider on web	Where do spiders sit on the web?
Child #2	Places fly on web	Why doesn't the spider get stuck in the web?
Child #3	Makes spider pounce on fly	Why pounce on a trapped insect?
Child #4	Wraps fly in yarn	Why do spiders wrap their prey?
Child #5	Places second fly on web	
Child #6	Helps second fly escape from web	
Child #7	Acts out spider eating first fly.	How do spiders eat their prey?

What challenges do some spiders face as predators that the use of a web helps them overcome? (small, wingless, poor eyesight, relatively slow compared to winged prey) What about the spiders that don't build webs – how do they capture prey?

Spider Scavenger Hunt

Objective: To observe a variety of spiders and webs in the natural world.

Introduce the different web types using illustrations or photos. Divide older children into pairs, younger children into small groups with a leader. Each group receives a Spider Hunt card. The children should search for as many of these items as possible, within set boundaries and a given time limit. On dry days, misting the webs with water makes them more visible.

Spider Hunt Card

See if you can find:

A messy web (a cobweb)

A web shaped like a sheet

A web shaped like a sheet with a funnel on one side

A web shaped like a wheel

A spider on a web

An insect caught in a web

A captured insect wrapped in silk

Part of a web that is sticky (look for liquid beads on the web, or lightly touch it)

Part of a web that is not sticky

A web near the ground

A web in a tree or bush

A spider egg case

A spider not in a web

Materials:
- illustrations of different web types
- Spider Hunt cards
- clipboards
- (optional) water mister

Weaving a Sharing Circle

Objective: To allow children to reflect on and share information about spiders.

Have the children sit in a large circle. Give a ball of yarn to one child who then completes this sentence: "One special thing about spiders is _____." That child holds on to the end of the yarn and rolls the ball to another child. This second child then shares one special thing about spiders, holds on to the string of yarn, and rolls the yarn ball to another child. Continue until you have created a messy sheet web on the ground and all have had a chance to speak.

Materials:
- ball of yarn

Activity Stations:
1) Felt Board Spider & Spider Spying
2) Spider Sensations
3) Spider Supper

Extensions

Diary of a Spider: Ask each child to find and watch one spider over a period of days or weeks and keep a journal of its activities. Suggest they go out at night with a flashlight to watch their spider constructing a web. Encourage the children to record their observations through illustrations, photographs, or videos.

Model Spiders: Provide recycled materials (foam balls, pipe cleaners, egg cartons, map tacks) for the children to use to construct a model of a spider. Younger children could hang theirs on a large bulletin board web.

Spider Sketching: Give each child a live spider to draw as if it were magnified to fill the paper. Use field guides to identify and research the spider.

Spider Folktales: Spiders are characters in the folktales of many cultures. Have the children read a variety of these stories and then create one of their own.

Web Design: Have the children construct pictures of different web types (orb, sheet, funnel, tangled) on paper. They should first draw the design in pencil, then cover the lines with glue and yarn or just colored glue.

Spiders and Webs

Characters: Benjy Bear, Charlotte Spider, Rhonda Robin

Prop: suitcase

Benjy Bear

Aren't you excited about tomorrow's big celebration, Charlotte? I can hardly wait!

Charlotte Spider

What celebration, Benjy Bear? The news hasn't made it to my corner of the woods.

Bear

We're having a big party to thank the Earth Fairy for this wonderful season. All the animals of the woods are doing something special for her.

Spider

Great! What are the spiders doing?

Bear

Well, all the insects are going to fly in a big parade in the air!

Spider

Fly?! But spiders can't fly.

Bear

Oh my, you can't? I don't understand it. I thought all insects could fly!

Spider

Well, most insects can fly, but spiders aren't insects. Everyone thinks we are, but we belong to a different class all together. We're called arachnids.

Bear

I never knew that. I always thought you were insects.

Spider

No, no. We're really very different. Spiders have no antennae and no wings, and we have eight legs. Insects only have six legs.

Bear

Eight legs! Let's see if that's right. *[Bear counts to seven]*

Spider

No, no, Benjy. You missed one, I'm sure. Now count again.

Bear

OK, OK, but I could use some help. Will someone please help me count? *[all count to eight]* Goodness. Eight legs and no wings! You really are very different from insects. I don't know what you spiders are supposed to do. If you don't have any wings, you can't very well be in the air show.

Spider

You mean no one has come up with anything special for the spiders to do! *[whimpering]*

Bear

Well, uh . . .

Spider

Why does everyone forget about us spiders? *[crying]*

Bear

Now don't cry, Charlotte. It's not that we forgot about the spiders; we just thought you were insects. I'm sure we can think of something special for you spiders to do. What else can you tell me about spiders?

Spider

Well, most of us have eight eyes. How many animals do you know with eight eyes?

Bear

Eight eyes! I hope you never need glasses.

Spider

Well, it wouldn't matter so much if I did. I can always hunt for my food by feeling it.

Bear

Hunt for food by feeling it? But you must have to catch it before you feel it.

Spider

Oh, we do. We catch our food and we know we've caught it when we can feel it.

Bear

Charlotte, I'm confused! What **are** you talking about?

Spider

Many of us spiders spin a web – a sticky silk web. Mine is shaped like a wheel, but other spiders spin different designs. Then we sit and wait.

Bear

Wait? For what?

Spider

We wait for dinner time. When little insects like flies and moths fly into the web and get caught, we feel something moving. Then we rush out, wrap some silk around whatever is caught, and eat it.

Bear

Boy, you sure use a lot of silk. That must get expensive.

Spider

Expensive! Ha, ha, ha! We don't buy the silk, Benjy. We make it.

Bear

Spiders can make silk? Wow, I wish I were a spider!

Spider

But don't forget. If you were a spider, everyone would be calling you an insect. Oh, it's so depressing. I guess I'll just stay at home in my web during the big celebration. *[Spider walks off]*

Bear

Poor Charlotte! It's tough being a spider. *[Robin enters]*

Rhonda Robin

Oh, Benjy, I'm glad I found you. I need your advice.

Bear

What's the problem, Rhonda?

Robin

I've been trying to find someone to decorate the forest for our celebration, but everyone is too busy.

Bear

Decorate the forest? What kind of decorations do you want?

Robin

Oh, you know, we need some white doilies and lacy garlands to hang in the trees.

Bear

Gosh, I wonder where we're going to get doilies and lacy garlands. Maybe somebody could make them for us. But who? *[pause; hopefully audience will suggest Charlotte the Spider]* Charlotte Spider? Of course! The spiders could weave the decorations with their silk. It will solve our problem and make Charlotte so happy. The spiders will have something special to do for the big celebration.

Robin

What a great idea, Benjy. I'll go tell the Organizing Committee right now! *[Robin leaves; Spider re-enters carrying a suitcase]*

Bear

Charlotte, Charlotte, where are you going with that suitcase?

Spider

The spiders just had a meeting. We can't stand the thought of a big celebration in which we have no part. So we've decided to move on to another forest. No one here will even miss us spiders.

Bear

But Charlotte, you can't leave. We **will** miss you, and besides, we **do** need you spiders. The spiders must make the decorations for our celebration! We need beautiful white doilies and lacy garlands to hang in the trees. Can you do that in time for tomorrow's festivities?

Spider

Of course! We spiders often spend our evenings spinning silk. Tonight we'll all work on spinning and weaving decorations. We could even string glistening beads of dew on each strand! Oh, Benjy, thanks so much for thinking of us.

Bear

Oh, no need to thank me, Charlotte. I'm just happy you'll be here to help us celebrate.

Spider

Well, you've made a lot of spiders very happy. I'd better go tell the others, Benjy, and then get right to work. We'll want to make sure this celebration is one that the Earth Fairy will never forget. Bye, bye, Benjy.

Bear

Bye, Charlotte. See you at the grand celebration.

Variations on a Leaf

THE GREAT PRODUCERS

Leaves come in a variety of sizes and shapes. They may have smooth or ragged edges; long or short stalks; a waxy, rough, or hairy coating. No matter their design, all leaves share a common function: they manufacture food for the plant. In an amazing process called photosynthesis, plants use the sun's energy to make the sugars and starches they need to grow and reproduce. Plants are the basis of the food chain, and all animal life ultimately depends on the ability of plants to produce food.

For **photosynthesis**, plants need water, carbon dioxide, sunlight, and a green pigment called **chlorophyll** that is found in their leaves. Plants obtain water through their roots. The water then rises to the leaves where, in the presence of chlorophyll, it is split into hydrogen and oxygen using energy from the sun. Carbon dioxide enters leaves through leaf pores called **stomata**. The hydrogen gas available from the splitting of water combines with the carbon dioxide to form the sugar glucose, a compound containing carbon, hydrogen, and oxygen. The leftover oxygen is released to the air, replenishing what we and all other animals need for breathing.

There are many adaptive designs that contribute to the efficiency of this photosynthetic process. Leaf size, even on the same plant, varies considerably. In general, leaves exposed to the full rays of the sun are smaller than those in the shade. The leaves in the crown of a red oak that bask in abundant sunlight have deeply cut lobes, while lower leaves that grow in less light are much broader. Birches and poplars that grow in sunny, dry areas, tend to have small, fluttery leaves that allow sunlight to reach their lower layers and that won't be overheated and dried out.

Leaf shape also varies considerably, from the needle-like leaves of the pines to the broad leaves of the maple. Leaves can be nearly round, star-shaped, or linear, but whatever their specific shape, the primary objective of leaves is to capture as much sunlight as necessary in order to carry out photosynthesis as efficiently as possible. The total leaf surface exposed for light absorption is often amazing. An American beech tree 15 inches in diameter was found to have 119,000 leaves with a total surface area of about 3,000 square feet.

Leaves are designed not only to capture sunlight but also to either retain or remove water, depending on environmental conditions. In regions where rainfall is plentiful, leaves often display water-shedding designs. Some leaves, such as those of the American elm tree, are asymmetrical, meaning that the two sides are

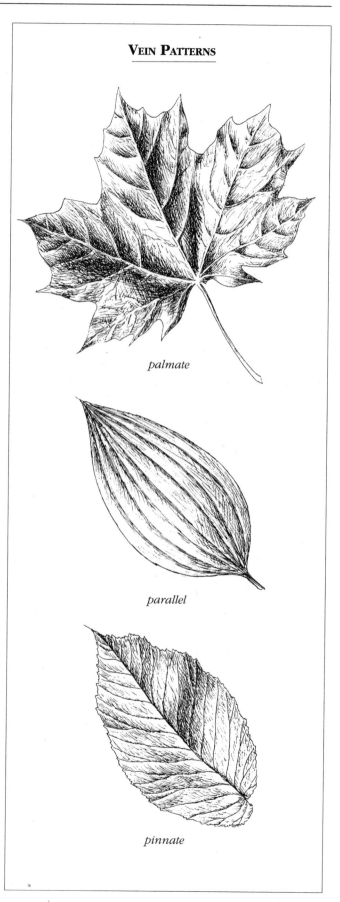

VEIN PATTERNS

palmate

parallel

pinnate

unequal. It is thought that this causes the leaves to tilt sideways in the rain, allowing water to drain off quickly, thereby preventing bacterial infection. Many leaves have sharp points at their outer ends or along their margins. These "drip tips" also cause the water to flow off the leaf quickly. Rainwater in drier areas evaporates rapidly, making these sharp tips unnecessary.

Many leaves are lobed or divided, such as those of strawberry plants and ash trees. Divided leaves are able to **transpire** more water and take in carbon dioxide more efficiently. The lobes also permit wind to pass through the leaf without the leaf blade being shredded. Similarly, the stalks, or **petioles**, hold the leaf and allow the leaf blade to twist in the wind and rain without injury.

While green plants must have sunlight to live and manufacture food, the heat of the sun often causes them to lose too much moisture for proper functioning. Much of the water taken up by a plant for photosynthesis and for mineral nourishment is eventually transpired through the leaves. Plants compensate for this with great efficiency. Leaves have many stomata, through which air can pass, allowing plants to take in carbon dioxide and release water vapor and oxygen. Most stomata occur on the underside of leaves. Cells on either side of each stoma expand and contract, controlling the water loss of the leaf. Desert plants have few stomata as they have to conserve what water is available. Willows, on the other hand, generally grow in moist habitats because they are unable to close their stomata completely, and thus they conserve water less efficiently. Other adaptations that effectively conserve the water supply of plants include leaves that turn on edge to avoid the hot sun, hairs that diffuse the sun's heat, thick outer leaf coverings, and temporary wilting that causes stomata to close up.

Some leaves have specific design adaptations that discourage animals, like deer, insects and rodents, from eating them. Fuzzy, rough, strong smelling, or poisonous leaves help provide this protection.

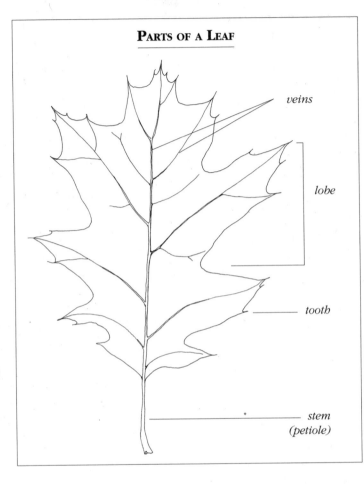

PARTS OF A LEAF

veins

lobe

tooth

stem (petiole)

Any discussion of leaf design is not complete without mentioning the veins that extend through each leaf. Inside these veins are the ducts that carry water and minerals to the various parts of the leaf and carry manufactured sugars away. There are three common vein patterns: **parallel**, **pinnate**, and **palmate**. Each species of plant has its own distinctive pattern of veins, and most fit into one of these three categories.

Leaves are shed and replaced on a regular basis. Where evergreen trees lose some of their leaves each year, **deciduous** trees lose all of theirs once a year in the fall. With the shorter days and cooler temperatures of late summer, chlorophyll starts to disintegrate, and hidden pigments such as xanthophyll (yellow) or carotene (orange or red) become visible. Anthocyanin, another red pigment, is produced by the leaves when nights are cold and days are sunny. In northern temperate climates this results in brilliant autumn foliage. As the chlorophyll disappears, the leaves are no longer able to manufacture food for the tree. At this time about 90 percent of the minerals in the leaves are transported out of them and stored in the tree's tissues. A layer of cork-like cells, the **abscission layer**, then gradually forms across the base of the leaf stems.

Soon, the leaves break away and fall to the ground while the buds that contain the beginnings of next year's leaves remain on the twigs. With the warmth of spring, leaf buds everywhere will unfurl into a multitude of designs, bringing bright shades of green to the landscape and a reminder that the great producers are at work.

Suggested References:

Brockman, C. Frank. *Trees of North America*. New York: Golden Press, 1968.

Ingrouille, Martin. *Diversity and Evolution of Land Plants*. New York: Chapman & Hall, 1992.

Knobel, Edward. *Identify Trees and Shrubs by Their Leaves*. New York: Dover, 1972.

Symonds, George W. D. *The Tree Identification Book*. New York: William Morrow, 1958.

Tolmie, Dr. Ghillean. *Leaves*. New York: Crown, 1985.

Variations on a Leaf

FOCUS: Leaves may vary in appearance and texture, but they are all designed to function as food producers for their plants.

OPENING QUESTION: *Why do plants have leaves?*

FIND YOUR OWN

Objective: To observe some differences among leaves.

Arrange the children in a large circle. Give each child a leaf to look at and to study using a hand lens. Allow the children time to get to know their leaf, then collect the leaves, and put them in a pile. Ask the children to find their own leaf in the pile and place it in front of them to display the variety of designs. Have the children share the unique characteristics that distinguish their leaf from the others. Point out the variety of shapes, sizes, textures, and patterns, and explain that these help serve important functions. Ask the children to hold up the leaves that:

~ catch the most sunlight. (Some trees have bigger leaves than others. When the leaves are small, there are usually more of them.)

~ are best protected from the wind. (Some leaves, like those of the poplar, have flat stems so they move easily in the wind rather than break; other leaves have deep lobes that allow wind to blow by without shredding the leaf.)

~ are designed so they won't become either too wet or too dry. (Some leaves have points on them or waxy coatings so the water will drip off; others have rough or fuzzy coats to help keep moisture in.)

Materials:
- a variety of leaves, one per child
- hand lenses

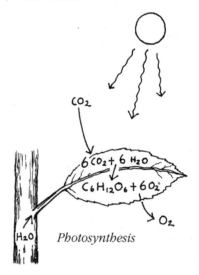

Photosynthesis

PUPPET SHOW

Objective: To understand the function of green leaves.

Ahead of time, hang up a poster with the words to the song from the puppet show. Perform, or have the children perform, the puppet show. Afterward, use a diagram to review the process of photosynthesis. Older children may be interested in the chemical formula for this process. How does the food that is manufactured in leaves nourish the entire plant?

Materials:
- script
- puppets
- props
- poster with song words
- photosynthesis diagram

LEAF RUBBING

Objective: To observe the different patterns of veins in a variety of leaves.

Each child selects one leaf. With the underside of the leaf up, place a piece of paper over the leaf and rub gently with a crayon to get a leaf rubbing. Examine the shapes and patterns that emerge. Allow the children to do this again with other leaves. Discuss the three common veining patterns: parallel, palmate, and pinnate. What function do the veins perform?

Materials:
- paper
- crayons
- leaves
- diagram of three patterns of veins

Leaf Look

Materials:
- Leaf Look Tasks sheets
- (optional) hand lenses

Objective: To discover the variety of leaves outdoors.

Divide the children into pairs, or for younger children, into small groups with an adult. Give each team a Leaf Look Tasks sheet. Within set boundaries and a set amount of time, have the teams go out and collect one leaf per task. (If you'd prefer that children not collect, they can use clipboard, paper, and crayons to do a rubbing of the leaf while it's on the plant.) Let the children know that all kinds of leaves are included in this activity, from blades of grass to palm fronds! When time is up, gather all groups together in a large circle. Read through the tasks one at a time and have each group display the leaf they collected to match that task.

Leaf Look Tasks

Find a leaf that is bigger than your hand.

Find a leaf that is smaller than your thumbnail.

Find a leaf with 5 lobes (fingers).

Find a leaf with smooth edges.

Find a leaf with jagged or toothed edges.

Find a leaf that is an unusual color.

Find a leaf that is lacy.

Find a leaf that is an unusual texture.

Find a leaf that has been partly eaten by something.

Back to Back Leaf Drawing

Materials:
- an assortment of leaves
- paper
- pencils

Objective: To make detailed observations in order to accurately describe a leaf.

Divide the children into pairs. In each pair, one child gets pencil and paper and the other selects a leaf, keeping it hidden from the partner. Sitting with their backs to each other, the child with the leaf describes it in detail while the other child draws the leaf according to that description. Reverse roles and repeat activity. Afterward, allow the children to compare the drawings with the leaves described.

Encourage older children to use leaf structure vocabulary in describing their leaves.

Have You Thanked a Green Plant Today?

ACTIVITY STATIONS:
1) Leaf Rubbing
2) Back to Back Leaf Drawing
3) Leaf Look

Objective: To become aware of our dependency on green plants.

Each child completes the sentence, "I'd like to thank a green plant for my favorite food: _____."

As each child shares, the leader writes the foods on a blackboard or newsprint. When all have finished, trace a few of the foods back to their plant origins.

EXTENSIONS:

Photosynthesis Skit: Give each child a neck tag with CO_2, H_2O, sunlight, or O_2 written on it. The whole group sings the Photosynthesis song from the puppet show, and as the children's parts are mentioned, they move into a center circle designated the "food factory" and hold hands. At the conclusion of the song, give each child a slice of apple or sugar cane to represent the food produced. Then eat it!

Crazy Leaves: Children collect 10 leaves each to play this game like Crazy Eights. In groups of 3-5, one child starts by putting a leaf in the middle. The child to his left must select a leaf from his hand that matches the top leaf in color, shape, or veining pattern and place it on the pile. Red leaves are wild cards and can be used anytime, but the player must declare what characteristic must be matched next. Play continues until one player has no more leaves left in hand.

Autumnal Tees: Decorate T-shirts with the pigments of autumn leaves by gathering a variety of leaves in the fall. Place each leaf stomata-down on a T-shirt and cover with wax paper. Rub a rolling pin or wooden fork over the leaf, crushing it onto the shirt. Peel off the leaf and admire the leaf stains (pigments) on the shirt – it's even cold-water washable.

Identification: After the children do a leaf rubbing, encourage them to use field guides to identify the leaf.

Stained Glass: Design pictures with leaves ironed between waxed paper. Cut a frame out of construction paper to tape around the wax paper. Then hang the pictures in a window.

Smooth rose. A compound leaf is made up of small leaflets. Note that a bud forms only at the base of the whole leaf.

Variations on a Leaf

Characters: Henry, Goldie Goldenrod, Chlorophyll (green circle with pipe cleaner arms)

Props: CO_2 (blob with CO_2 label), Water Drop (blue drop), Sun

Henry

[examining Goldenrod puppet] Nope, not here.
[examines Goldenrod from the other side]
Not here either.

Goldie Goldenrod

Have you lost something?

Henry

Huh! Who said that?

Goldenrod

I did. You seem to be looking for something.

Henry

Well, I was, but I...I didn't lose anything. I was just...well, oh dear, I uh, I was just looking for your mouth.

Goldenrod

For my mouth! But why?

Henry

Well, I wanted to know how you eat. I mean, you're a living thing, right?

Goldenrod

Right.

Henry

Well, I know living things have to get energy from food to live and grow, but I just can't figure out how **you** eat your food.

Goldenrod

We plants don't **eat** food. We **make** food. We're the Producers.

Henry

Producers? You make food? Well, what do you use to make it?

Goldenrod

Oh, it takes a little sunshine and a little water. And I couldn't make food without the air around me and the chlorophyll inside me.

Henry

The **what** inside you?

Goldenrod

Chlorophyll. Why, chlorophyll is the most important thing we green plants have. It's what colors my leaves this gorgeous green. And, it's the secret to our food-making success.

Henry

Well, I wondered what your secret was. But I'm still not sure I understand.

Goldenrod

Well then, maybe this little song and dance I wrote with a friend of mine will help. You ready, Chlorophyll?

Chlorophyll

[offstage] I'm ready!

Goldenrod

A one, and a two, and hit it.
[Chlorophyll appears. Goldenrod exits]

Chlorophyll

Chlorophyll's the name, and making food's the game.

[To the tune of "Oh My Darling, Clementine"]
First I take some
carbon dioxide
which I get
straight from the air.
[CO_2 pops up]
And I mix it
with some water
for the food I
will prepare.
[Water pops up]
Then I capture
me some sunshine
[Sun pops up]
and I mix
the whole thing up.
And Presto,
there is plant food
good for breakfast,
lunch, or sup.
So remember
how important
these three friends
all are to me.
With their help I
find I'm able
to make plant food easily.

Henry

Bravo, bravo! I'd like to hear that again.

Chlorophyll

Sure. But this time I'd like to have everyone join in. How about it?

Henry

Great. I'll get us started. With a one and a two, and hit it. *[Repeat song; display words on a large poster]*

Chlorophyll

[following song] Now I'd better get back in Goldie's leaves where I belong.

Henry

Thanks for the show.

Chlorophyll

Anytime, anytime. Hollywood, here I come! *[Chlorophyll exits; Goldenrod re-enters]*

Henry

That was great, Goldie. But now that I know **how** you green plants make food, tell me **where** you make it. Where is this food factory?

Goldenrod

Food factory? I like that! I guess you could say my leaves are the food factories. I gather water through my roots and bring it to my leaves. My leaves are specially designed to gather the sunlight and air I need to make my food, and that's where most of my chlorophyll is, too. I couldn't do it without my leaves and my chlorophyll.

Henry

So, I guess if you have chlorophyll, you can get your "fill" without even eating.

Goldenrod

Right! And now I'd better get back to producing food before I get hungry.

Henry

And since I'm not a producer, I need to get home for lunch! Thanks for everything, Goldie. Bye. *[both exit]*

Cones

CRADLES FOR THE CONIFERS

Cones are among the beautiful designs of nature that often go unnoticed, hanging as they usually do high in trees, out of reach. Cones from different species of **conifers** vary greatly in size and shape, in color and texture. Yet, they all serve the same function, as cradles for maturing seeds.

For many people the term pine cone is the generic name for a cone from any evergreen tree. However, there are many different kinds of cones. Trees that bear cones, such as pines, spruces, firs, hemlocks, and cedars, are called conifers, and their cones take on the name of the parent tree, such as spruce cone or hemlock cone. In general, the leaves of conifers are either long and slender (needle-like) or small and overlapping (scale-like). A careful examination of the branches and leaves of the cone-bearing tree helps us in the identification of the cone; likewise, studying the design of the cone can be a help in identifying the tree.

eastern hemlock in May

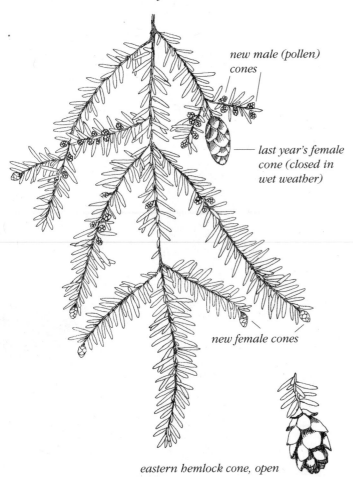

new male (pollen) cones

— last year's female cone (closed in wet weather)

new female cones

eastern hemlock cone, open

A cone consists of a woody stalk with stiff scales growing from it. Each scale is designed to cradle its seeds during infancy and to shield those seeds until the opportune time for dispersal. Each cone is composed of many overlapping scales, like shingles on a roof. If you look closely, you can see that the arrangement of scales makes a pattern of parallel spirals going up the sides of the cone. This architecture channels air and wind around the cone, aiding in **pollination**. Most ripened cones hang upside down from the tree, and when the seeds are fully mature, the cone scales separate slightly allowing the seeds to fall out. If you've ever made a pine-cone wreath, you may have started with cones whose scales were closed. After a few days in a heated house, however, the scales begin to separate, and the cones seem to enlarge. This is part of the natural process for releasing seeds.

blue spruce cone

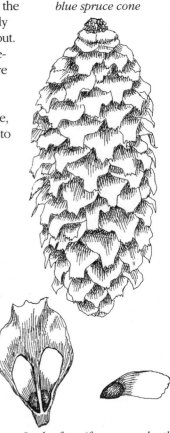

The production of seeds by conifers is considered a somewhat primitive process compared with the process found in flowering plants **(angiosperms).** Unlike the seeds of flowering plants, conifer seeds are not enclosed within an ovary. **Gymnosperm,** the name given to the class of plants that includes conifers, means "naked seed." Conifer seeds develop in an exposed position on top of the cone scales. With a wing-like membrane attached, the seeds rest in grooves on the scales until they fall out or are removed. Some coniferous trees, however, such as junipers and yews, have berry-like seed-bearing structures instead of cones, but even these develop from naked ovules.

Seeds of conifers are under the scales, ready to drop out as soon as the cone opens.

Nevertheless, cones are the most common seed cradles for the conifers, and some of the most familiar of the cone producers are the pines. A description of the development of pine cones and seeds will serve to illustrate the process, although not necessarily the timetable, by which most conifer cones and seeds are formed. In the spring when pine buds begin to swell, the first parts to emerge are the new pollen "cones," clustered at the base of the growing twig. Inside these are pollen grains that, once mature, will fertilize the seeds. (Even in winter you can find dried clusters of last year's pollen cones, looking a little like papery shavings at the base of last year's new twig.) The twig elongates, sprouting needles along its length, until finally in late spring, the new seed cones appear on some of the most vigorous branches, standing straight up at the tips of the twigs. These gumdrop-sized cones are soft and green when they first appear.

Once the **pollen** grains have ripened, the pollen cones elongate, and the sacs containing the pollen are exposed to air. Drying causes the sacs to split open and release clouds of yellow pollen that gets carried by the wind. This is often seen as yellow dust floating in ponds and lakes or covering cars, usually in late May or early June. At this time, the scales of the seed cones separate slightly to allow pollen grains to reach the two **ovules** that lie near the base of each scale. To further insure successful pollination, each ovule secretes a small drop of sticky fluid that traps the landed pollen grains and draws them into the pollination chamber. Once the seed cone has been pollinated, the scales close up again, the stalk of the cone begins to bend downward, and the scales enlarge and harden, so that by the end of the first summer the cones are tightly sealed and hanging upside down. With pollen in place and ovules enlarged, the seed cone rests through the winter.

In pines, it is not until the second spring that fertilization actually takes place, when the sperm from the pollen reaches the ovule. Then development and growth begin in earnest. A white pine seed cone that is three-quarters of an inch long at the end of its first summer grows to a length of four to eight inches by the end of its second summer. After being fertilized, the two ovules at the base of the once tiny scales enlarge and mature to become two winged seeds resting on inch-long scales.

Some seeds disperse from the hanging cone; others hitch a ride to the ground with the falling cone where they may be eaten by squirrels. Some cones, like those of jack pines and western lodgepole pines, may rest for years until fire opens the closed cones to release the seeds. This remarkable adaptation allows these trees to colonize newly opened areas.

Since pine cones take two seasons to mature, pine trees may have two sizes of seed cones present at one time. Cones from other conifers take only one season to mature, even though there is still an interval between pollination (when the pollen lands) and fertilization (when the sperm reaches the ovule). In some species, mature cones may remain on the parent tree for up to 20 years.

Cones have been successful for millions of years in fulfilling their roles as efficient seed nurseries and effective seed dispensers. With their spiral arrangement of overlap-

white pine

white pine cone

ping scales, their many variations of scale size and shape, and the delicate color pattern and shading of each individual scale, cones are among the most artistic and functional designs of nature.

Suggested References:

Eastman, John. *The Book of Forest and Thicket: Trees, Shrubs, and Wildflowers of Eastern North America.* Harrisburg, PA: Stackpole, 1992.

Harlow, William. *Trees of the Eastern and Central United States and Canada.* New York: Dover, 1957.

Petrides, George A. *A Field Guide to Trees and Shrubs: Northeastern and North-central United States and Southeastern and South-central Canada.* Boston: Houghton Mifflin, 1986.

Pielou, E. C. *The World of Northern Evergreens.* Ithaca, NY: Comstock, 1988.

Cones

FOCUS: Cones are beautifully designed with layers of overlapping scales to hold and protect their seeds until dispersal.

OPENING QUESTION: *Why do some trees have cones?*

CONE HUNT

Objective: To find a variety of cones outdoors.

Divide into small groups. Give each group a pencil and a Cone Hunt card. Within set boundaries and a given time limit, challenge the children to find as many items on the card as possible. Record any questions the group has about cones. Afterward, gather groups together and ask each group to select and share one thing they wondered about cones.

CONE HUNT CARD

Find:

A cone that hangs upside down

A cone that sticks straight up

A cone that is twice as long as it is wide

A cone smaller than your thumb

A tightly closed cone

A cone with open scales

A cone with sap on it

A cone that has fallen to the ground

A cone that's been partly eaten

A part of a cone

A tree with long needles

A tree with short needles

A tree with overlapping scales instead of needles

Materials:
- Cone Hunt cards
- clipboards
- pencils

PUPPET SHOW

Objective: To learn about the function of cones as seed-bearing organs.

Perform, or have the children perform, the puppet show. Afterward, discuss the importance of seeds and how cones are involved in seed production. Use a diagram *(see illustration of eastern hemlock in May)* to illustrate the growth and development of a cone.

Materials:
- script
- puppets
- diagram showing cone development

*Penny and Paula,
young cone puppets*

SPIN THE CONE

Objective: To examine differences in branch, leaf, and cone design among conifer species.

Materials:
• cones and branches from 5-6 different kinds of conifers
• cards describing branches, wired onto appropriate cones
• large cone (such as Norway spruce) for spinner

Ahead of time, gather five or six different kinds of cones and matching branches. Use wire to attach a card to each cone that describes its branch.

Have the children examine the branches, noting characteristics such as number of needles per bundle, length of needles, texture (soft, prickly, scaly, or smooth). Explain that each kind of conifer has a different cone.

Now, arrange the labeled cones in a circle and place a spinner in the center. At a signal from the leader, one child turns the spinner, reads the card attached to the designated cone, and matches the cone with the appropriate branch. Repeat the process, allowing every child to have a turn. For older children, provide resources so they can identify the branches. The following cone clues can serve as examples:

> Clusters of 5 needles (white pine).
> Clusters of 2 needles, over 4" long (red pine).
> Short (about 1/2" long), flat needles with tiny stems, attached only to the sides of the twigs (hemlock).
> Short (about 3/4" long), 4-sided, somewhat prickly needles that grow all around the branch (spruce).
> Flattened needles with overlapping scales, in fan-like sprays (white cedar or arborvitae).
> Bundles of short (3/4"-1 1/4"), narrow needles arranged in circular clusters; sheds needles in winter (tamarack).

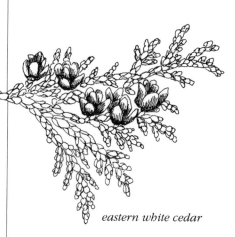

eastern white cedar

PINE CONE PATTERNS

Objective: To observe how cones are designed to hold and protect seeds.

Materials:
• pine cones
• craft paint or markers
• small paintbrushes

In small groups, give each child a cone to examine carefully. Ask them to look for seeds on top of the opened scales. Shake out or take out a seed and examine it. Can they see the indentations where the seeds rested? Now ask the children to examine their cones from the following different angles, and in turn, to tell the colors, shapes, and patterns they see.

> With the bottom or stem end toward them.
>
> Horizontally, with the cone on its side.
>
> With the top pointing diagonally up so that the bases of the scales are visible.
>
> With the top pointing straight toward them.

Note that the scales of the cones form two sets of spirals, one circling steeply, the other more gradually. Have the children paint the scales of a steep spiral to highlight it. Older children could use another cone to paint the scales forming one of the gradual spirals. What are some of the advantages of this spiral design?

loblolly pine cone (closed) with spirals indicated

Where's My Twin?

Objective: To use the sense of touch to match cones of the same species.

Arrange the children in small groups. Give each group a bag containing five or six different cones and another bag containing a duplicate set. Each child in turn is handed one cone from the first bag to examine with eyes closed. The child then tries to find the matching cone by feel in the second bag. Repeat until all have had a turn.

Sharing Circle

Objective: To review and share new information about cones.

Have the children gather in a circle. Pass a cone around the circle, and as each child receives it, he or she completes the sentence, "One new thing I learned about cones is _____."

Materials:
- duplicate sets of 5-6 different kinds of cones
- bags
- (optional) blindfolds

Materials:
- cone

Activity Stations:
1) Pine Cone Patterns
2) Spin the Cone
3) Where's My Twin?

Extensions

Cone Creatures: Let each child choose a cone and use it to make a cone creature by adding material scraps, pipe cleaners, seed pods, acorns, etc. Use florist's clay as an adhesive. Place the creatures in a predesigned diorama.

Cone Bird Feeders: Prepare a mixture of melted shortening and bird seed. Have children dip pine cones in the mixture. Hang the coated cones where the birds will find them.

Cone Display: Invite the children to bring in different kinds of cones and branches; identify them using field guide books. Then label and set up a display.

Math in Nature: Have the children count the number of steep and gradual spirals on their cones. Discuss the Fibonacci sequence and how the number of spirals on the cones relates to it. Try other types of cones and compare. What other spirals are found in nature? (seed heads, pineapples, etc.)

balsam fir

Cones

Characters: Penny Pine Cone (upright pine cone with closed scales), Paula Pine Cone (upright pine cone with closed scales), Carol Cone (pine cone with open scales hanging upside down so the openings face downward)

[Try decorating the cones by painting faces or adding yarn hair, hats, or other accessories.]

Penny Pine Cone

Oh my, I'm so upset! Summer has come and gone, winter is here, and still not a flower or fruit on this whole pine tree.

Paula Pine Cone

But Penny, you know pine trees don't have flowers or fruits. At least I've never seen any. We're different from some other trees. Why just look at us – we have hard waxy needles instead of broad, flat leaves.

Penny

That's true but...

Paula

And we don't lose our leaves all at once the way some trees like maples and birches do.

Penny

Yes, I know we're different in those ways, but what I'm worried about is making seeds.

Paula

What do you mean?

Penny

Well, apple trees have beautiful blossoms in the spring and apples in the fall. The apples hold the seeds to make more apple trees. But what about us pines? If we really don't have any fruit or flowers, we must not have any seeds.

Paula

And without seeds, no more pine trees can grow.

Penny

Exactly. Why, we might be the last grove of pine trees ever!

Paula

Oh my, you're right. We do have something to be upset about.

Penny

Look, here comes Carol Cone. She's older than we are. Maybe she'll know the answer.

Paula

I don't know. I'm afraid she's not feeling well. She's been hanging from her branch head down. And look how her scales are sticking out, not tightly closed like ours. I'm afraid to ask her anything. I don't want to bother her.

Penny

Well, **I'll** ask her then. You hide here and listen. *[Paula exits]* Carol Cone, Carol Cone. I've got some questions for you! *[Carol Cone enters]*

Carol Cone

Hello, Penny. What's the problem?

Penny

Are you feeling all right, Carol?

Carol

I'm feeling just fine.

Penny

Well then, how come you've been hanging upside down for so long? And why are your scales open and my scales are closed? And, oh Carol, I'm worried about the future. How come pine trees don't make seeds?

Carol

Now, now, my dear, you don't need to worry. You've got lots of good questions, and there's one very simple answer to them all.

Penny

There is?!

Carol

The answer is: pine trees do make seeds.

Penny

They do!? Where do they make them? I've never seen any!

Carol

Well, if you can figure out where pine trees make seeds, you'll have figured out the answers to all your questions.

Penny

I know! They make seeds in their roots.

Carol

No.

Penny

They make them in their twigs?

Carol

No.

Penny

They make them in their needles?

Carol

No.

Penny

Well, the only part I haven't guessed yet is the cone...hmmm, I wonder. Could there be seeds inside of us cones? That's a little hard to believe.

Carol

Well, there's only one way to find out. Why don't you take a closer look at me? *[Penny moves closer to Carol]*

Penny

OK...Wow! Wow, wow! There are seeds in there! Right there on the scales!

Carol

Yes! And now do you know why you're not opened up and hanging upside down?

Penny

Maybe I'm not old enough, and my seeds aren't ready to come out.

Carol

That's right. *[starts to wave up and down]* Oh dear, I can't chat any longer – the wind is blowing me around too much. Luckily, it'll help my seeds fly on their way. Goodbye, Penny.

Penny

Bye, Carol. *[Paula appears again]*. Oh, there you are Paula. Did you hear that? The seeds come from us pine cones.

Paula

Gee, it's hard to imagine that I'm holding dozens of seeds inside my scales. I wonder what it will be like hanging upside down.

Penny

We'll know soon enough. For right now, I'm happy just knowing that this isn't the last pine grove on earth!

Paula

Me, too! Hurrah for the pines!

Penny

You mean, hurrah for the cones!

Paula

Yes, yes, hurrah for us!! We should take a bow.

Penny

Yes, a pine bough! Ha ha ha. *[both bow]*

Snow and More

CRYSTALS IN THE CLOUDS

Water is a remarkable compound, and when water in the air meets with freezing temperatures, the result can be a surprising array of sparkling frozen forms. From beautiful crystalline snowflakes to hailstones two inches in diameter, the design of the frozen precipitation can tell you much about its creation and the weather conditions that shaped it.

If you look very closely at a snowflake, you'll quickly discover it is far from just a small round speck of ice. Snowflakes consist of one or more tiny and intricate snow crystals that have joined together in their descent to earth. Although they begin simply as water vapor freezing around tiny, solid particles in the air, such as dust or salt, snow crystals take on beautiful shapes as they form. A crystal, by definition, is a regular and repeated arrangement of atomic particles. All crystals can be grouped based on similar geometric features, and snow crystals are part of the hexagonal system, consisting of six-sided vertical or horizontal prisms. An individual snow crystal can then be categorized according to its particular type of hexagonal crystal growth and its modifications. Some of the more common classifications for snow crystals are:

Hexagonal Plates: six-sided flat crystals with varying internal designs

Hexagonal Columns: six-sided cylinders with flat ends

Bullets: hexagonal columns with one conical end; sometimes a number of them grow outward from a common point to form a rosette

Capped Columns: hexagonal columns with hexagonal plates on either end

Needles: long, slender six-sided columns looking like tiny bolts of lightning

Stellar Crystals or Dendrites: star-shaped with six branches having simple to elaborate designs radiating from the center

Spatial Dendrite: feathery stellar crystals with other branches projecting, usually at 90° angles, from each of the six original branches.

What determines which of these crystal formations will develop? The differences are due chiefly to the temperature and moisture content of the air in which the crystal grows and through which the crystal falls as a snowflake. When the temperature is very low, there is relatively little water vapor and the crystal growth will be relatively slow. Slow crystal growth results in small, simple snow crystals such as the columns and bullets that form in high, wispy-looking cirrus clouds. The low-lying, heavier-looking clouds, on the other hand, have a higher temperature, more water vapor, and thus more rapid crystal growth. This yields the more complex varieties such as stellar crystals and spatial dendrites.

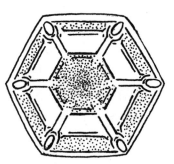

hexagonal plates

In addition to the varied crystal shapes, the designs within the crystals, formed by ridges, grooves, and cavities, are also splendid to examine. Ridges are thicker portions within the crystal, occurring along junctions of crystal segments. Grooves result where crystal segments have only partially welded together. Most of the lines and dots on snow crystals, however, are due to cavities, usually empty, but sometimes partially filled with water. These create lovely patterns and designs within the crystals by dispersing light in different ways.

bullets

Snow usually appears white because the crystals reflect and scatter all colors of the spectrum. Occasionally, however, freshly fallen snow may show blue shadows as it reflects the blue sky above. You may also see this blue tint in depressions or holes in the snow; the deeper the hole, the bluer it appears. The snow crystals absorb many of the colors in the sun's spectrum (the rainbow) except blue light, which then bounces around.

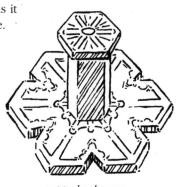

capped columns

Each snowflake experiences different events and weather conditions during its formation and descent. Often as they fall from the sky, crystals collide and the branches may interlock or break off, destroying the symmetry and altering the design. It is rare to find an example of a perfectly symmetrical snow crystal. Once snowflakes reach the earth, the crystals quickly lose their delicate form and become small, irregular ice granules.

Under diverse weather conditions, frozen precipitation may take on other forms, such as sleet, hail, freezing rain, or graupel. Like snow crystals, each of these forms is made of water exposed to freezing temperatures. However, the formation processes are quite different. Sleet, also known as ice pellets, occurs when rain passes through a deep layer of cold air, freezing into ice particles before it hits the ground. Freezing rain, on the other hand, falls through a more shallow layer of below-freezing air near the ground. The liquid drops then freeze on contact with the colder surfaces. Hail, more common in spring and summer than in the winter months, begins as raindrops that are tossed upward by strong winds into the cold upper air. There they freeze, fall back into the rainy layer, and are coated by more water, only to be blown upward again where this next water layer freezes. This process continues until the hailstone becomes too heavy to remain airborne. Graupel, also called soft hail or snow pellets, is made primarily of frozen water droplets that appear as miniature snowballs. These form in local snow showers, or sometimes in intense snow storms, where updrafts in the clouds help their formation.

Each of these forms of frozen precipitation can transform a landscape. Snow softens contours with its insulating blanket of white. Hail often damages objects as it crashes down, covering the ground with frozen pellets. And sleet or freezing rain may coat a scene in a glaze of ice. A close-up look at any of these frozen forms, however, reveals the glistening beauty of one of our most precious natural resources, water.

Suggested References:

Bentley, W. A., and W. J. Humphreys. *Snow Crystals*.
New York: Dover, 1962.
Halfpenny, James C., and Roy Douglas Ozanne. *Winter:
An Ecological Handbook*. Boulder, CO: Johnson, 1989.
Marchand, Peter. *Life in the Cold*. Hanover, NH:
University Press of New England, 1996.
Stokes, Donald W. *A Guide to Nature in Winter*. Boston:
Little, Brown, 1976.

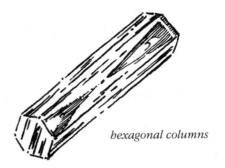

hexagonal columns

stellar crystal

needles

spatial dendrite

Snow and More

FOCUS: In freezing temperatures, water may take many forms, each with its own design and special beauty.

OPENING QUESTION: *How many different forms of frozen water can you think of?*

SNOW MELT

Objective: To investigate how much of the volume of snow is occupied by air.

Divide the children into groups of two or three. Give each group a clear plastic cup full of loose snow. Ask them to decide where they think the water level will be when the snow melts, and have them draw a line at this spot with a waterproof marker. Set the cups in a warm place.

Later, when the snow has melted, ask the groups to compare their results with their predictions. How close were their guesses? How is frozen water different from liquid water?

Materials:
- clear plastic cups
- snow
- waterproof markers

PUPPET SHOW

Objective: To learn about and compare the designs of five different kinds of snow crystals.

Perform, or have the children perform, the puppet show. Afterward, use the puppets to compare the different kinds of snow crystals. In what ways are they similar?

Materials:
- puppets
- script

MAKE A FLAKE

Objective: To reproduce the six-sided structure of snowflakes.

Give each child a **square** of paper. Explain the steps illustrated in the "Make a Flake" diagram and have the children fold and cut as instructed. Open up the folded and cut paper, and you have beautiful six-pointed snowflakes, each design different from all the others. To create a wintry display, hang the flakes by threads of varying lengths from the ceiling.

For younger children, the folding can be done beforehand, with just the cutting for them to do.

Materials:
- white paper squares
- scissors
- thread
- Make a Flake diagram

FLAKES ON FILM

Objective: To observe the different types of snow crystals and other forms of frozen precipitation.

Show pictures of snowflakes, sleet, hail, and other frozen precipitation. Discuss how the temperature and moisture content of the air influence the formation of different forms of frozen precipitation.

Materials:
- slide show or photos and drawings of snowflakes and frozen precipitation

MAKE A FLAKE

How to Make Six-Pointed Paper Snowflakes

Start with a square of paper.

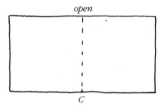

Fold in half,

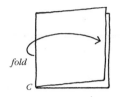

*and in half again.
Keep track of the center (C).*

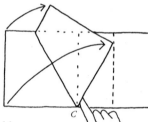

Open the last fold,

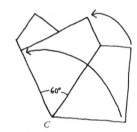

fold one side to the center,

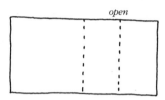

and open again.

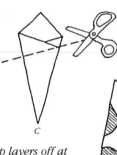

Fold lower corner to meet the fold line as shown, keeping center point.

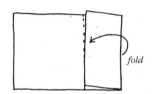

Fold the other lower corner as shown, keeping center crisp and edges lined up.

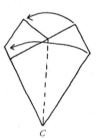

Fold in half.

Trim top layers off at a slant as shown.

Now the fun part – snip out bits along the edges, and open up to see your snowflake.

Snowflake Fantasy (grades K-2)

Objective: To act out the process that results in a snowflake.

Have the children find a quiet place where they can curl up and close their eyes. Read the fantasy and have them imagine that they are living it.

Materials:
• Snowflake Fantasy script

Icy Acts (grades 3-6)

Objective: To examine the different processes that form various types of frozen precipitation.

Arrange the children in three or four groups. Give each group an information card (with diagram) that describes how one type of frozen precipitation (snow, sleet, freezing rain, hail) is formed.

Materials:
• Icy Acts information cards and diagrams
• (optional) role nametags and assorted props, such as ping-pong balls, cotton balls, snow shovels, plastic wrap, windshield scraper

Icy Acts Information Cards

(Words in bold may be made into nametags and assigned as roles for the children to act out.)

<u>Card #1 Snow:</u> All snowflakes start with **water vapor** freezing around a **small speck of dust**, salt, or even pollen. When the temperature is just right (anywhere from **32ºF to -39ºF**), water vapor cools very rapidly and crystallizes, forming a snow crystal. A **snowflake** may be one snow crystal or a cluster of many crystals joined together.

<u>Card #2 Freezing Rain:</u> Freezing rain isn't frozen when it falls from the sky. It comes down as liquid **raindrops**, passing through relatively **warm** (above-freezing) **air temperatures**. It passes through a shallow layer of below-freezing air near the ground, and the raindrops freeze as they touch the **frozen ground** or other cold surfaces.

<u>Card #3 Hail:</u> Hail starts out as **raindrops**, but **strong winds** toss the raindrops up into the very cold upper air where **freezing temperatures** can be found even in spring or summer. As the wind blows the raindrops upward, they freeze and then fall down, only to be blown back upward again and again. This continues until the original rain drops, now coated with many layers of ice, become so heavy they fall to the ground as **hail**. Hailstones can be very large – even the size of golf balls!

<u>Card #4 Sleet:</u> Sleet starts out as liquid **raindrops**. However, as this rain passes through a deep layer of **very cold air**, it freezes into the tiny ice particles of **sleet** before hitting the ground.

Each group presents a skit that shows how one type of frozen precipitation is formed and how that precipitation affects our lives (one example will be sufficient). Encourage the children to use props in their skits. Other children guess which type of weather is being dramatized. After each performance have the children use the diagram to review the steps.

Flakes Up Close

Objective: To notice firsthand the intricate designs of real snowflakes.

If it is snowing, take the children outside to examine the falling flakes. Tape a small scrap of chilled dark material or dark construction paper on the arm of each child, and give everyone a hand lens. Look closely at the snowflakes that fall on the material. Share observations: who sees a star-shaped flake? a six-sided box? a six-sided plate?

Materials:
• black cloth or black paper scraps
• tape
• hand lenses

Snow Snooping

Objective: To observe water in some of its many frozen forms.

Materials:
- Snow Snoopers cards, one for each small group
- hand lenses

Divide into small groups. Each group receives a Snow Snoopers card to guide them in their exploration. Before heading out to snoop, decide when and where the groups will reassemble outside.

> **Snow Snoopers Card**
>
> *Snow Snoopers, try to find:*
>
> The longest icicle
>
> A patch of ice with spots or lines in it
>
> The widest melt circle around a tree
>
> The deepest melt circle around a tree
>
> Evidence of damage caused by ice or snow
>
> A place where the snow is melting and water is dripping
>
> A place where a dark object has melted into the snow
>
> One snow granule. Look at it through a hand lens
>
> Snow that looks clean
>
> Snow that looks dirty – use a hand lens to find out why
>
> Snow you can blow
>
> Snow you can make into a snowball
>
> An imprint – other than a footprint
>
> The biggest piece of crust you can carry. Take it back to the meeting place.

Once groups have reassembled, compare crust samples. Build a snow crust cairn by stacking crust pieces on top of each other.

Hidden Images

Objective: To examine the designs within a snowflake and think about what causes these.

Materials:
- Hidden Images snowflake pictures, enough copies for each pair of children

Give each pair of children a sheet of snowflake pictures. Ask the children what they think might cause the different designs within a snowflake. Discuss how these patterns are formed by ridges, grooves, pits, and water films on the surfaces of the crystals. Have the children search for familiar images in the snowflake pictures and share these with their partner (e.g., I see something that looks like birds flying, or I see something that looks like six horses' heads). Partners look for the snowflake with that image. Afterward, let each pair share their favorite image for the whole group to find.

Extensions

Snow Shakers: Children each create a winter snowstorm in a small glass jar. Use florist's clay to attach a piece of Styrofoam to the bottom of the jar and then stick a small evergreen twig into the Styrofoam. Fill each jar with water, add some silver glitter, and seal well. Shake the jar and watch it snow!

Snow History: Have the children research snowfall records over the past 100 years in the local paper. What is the record snowfall? What other frozen precipitation events made headlines?

Silver Stellars (K-2)**:** Cut a 12" silver pipe cleaner into 3 four-inch pieces and twist these together at the center to form a 6-pointed star. Cut smaller (1/2") pieces to add branches to each arm. Attach black thread to hang this sparkling snowflake ornament.

3-D Flake Construction (3-6)**:** In small groups, have the children make 3-dimensional models of various types of snow crystals out of card-stock paper. Label and display the different crystal types.

Icy Acts

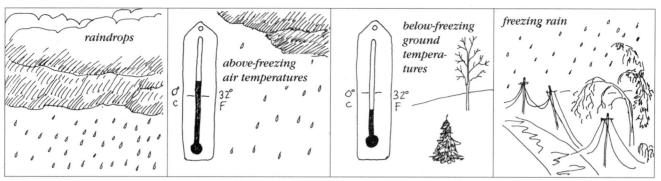

FREEZING RAIN

HAIL

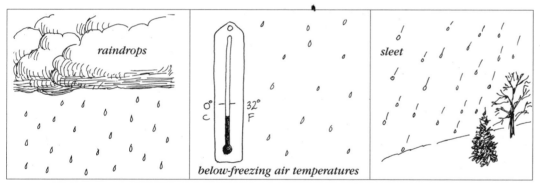

SLEET

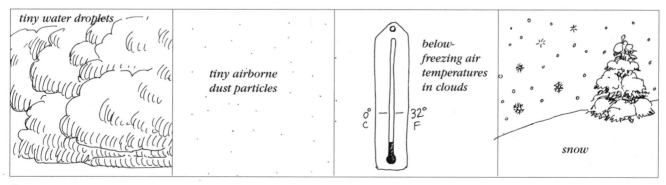

SNOW

Snow and More

You are a tiny speck of dust, sitting on top of a dried-up weed in the middle of a big field. It is January, and a cold, strong wind blows. Last fall, just when the children were going back to school, you landed on the weed while it was alive and green. Now you are wondering, will I ever become unstuck from this dried-up old weed?

The cold wind blows longer and harder, causing your weed stalk to shake back and forth. Suddenly, you are thrown off the weed, and you are headed right for a grove of trees. The bare branches of the trees look closer and closer as the wind carries you toward them. Just at the last instant before you crash, you are lifted up over the treetops and into the open sky.

As you rise higher and higher, you feel light like a feather. Down below, the field that you came from looks like a tiny speck on the earth. Just as you look up, the wind carries you into a dark gray cloud. Here, in the cloud, there are millions and billions of dust particles rushing around and bumping into one another. "Hey, watch it," you yell as a careless piece of dust bumps into you. "Ouch! It's too crowded in here!"

It is also very wet and cold in the cloud, and some water vapor begins to freeze on you, forming tiny ice crystals on your side. More and more water vapor freezes on you, and you begin to form little arms of feathery ice. You feel like a cold piece of white lace.

Now you have six beautiful arms growing longer and wider, so big that you become too heavy and start to fall. All around you, thousands of other crystals are falling and floating down. Lower and lower you sink. Your arms stop growing, and you look at them. You are now a tiny snowflake, a white, shining star falling to the earth. In every direction you look are other snowflakes, and the whole world seems to be white.

You can't tell where you are going. You begin asking yourself, when will I ever land? Where will I be?

Afterward, have the children sit up. Ask them:

Where do you want to land? Why?

What do you want to become after you hit the ground?

What happens to snowflakes?

HIDDEN IMAGES

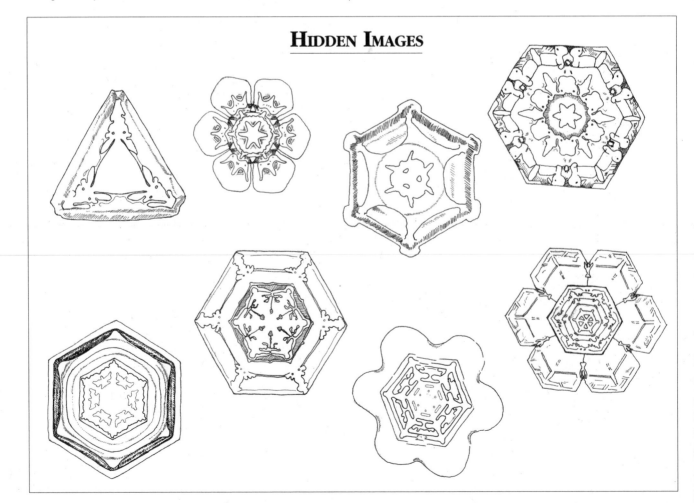

Snow and More

Characters: Needle Crystal, Plate Crystal, Column Crystal, Capped Column Crystal, Stellar Crystal
(for plate and stellar puppets: mount pictures of crystals on poster board, then attach to paint sticks;
for needle, column, and capped column puppets: create 3-D crystals by folding poster board into
appropriate hexagonal shapes; then mount on paint sticks)

Needle Crystal

Hi, who are you?

Plate Crystal

I'm a snow crystal. Who are you?

Needle

I'm a snow crystal, too.

Plate

But you don't look at all like me. You look like a sharp needle made of glass.

Needle

That's why I'm called a needle crystal. You're flat, like a plate.

Plate

That's why I'm called a plate crystal. Do you think we could both be snow crystals even though we look so different from each other?

Needle

Well, I don't know. I'm made out of frozen water. What about you?

Plate

I'm made of frozen water, too. I was made in a cloud high up in the sky.

Needle

Me, too! This is amazing. We have so much in common, but we don't look at all the same. Crystals like me are made when the air temperature is just below freezing.

Plate

Well, that's where we're a bit different because plate crystals like me are made when the air is a little colder than that.

Needle

What beautiful decorations you have all over you. I'm just plain.

Plate

These aren't decorations. They're places where I'm not perfectly flat – little bumps and ridges, grooves and pits. Each of us plate crystals looks a little different from the others.

Needle

You're lucky to be so pretty.

Plate

Aw gee, thanks. Do all needle crystals look the same?

Needle

No, we can be big or medium or small, and sometimes we get stuck to each other and come to the ground in little bunches and then we look pretty fancy. Uh oh, looks as if it's time for me to prepare for landing.

Plate

Goodbye, Needle Crystal. Hope we meet again in the clouds.

Needle

Goodbye. *[voice fading as Needle exits; Column enters]*

Column Crystal

Hello there!

Plate

Why, hello. Are you a snow crystal, too?

Column

I sure am. I'm called a column crystal, because I look like, well, a column! And I'm hollow.

Plate

Oh, I see. You're like a tube.

Column

Well, not exactly. After all, a tube is round, but I have six sides. See? You can start counting here and go around: one, two, three, four, five, six.

Plate

Well, that's amazing because I have six edges. Count them.

Column

One, two, three, four, five, six. Hey, you're right. We both have six sides, even though you're flat and I'm a column.

Plate

I wonder if we're made in the same kind of cloud.

Column

Well, I'm made in the highest, coldest clouds way up in the sky where there's not much moisture. The temperature can be thirty below zero! How about you?

Plate

Oh my, that sounds miserable! I'm made in warmer clouds where there's lots of moisture. I guess we're not from the same place in the sky.

Column

I guess not. But we're both going to the same place. See you in the snow bank! *[exits]*

Plate

Bye, Column Crystal. See you later. *[Capped Column enters]* Hey, aren't you...didn't you just...didn't we just meet?

Capped Column

It's me all right, but I just picked up a couple of friends.

Plate

You've got a plate crystal stuck to each end!

Capped Column

Yeah. I was feeling kind of plain, and these two needed a ride. So now I'm not just a plain column crystal. I'm a capped column! So long. *[Capped Column exits and Stellar enters]*

Plate

What a **cap**ital idea. Oh, here comes another snow crystal. Hello there.

Stellar Crystal

Hello. Bonjour.

Plate

Why, you're beautiful. You look as if you're made of lace. You couldn't be a snow crystal, too.

Stellar

Yes, of course I am, darling. I'm a stellar crystal.

Plate

Stellar means star. I wonder why you're called star?

Stellar

Perhaps because I look like a star.

Plate

I think it's because you're the star of the show! Where do beautiful stellar crystals like you come from?

Stellar

We form in the low clouds where there is lots of moisture and the temperature is just right. It must not be so cold that it makes columns and not so warm that it makes needles.

Plate

Just like me. We could be from the same cloud.

Stellar

It's possible. Now take my arm, and we will fall to earth together.

Plate

Um, ah, which arm? You've got six.

Stellar

It does not matter. They are all the same. Are you ready?

Plate

Sure. I'll count to three. One, two, three! Here we go!

Stellar

Look out below!

Plate

Here comes the snow!

All

Yippee, hurrah! *[all five crystals appear and dance around, then exit]*

Tracks and Traces

CLUES THAT TELL A TALE

To discover animal tracks in the snow or to come across footprints along the muddy edge of a pond is almost as exciting as seeing the creatures that made them. In fact, it may be more so, because unless you are lucky, or very well concealed, an animal will flee at the first hint of your presence. Tracks, on the other hand, remain waiting to be examined, measured, and followed. Infrequent glimpses of an animal tell you little about its habits and behavior, but the tracks and traces it leaves behind can give you a wonderful window into its life.

The immediate impulse upon finding a track is to ask what made it. There are many questions to pose that can help to identify the animal. One of the first questions has to do with habitat – where is the track? Is it near a pond? a stream? in the middle of a field? deep in the woods? The answer eliminates some track-maker possibilities. You would not expect to find a beaver track in the center of a field any more than a rabbit track in the middle of an iced-over lake.

Additional questions go beyond a consideration of habitat. How big and how deeply embedded is the print? Sometimes, in deep snow especially, it is unclear whether a print was made by a single foot, or by all four feet landing together. At this point, it is necessary to decide whether you are looking at one large footprint, like that of a dog, or a cluster of smaller footprints, like those of a squirrel. Once the relative size of the footprints is determined, one can guess the approximate size of their owner.

If the prints are clear, individual tracks can help solve the mystery of who made them. They rarely are though, because melting, freezing, rain, snow, and wind tend to blur distinct features. Sometimes the prints are obliterated by excited trackers; remind children to step carefully. Details to notice that will help identify the print include shape, length, width, number of toes, presence or absence of toenail marks, and even the shape and number of pads.

Just as habitat and details of the footprint provide us with information about the animal, so too does the pattern of the trail. For many trackers, the design or pattern made by a series of footprints is the single most important clue. Most animals typically move in one of four distinct ways: walking/trotting, galloping, bounding, or waddling.

In the walking/trotting pattern, the animal alternates right and left feet, placing the hind feet in the prints made by the front feet. The result is a straight or nearly straight line of single prints. Members of the dog, cat, and deer families typically walk in this fashion. Gallopers such as mice, squirrels, and rabbits push off with their front feet and then swing their two back feet around so that they land in front and a little outside of the prints made by the smaller front feet, creating a cluster of prints. Stocky, short-legged animals like raccoons and porcupines waddle from side to side, creating an alternating big-little pattern as they place their large hind feet next to their smaller front footprints. The bounding pattern is typical of animals like the weasel, mink, and otter. These long-bodied animals land with their front feet planted nearly side by side, and then leap forward so that the two back feet fall into the prints just vacated by the front feet.

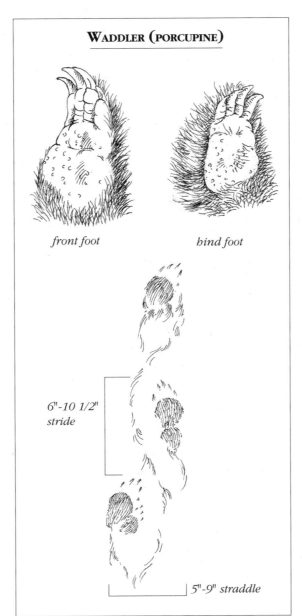

WADDLER (PORCUPINE)

front foot   *hind foot*

6"-10 1/2" stride

5"-9" *straddle*

Once you have determined the pattern, it's time to take measurements. This can help you to distinguish among animals that create similar patterns. The distance between footprints (as in the walkers) or between sets of footprints (as in the gallopers) is called the **stride**. The width of the track from the outer edge of one print to the outer edge of the next print or across the set of prints is called the **straddle**, or trail width. The size of animals of the same species can vary, as can their speed of movement, and both of these can affect the size of their leaps. But if you consider all the clues available – the habitat, the pattern, the stride and straddle, and the print size – you will have a better chance of figuring out what made the track. Of course, a good field guide is indispensable.

Identifying the tracks gives great satisfaction, but following them reveals a chapter in the animal's life otherwise closed to most of us. Knowledge about any animal increases with observations about where it's been, what it's been doing, and where it's going. Does the path follow a fairly straight route toward some seen or unseen destination, and are the tracks evenly spaced? If so, the animal was probably neither pursuing nor being pursued. Does the animal go under low branches, around them, or step over them? (A hint as to its height.) Do the tracks end at a tree or at a hole?

Animals move about for three main reasons. They are probably looking for food, for shelter, or for mates. Nibbled branches, cone scales, bits of fur or feathers, and blood are traces left by very different animals. Deer or rabbits nibble branches; squirrels peel off cone scales to get at the seeds; and **predators**, like foxes or hawks, account for remnants of hair or feathers. Holes in the snow or well-trodden paths leading to hollow trees or evergreen groves can indicate a shelter or home. Porcupines and deer reuse paths to and from their sheltered spots. Ruffed grouse will dive into soft snow on cold nights to take advantage of the snow's insulation. In the spring, tracks of animals that usually appear alone, like fox tracks, are often paired. Such signs inform us of the habits and activities of the animals who left them.

Tracking may leave its followers with more questions asked than answers found, but, without interfering in their lives, there is no better way to learn about the secretive world of wild creatures.

Suggested References:

Ennion, E. A. R., and N. Tinbergen. *Tracks*. London: Oxford University Press, 1967.

Headstrom, Richard. *Whose Track Is It?* New York: Ives Washburn, 1971.

Murie, Olaus J. *Field Guide to Animal Tracks*. Cambridge, MA: Houghton Mifflin, 1954.

Rezendes, Paul. *Tracking and the Art of Seeing*. 2nd edition. New York: Harper Collins, 1999.

Stokes, Donald W. *A Guide to Nature in Winter*. Boston: Little, Brown, 1976.

Stokes, Donald W., and Lillian Q. Stokes. *Guide to Animal Tracking and Behavior*. Boston: Little, Brown, 1986.

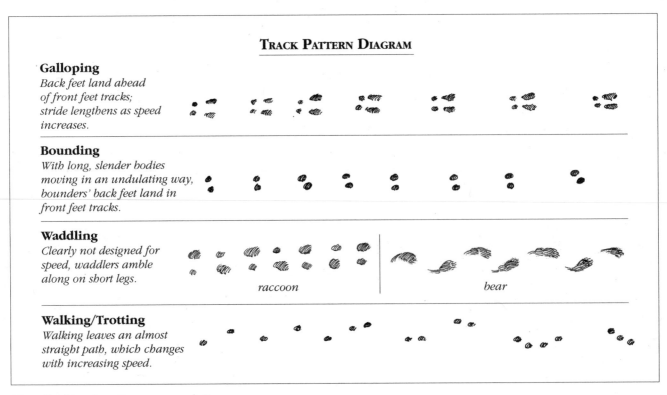

TRACK PATTERN DIAGRAM

Galloping
Back feet land ahead of front feet tracks; stride lengthens as speed increases.

Bounding
With long, slender bodies moving in an undulating way, bounders' back feet land in front feet tracks.

Waddling
Clearly not designed for speed, waddlers amble along on short legs.

raccoon   bear

Walking/Trotting
Walking leaves an almost straight path, which changes with increasing speed.

Note: *Real track patterns vary a lot!*

WALKER (WHITE-TAILED DEER)   GALLOPER (COTTONTAIL RABBIT)   BOUNDER (FISHER)

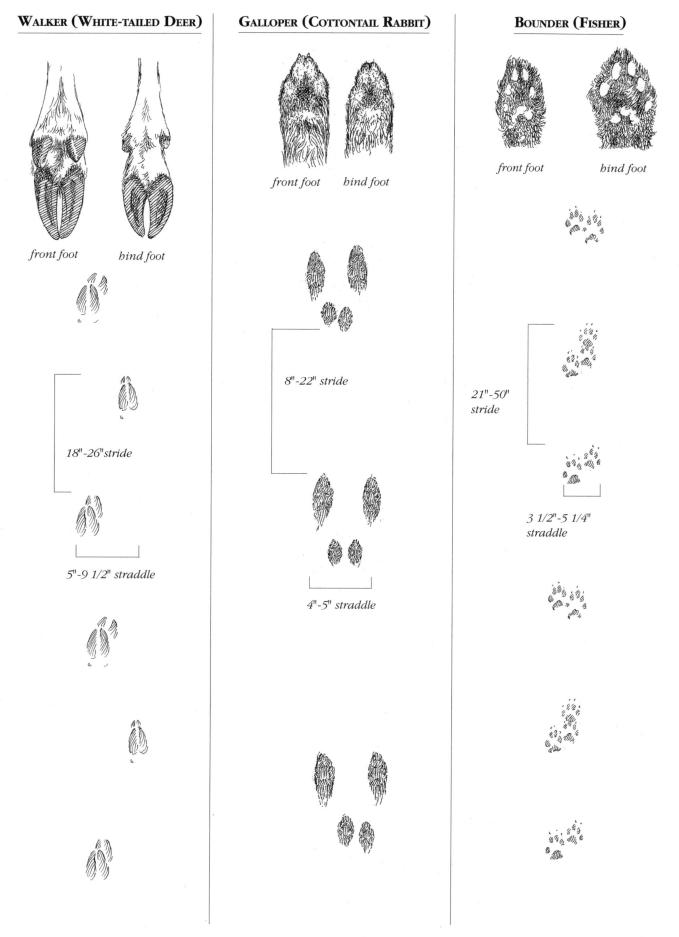

front foot   *hind foot*

front foot   *hind foot*

front foot   *hind foot*

18"-26" stride

8"-22" stride

21"-50" stride

5"-9 1/2" straddle

4"-5" straddle

3 1/2"-5 1/4" straddle

Tracks and Traces

FOCUS: Tracks and traces can provide a glimpse into the lives of animals whose actions are otherwise hidden from us.

OPENING QUESTION: *What can we learn about animals from their tracks?*

PUPPET SHOW

Objective: To compare some different animal track patterns.

Perform, or have the children perform, the puppet show. Afterward, review the animals in the puppet show, how they move, and what their track patterns look like.

Materials:
- script
- puppets
- props

FOLLOW THE FOOTPRINTS

Objective: To learn how track patterns can help us identify the animals that made them.

Using a track pattern diagram, discuss the four main types of patterns and which groups of animals make them. Tape oval pieces of cardboard to the floor in each of the four track patterns (adjusting the scale to the size of the children). Then have the children try to follow each set of footprints on hands and feet. (It's not easy!) Discuss how each animal's shape and way of moving relate to the track pattern it leaves. Explain how to measure stride and straddle, and show how these give clues about an animal's size and shape.

Materials:
- track pattern diagram
- cardboard ovals, in two sizes when need to show front and back feet tracks
- tape

WINDOW SHADE STORY

Objective: To practice interpreting tracks to discover the story they tell.

Ahead of time, copy or illustrate a "tracks and traces" story on an old window shade. Gradually unroll the window shade to show the picture story in stages and have the children figure out what happened. Discuss how the location of the tracks, as well as the tracks themselves, give clues about the tracks' makers.

Story: A fox came out of the woods, trotted to a hen house, and grabbed a chicken. Just then, someone let the dog out of the house. The dog chased the fox, and the fox ran off, dropping the hen. The hen hurried back to the hen house. The dog walked along until it encountered a skunk, which turned and sprayed the dog. The dog ran home to its doghouse. Elsewhere that same evening, a mouse emerged from a hole near the base of a tree. An owl swooped down and captured the mouse.

Materials:
- shelf paper or window shade
- permanent markers

PRINT MATCH

Objective: To practice distinguishing among prints.

Divide the children into groups with even numbers. Assemble each group near a different patch of clear snow or dirt. (Use a broom to clear away footprints, if necessary.) This could also be done on a dry cement walk with wet footprints. Ask the children to form pairs, one a tracker, the other a print-maker. Have the trackers move away from their partners, face the other direction, and close their eyes. Each of the print-makers makes a footprint in the fresh snow, and then shows the sole of their boot to their partner. Trackers now try to distinguish their partner's prints from the others. Have the children switch roles and try it again.

PATTERN PRACTICE

Objective: To practice making and recognizing the four basic track patterns.

Divide the children into small groups. Assemble each group near a clear patch of snow or dirt. One child in the group is selected to make one of the four basic track patterns (galloper, walker/trotter, waddler, or bounder) while the other children hide their eyes. They then look at the pattern and try to guess what movement was used. Now let other children try making different track patterns. Bring out the track pattern diagram for children to refer to. Children may also have fun making up their own movements (e.g., skipping, rolling, hopping) for the other children to try to interpret.

For younger children, play "Follow the Leader" in pairs with one child recreating the movement of the leader by following in his/her tracks.

Materials:
• broom
• track pattern diagram

PICTURE PARADE

Objective: To recognize the tracks and traces of some common animals.

Show pictures or slides of different animals' tracks, including representatives of the dog, cat, rodent, rabbit, weasel, and deer families, and have the children describe the differences they observe. Also include pictures of scat, browse or chew marks, and denning or nesting evidence of animals common to your area.

Materials:
• slides or pictures of animal tracks and sign

TRACK DETECTIVES

Objective: To find tracks and traces of animals, and, by noticing the location, size, shape, and pattern, to try to determine what animals made them and what they were doing.

Lead the children in small groups to areas where you have previously found animal tracks or sign. Have them comment on all the characteristics they notice about the tracks, and together try to discover as much as possible about the track-maker and what it was doing. With older children, measure the prints and patterns, and refer to track field guides for identification.

Materials:
• track and sign field guides
• track pattern diagram
• tape measures or yardsticks

TRACK STORIES

Objective: To reconstruct a story they "read" outside or to invent a story and tell it in tracks.

Divide the children into groups of 3 or 4. Each group gets a large piece of paper. First ask each group to decide on an animal story – either one they have seen outside or a made-up one. After sketching and coloring in the habitat, children use sponge tracks or stencils you've provided to print their story. It works well to have each child in charge of printing one set of tracks. Groups take turns holding up their track story while the other children try to interpret it.

Materials:
• sponge cushion tracks or stencil tracks of representatives of the dog, cat, deer, rodent, rabbit, and weasel families
• markers or ink pads
• shelf paper, crayons

ACTIVITY STATIONS:
1) Print Match, Pattern Practice, and Track Detectives (outside)
2) Track Stories (inside)

EXTENSIONS

Home Track Detectives: Ask the children to check outside their house every morning for tracks. Can they tell which person, car, or animal was there?

Track Scouts: Suggest to the children that they go for a walk with their parents to look for tracks. Follow some and try to figure out what the animal was doing.

Creative Writing: Have the children write a story that involves following tracks as an important part of the plot.

Tracks and Traces

Characters: Henry (dressed in coat and hat), Mother, Mouse, Mink, Porcupine
Props: 4 track pattern signs – Walker/Trotter, Galloper, Waddler, and Bounder

Henry

Mom, have you seen Woof? I've called and called, and he hasn't come home. It's his dinnertime.

Mother

That's odd. He's always hungry for his supper. Let's take a walk and see if we can find him.

Henry

With all this snow, we should be able to follow his tracks. Those must be his over there. *[walk/ trot track pattern sign held up briefly]* They're in a pretty straight line, and I can see his toenail marks.

Mother

Well, let's see where he went. You can go first, if you'd like. I know you want to run, and I'll follow along behind.

Henry

OK, Mom, good idea! I'll follow Woof's tracks, and you can follow mine! *[both exit; Henry re-enters alone]* Hmmm. Woof's tracks are going right over to this stone wall. *[galloper track pattern sign appears]* Looks as if he's following these other little tracks. *[Mouse enters]*

Mouse

[out of breath] You bet he was following those tracks. They're mine, and I was scared to death hearing that loud, snuffling nose and feeling the snow shake under those big, clumsy feet.

Henry

I'm sorry, Mouse. Your tracks are so tiny; Woof probably had to use his nose to follow you. But I can see your tracks clearly – four prints, then a space, then four more prints.

Mouse

Yes, I may be little, but I can move pretty fast. I'm a galloper. My strong back legs push off, and I land on my front feet. Then, my back feet land ahead of my front feet and push off again. I have to move along quickly so I won't get caught and eaten by dogs, or owls, or other predators. Speaking of which, I'd better be galloping. Goodbye. *[galloper prints down; Mouse gallops off]*

Henry

I'm glad that mouse went back under the snow before Woof got him. *[walks along, then stops; waddler track pattern sign appears]* What tracks did Woof follow here? That's a pretty wide path, and deep, too. I'd say the animal wasn't in much of a hurry. *[Porcupine appears]*

Porcupine

No need to hurry with these quills of mine to protect me.

Henry

Oh, you must be a porcupine. Do you leave a wide path in the snow when you walk?

Porcupine

Yes, I do. Good thing too, makes it a lot easier to get around, following my own trail in the snow.

Henry

And I see a big print next to a little print.

Porcupine

Yup, I put my back paw next to my front paw. Some folks call me a waddler because I walk along flat-footed and slowly, but I usually get where I'm going. Now, if you'll excuse me, I'm going that way. *[waddles straight at Henry, who moves aside]*

Henry

[backing up] Certainly, certainly. Don't let me hold you up. *[waddler pattern down; Porcupine leaves]* Whew! When I find Woof, I'm going to tell him to stay away from **that** track. It could mean prickly trouble. Hmm, here it looks as if Woof headed down to the stream. *[bounder track pattern sign appears]* Now, what are these tracks Woof found? Looks like somebody doing a whole bunch of broad jumps. *[Mink appears]*

Mink

[cheerfully] Did you say broad jumps? Ha, ha, ha, I never thought of it that way. I think of myself as a bounder.

Henry

Hi, Mink. So you're a bounder? Well, how do you make those tracks?

Mink

Just look how I'm built, long and skinny, with short legs. I spring forward and land on my front feet. Then my back feet follow into my front footsteps. My back arches up when I run that way.

Henry

You mean, sort of like a slinky coming down the stairs?

Mink

Slinky minky, that's me. And now I'm in need of some dinner, so I'll just bound off to where the crayfish abound! *[bounder pattern down; Mink exits]*

Henry

Speaking of dinner, I'm hungry. From Woof's tracks, I'd guess he is, too. *[walk/trot pattern sign appears briefly]* He's trotting right along toward home. *[hurries across stage and exits]*

Mother

[Henry and Mom enter] There you are, Henry. I've finally caught up with you, and here we are back home again.

Henry

Hi, Mom. I ran a lot. And when I got back home, there was Woof waiting for me on the front stoop!

Mother

That dog led us on a wild goose chase all through the fields and the woods, and then he beat us home!

Henry

Well, there weren't any wild geese, but there was a galloping mouse, and a waddling porcupine, and a bounding mink. Woof's a great tracker.

Mother

He sure is. And I can see that he's still tracking.

Henry

Gee? He is? What can he be tracking in here?

Mother

He's tracking mud all over the kitchen floor! And you'd better go get those wet boots off or you will be, too.

Henry

Oops! OK, Mom, I'll go do that right now!

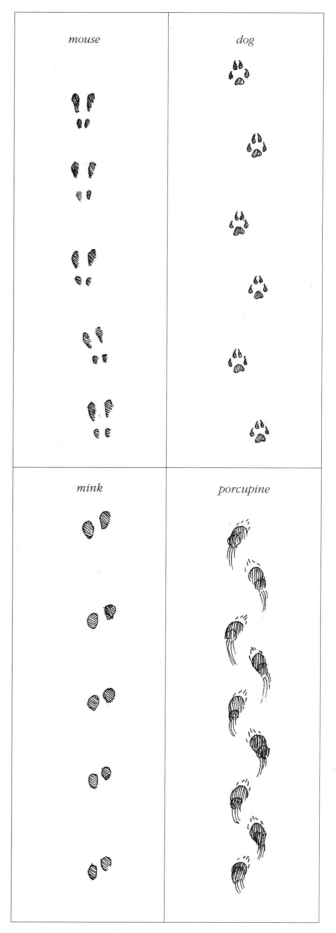

mouse   dog

mink   porcupine

Winter Weeds

RUGGED REMNANTS OF SUMMER FLOWERS

Wildflowers are a cherished part of many landscapes, from those that bloom each springtime in the desert or the forest, to those that fill the fields in summer and fall. When wildflowers grow where they are not wanted, we call them weeds. Winter weeds are merely the remnants of summer's growth, their sturdy stalks displaying a wide variety of colors and shapes. Their delicate patterns and subtle shades of browns and grays belie the fact that these are rugged, resilient plants that stand up through harsh weather until they crumble from exposure or are buried by the snow. Winter weeds interrupt the smooth, white fields of snowy landscapes, adding to the beauty of many a winter scene.

These winter skeletons each reveal the arrangement of the leaves and flowers they once held. Designed to last into or through the winter, winter weeds have extra time to disperse their seeds, and they do so in many different ways. Plants with stiff but not easily broken stalks, such as evening primrose or St. John's-wort, may catapult their seeds when whipped about in strong winds or brushed against by animals. Atop sturdy stalks, milkweed plants keep their seedpods well into winter so that every fluff-tipped seed has the chance to emerge and be carried off by the wind. Hitchhiker seeds, like burdock, need the extra months to catch a ride on unwary passersby.

Often covered with a hard coat, weed seeds are able to survive very harsh conditions, sometimes germinating years after they were produced.

A close inspection of the dried flowers of winter weeds exposes the intricate structures that cradle the seeds, from compact heads, to umbrella-like umbels or elongated spikes. The starry **bracts** of asters, the four-parted capsules of the evening primrose, and the translucent pods of mustard plants are all designed to protect and release the plant's seeds when ripe and ready. The seeds themselves also show a surprising diversity of design.

Most winter weeds are **herbaceous** plants, which means the living parts of the plant above the ground die by the end of each growing season. Herbaceous plants have three possible life cycles. Annual plants grow, bloom, and die after just one season, leaving their seeds to make the next generation. Ragweed, wild mustard, and chickweed are some familiar annual weeds. Biennial

Queen Anne's lace in bloom

Queen Anne's lace, seed head developing

Queen Anne's lace, gone to seed

plants, such as Queen Anne's lace and mullein, black-eyed Susan and evening primrose, take two seasons to complete their growth cycle. In the first year they produce a flattened circle of leaves near the ground, and in the second year they send up a flowering shoot, which may remain standing through the winter. Perennial herbaceous weeds, like chicory and goldenrod, send up new shoots in the spring from surviving roots and set seed every year. Some perennial weeds, like steeplebush and meadowsweet, are **deciduous** rather than herbaceous, losing their leaves in the fall and budding new ones each spring. Thus perennials can be found growing in the same fields and lots for many years.

Whether annual, biennial, or perennial plants, winter weeds exhibit the hardiness characteristic of plants able to grow where they are not wanted. Many grow where other plants could not survive, where the soil is poor and/or the climate is harsh. Since weeds grow primarily on land that has been disturbed, perhaps by fire, flood, plow, or bull-dozer, these plants often hold the soil and prevent erosion. They succeed because they are such aggressive colonizers, producing thousands of seeds that travel great distances carried by wind or by animals. Their seeds are often the first to arrive in a newly disturbed area, and, taking advantage of this lack of competition, the weed-colonizers are quick to settle in.

Once their seeds have sprouted, weeds have remarkable physical characteristics that help them survive. These may include hairy or fuzzy outer layers of stems and leaves to reduce moisture loss; fibrous, almost woody, stems to hold the seed heads erect; long taproots to reach scarce water; or tenacious root systems that spread quickly, helping the plant compete for space. Effective designs such as these help many weeds thrive in poor conditions as well as stand up to the rigors of winter.

Most of the plants we consider weeds are not native to this country. Weeds can travel great distances by wind and water, and birds may transport seeds on their bodies for hundreds of miles. Many weeds were introduced by settlers, either on purpose or carried inadvertently with other plants and herbs. As people relocated to other countries, so did weeds – in baggage, animal feed, supplies, even in pockets or pant cuffs. Although the deep forests of New England would not have been hospitable to the seed newcomers, as soon as the settlers began to clear the land, the sun-filled fields and waste places provided perfect habitat for these weeds, with little competition from native shade-loving plants. Once established in the East, the seeds of these plants soon found their way to the West.

Winter weeds can be appreciated for the story they tell about the land on which they grow and the history of their arrival. Found in open spaces from vacant lots and railroad tracks to garden plots or country roadsides, these pioneer plants provide seeds for birds, homes for wintering insects, and food and building materials for small mammals. In addition, the subtle beauty of winter weeds provides pleasure for those who take the time to look closely at these plants, standing tall in rugged defiance of winter.

Suggested References:

Brown, Lauren. *Weeds in Winter*. New York: W.W. Norton, 1976.

Embertson, Jane. *Pods: Wildflowers and Weeds in Their Final Beauty*. New York: Charles Scribner and Sons, 1979.

Levine, Carol. *A Guide to Wildflowers in Winter*. New Haven: Yale University Press, 1995.

Stokes, Donald W. *A Guide to Nature in Winter*. Boston: Little, Brown, 1976.

evening primrose

Winter Weeds

FOCUS: Winter weeds beautify the landscape with their variety of colors, shapes, and structures. They are stalwart remnants of last summer's flowers, carrying the seeds for future plants.

OPENING QUESTION: *What is a winter weed?*

PUPPET SHOW

Objective: To identify some of the common characteristics of winter weeds.

Perform, or have the children perform, the puppet show. Afterward, hold up the weed puppets and compare the design of each, including leaf arrangement and flower and seed structures. How might these different structures contribute to the success of weeds?

Materials:
• puppets
• script
• props

CARD PARTNERS

Objective: To make detailed observations of some common weeds.

Give each child a weed card. Have the children examine their weed to notice its branching pattern, flower structure, and other special features. Ask them to look for its seeds. Where are they held? How might they disperse? Then have the children find another child who has the same kind of winter weed. When all have found a partner, each pair introduces their weed and tells what's distinctive about it. The leader may show pictures of each flower in bloom and provide information and name, if appropriate. Encourage pairs of older children to identify their plant using a field guide. What similarities and differences do the children notice in the weeds?

Materials:
• two identical sets of winter weeds o. their seed heads mounted on cards
• hand lenses
• wildflower and weed field guides

WINTER WEED WALK

Objective: To discover the variety of winter weeds growing nearby.

Divide the children into small groups. Give each group a hand lens and a Winter Weed Walk card. Ask the children to find the objects listed and examine some of their discoveries with a hand lens. If appropriate, the children can bring their Card Partners weed card outside and try to locate a matching weed.

Materials:
• Winter Weed Walk cards
• hand lenses
• pencils
• clipboards
• (optional) map of search area, surveyor's tape, and Card Partners weed cards

WINTER WEED WALK CARD
Find a winter weed:
 with a seed head that looks like an umbrella;
 with seeds in a compact head;
 with seeds that stick to your clothes;
 with seeds that float;
 with a stalk that looks rough and tough;
 with a stalk that looks fragile;
 growing in a dry, rocky place;
 with a fuzzy parachute.

Match each of these colors to a winter weed *(use brown, gold, tan, and gray color chips from paint sample cards)*.

Find three different weeds that have a lot of seeds.
 Why do you think the seeds are still there? How will the seeds finally escape?

Notice a weed that is all by itself.
Find a big patch of one kind of weed and count the number of plants.

meadowsweet

Afterward, if time allows, have each group choose one especially interesting weed to tag with surveyor's tape. Mark the location of each chosen weed on a map. Then in the springtime, have each group check to see if their weed or another of the same species is growing and flowering.

Snow Bouquet

Objective: To appreciate the varied colors and shapes of winter weeds while creating a three-dimensional work of art.

Give each child a chunk of white craft dough the size of a golf ball, and provide an assortment of winter weeds for the children to choose from. Each child should make a miniature bouquet arranged in the white dough using cuttings from the different weeds. Inscribe the children's initials into the dough and let it dry overnight.

To display the bouquets, drape a white cloth over books to create a snow-covered landscape and have the children place their creations on the landscape.

Weed Poems

Objective: To creatively express observations and feelings about winter weeds.

Divide the children into at least three groups and give each group a winter weed hidden in a paper bag. Each group takes its bag to a private corner and removes the weed. Ask each member of the group to look at the weed from a different angle, i.e., far away, close up, from above, from below. One person in the group acts as recorder while each child shares one or two adjectives describing the weed. Together the group members compose a poem that includes these adjectives. Groups place their weeds back in the bags for the leader to collect, and then the winter weeds are displayed centrally. With the children in a circle around the weed display, read each poem aloud and have the children guess which weed each poem describes. As an added challenge, include an extra weed or two in the display.

Materials:
- collection of winter weeds
- clippers or scissors
- white cloth
- craft dough

Craft dough recipe:
4 c. baking soda
2 c. cornstarch
2 1/2 c. cold water
Mix ingredients in a saucepan. Cook over medium heat, stirring constantly, for 10 minutes, until the dough is the consistency of mashed potatoes. Turn the mixture onto a plate and cover with a moist cloth until cool. When cool, knead until smooth and then seal in a plastic bag and keep in a refrigerator until ready to use.

Materials:
- variety of winter weeds
- paper bags or other hiding devices
- florist's clay as a base for each weed
- pencils
- paper

Extensions

Winter Seeds: With the children, gather some seeds from winter weeds collected outside. Plant the seeds to see if they will germinate. Some seeds need winter's cold temperatures in order to grow in the spring. Record what happens.

Who Grows There: Have the children design experiments to test the hardiness of weeds by planting some weed seeds indoors in various soils, light conditions, or using another controlled variable.

Weed Feed: Many animals depend on weeds for food or shelter through the winter. In pairs, have the children select one species of weed to research, through observation as well as searching in books and on the Internet, and report on ways in which the weed is valuable to animals.

Winter Weed Drama: In small groups, the children write skits about life as a weed, including the difficulty of finding a suitable place to grow, the challenges of weather, and the need to reproduce. Allow groups to create backdrops for their skits. Have them present their skits to each other.

Weed Cards: Press winter weeds and use these to create homemade note cards.

Winter Weeds

Characters: Mother, Polly, Burdock Weed (real), Milkweed (real), Queen Anne's Lace (real)

Props: tape, burdock bur, milkweed seeds

Polly

You know, Mom, I love our backyard, even in the winter. There's always so much to look at.

Mother

I suppose so, but I miss all the colors of summer when my flowers are blooming.

Polly

Gosh, I think it's pretty with all those dried plants in different shapes and colors.

Mother

Well, all I see are the weeds I should have yanked out before they went to seed. *[Burdock appears]* Yikes! Look at that big burdock standing smack dab in the middle of your dad's squash garden. I'm going to find the shovel and get rid of it before that dog of yours gets all those burs tangled in his coat. *[Mother exits]*

Burdock

No way. I'm not going anywhere!

Polly

[looking around] What? Huh? I thought I heard a voice!

Burdock

That was me, Burdock. And I'm not too happy about your mother's plan to pull me from this spot. I've been working for two years to produce these seeds, and I'm proud of them. Just look how neatly they're arranged along my branching stem. No, sir, it hasn't been easy, and I'm not leaving without a fight!

Polly

Well, where did you come from? You really don't belong in a garden.

Burdock

Who are you to tell me where I belong?! Your dog, Ruff, planted me here a couple years ago. And you helped! Remember? Ol' Ruff was covered in burs. You brushed him out and threw the fur, and me, onto the ground. Yep, I hitched a ride on him, just like some of my own seeds did this year.

Polly

So those burs are your seeds?!

Burdock

Not quite, kiddo. My burs protect the seeds inside. Not many animals want to nibble on these hooks. Plus, my seeds move around by hitching a ride on any animal that brushes past, like Ruff. Pretty clever design, huh?

Polly

Gee, I guess so! But why do your seeds hang on so long? How will they ever grow in the wintertime?

Burdock

They won't. But they'll have all winter to catch a ride and find a good place to start growing when springtime comes.

Polly

Well, with all those burs full of seeds, some are sure to grow.

Burdock

You bet! But, if you want to talk numbers, go check out that milkweed near the driveway. Each of her pods is packed with seeds. Before you go, do you mind giving someone a lift? *[stick bur to Polly's clothes]* Hey, kid, you're OK. I'll see you around. *[Burdock exits]*

Polly

Looks as if I'll be seeing **you** around. Bye! *[Milkweed appears across stage; Polly walks toward it, talking to herself]* Well, now, there's that milkweed, standing tall and stiff just like that old burdock. *[louder]* Hey, Milkweed, I'm looking for my mom. Did she come by here armed with a shovel?

Milkweed

Yes, dear, she did walk by. But I didn't see her carrying anything. She was headed into your home.

Polly

Speaking of homes, Milkweed, this isn't a great spot to grow, here in the sand near the driveway. What's a nice weed like you doing in a place like this?

Milkweed

Well, I'm a pioneer at heart. I floated here a few years ago on my silky plume. This dry, sandy spot might not have been right for some plants, but I'm quite hardy. I've been happy here. And look, now my own seeds are ready to fly away.

Polly

So that's why you're standing here with your pods split open like that. I can see you've got lots of seeds in each pod.

Milkweed

Yes. Isn't it nice? And I don't have to rush to get rid of my seeds in a hurry. I can take all winter, just waiting for the wind to carry them away one by one. You see, my stalk is so sturdy that even winter weather doesn't knock me down.

Polly

Well, it'd be awfully bare around here without a few weeds poking up through the snow. Say, who is that neighbor of yours? She looks as if she's sprouting bird nests!

Milkweed

Oh, yes, that's Queen Anne's Lace. *[whispering loudly]* She's quite a wild carrot! Those little bird nests are her flower heads! *[Queen Anne's Lace appears; Milkweed exits]*

Queen Anne's Lace

Are you two talking about me?! It really isn't polite to whisper.

Polly

Oh, no, your majesty, I was just noticing the interesting design of your dried flower stalks.

Queen Anne's Lace

Yes, aren't they lovely? They're holding up my many precious seeds. I like to keep them safe as long as I can. It isn't easy being a weed, you know.

Polly

Oh, you're not a weed. You're so beautiful. You, and Milkweed, and Burdock. Why, you're like a garden, a winter garden.

Queen Anne's Lace

Oh, my! A winter garden? How nice. Often people just think of us weeds as ugly nuisances. *[wind noises]* My my, it sounds as if the wind is picking up again. *[more wind; blow milkweed seeds into audience]* Oh, look, there go some of my neighbor's seeds.

Polly

Yes! Well, I'm going now, too, your highness. That cold wind is too much for me. Good luck out here this winter.

Queen Anne's Lace

Good luck to you, too, Polly. Goodbye. *[Queen Anne's Lace exits; Polly crosses stage]*

Polly

Hey, Mom, are you in here? *[Mother enters]* I thought you were coming back to dig up that burdock.

Mother

Well, I started thinking about what you said. You were right. Those weeds are just as pretty in their own way as my summer roses.

Polly

And they're a lot easier to grow! In fact, I'd better go get these burs off my socks before we have winter weeds growing right here in our living room!

Mother

Okay, dear. Bye! *[both exit]*

Camouflage

DESIGNED TO CONCEAL

The ability to go unnoticed has saved the lives of many animals, **predator** and **prey** alike. A predator that depends on its skill as a hunter can get much closer to its prey before attacking if it can blend in with the surrounding habitat. And prey have a safety advantage if their skin covering and shape render them inconspicuous. The coats and coverings of many animals are so well designed to match their surroundings that they are almost impossible to see: a motionless green frog at the edge of a pond, a moth with wings spread flat against the bark of a tree, a ruffed grouse nesting on the forest floor.

The term "camouflage" came into common usage during World War I when armies had to disguise their men and operations from aerial reconnaissance. In current usage, camouflage has broadened its meaning to describe disguises of color, pattern, and shape.

To escape notice is the primary adaptive function of camouflage. But no matter how effective the disguise of a creature may look, it is only successful if the creature is able to **freeze**, to remain absolutely motionless. A spotted fawn lies still when danger is near. Cats crouch motionless between advances toward their prey. Anyone who watches birds in the spring knows that spotting a sudden movement in the newly leafed trees is the only way to catch a glimpse of an elusive warbler.

Matching color is the most common and obvious disguise, where the color of skin covering approximates the color of the environment, such as a lion on the African plains or a polar bear on the Arctic ice cap.

Some animals have evolved to the point where they change the color of their skin covering seasonally in order to match their surroundings: the snowshoe hare becomes white in winter, as do weasels in the north; the deer's winter coat is darker with more brown-gray tones. The process of shedding and growing in different coats is triggered by the shortening and lengthening daylight hours.

Some creatures can change color quite quickly. Able to match its surroundings within minutes, the chameleon is perhaps the most famous camouflaged creature, although other animals have similar abilities. In fact, cuttlefish, squid, and octopuses can change color within fractions of a second! This color change occurs as the top layers of pigment cells expand or contract to reveal or conceal the under layers of pigment cells.

Unless observed in its own habitat, one would hardly think of a spotted giraffe or a copperhead snake with its distinctive hourglass pattern as camouflaged. But these coats exhibit **disruptive coloration** – a series of patterns, spots, or stripes that, against a partially sunlit background, eliminates the sharp outline of body shape and thus causes the animal to blend into the surroundings. Patterns combined with special colors have evolved on certain creatures, allowing them to blend with specific backgrounds. The birch moth's white and gray lined wings make it unnoticeable on a birch tree.

Shape and texture have also been copied to provide near invisibility. The walking stick, when motionless on a branch, is almost impossible to see, and its cousin, the water scorpion, looks just like a floating twig. An inchworm camouflages itself by using its hind appendages to grasp the twig on which it is climbing and then holding the rest of its body rigidly at an angle. In experimental tests, even hungry blue jays have overlooked this juicy morsel, mistaking it for just another twig.

The snowshoe hare's winter coat is white except for black ear-tips.

The effect of light and shadow has been used to advantage by some creatures interested in concealment. Many birds have light bellies that show up poorly against the light sky and dark backs which, from above, blend against the dark earth. This light against light and dark against dark is called **countershading**. It gives a two-dimensional effect that tends to flatten the animal's shape. Fish also are often light on their bellies and dark on their backs; a motionless trout is nearly impossible to spot. Shadows can also be a giveaway, and there are creatures that try to eliminate their shadow. Polar bears lie flat to avoid casting a shadow. Moths with their horizontal wings can lie almost flush against a surface. And one amazing bird, the Australian nightjar, turns with the sun so that its long tail casts the smallest possible shadow.

Just as humans do, animals sometimes devise a disguise from available materials. This is called **masking**. The caddisfly larva, for instance, creates a tube-like shelter from available twigs, pebbles, or reeds on the bottom of the pond or stream. By using materials from its own habitat, its shelter is unnoticeable to the untrained eye. Likewise, spider crabs stick bits of sea vegetation to their shells and end up looking just like the bottom of a tidal pool. Even bird nests are constructed with materials that blend with the shrubs, trees, or ground where they're built.

Not all coloration, however, contributes to disguise. In fact, certain colors and patterns are very noticeable, and this phenomenon is called **warning coloration**. The creatures that exhibit this either smell or taste bad or inflict pain. Skunks are the best-known mammals with warning coloration. A young, inexperienced predator might well attack a skunk, but once that spray has hit, the receiver is not likely to forget the black and white striped creature that emitted the odor. Monarch butterflies apparently carry a distasteful residue of toxin from the milkweed plants they ate as caterpillars. The butterflies' bright color insures that insect-eating birds recognize and avoid them. The viceroy butterfly takes advantage of this with its similar orange and black striped pattern. This **protective mimicry** helps to keep the birds away from the viceroys as well.

One other common protective coloration trick is to look scary. Among insects, an effective scare tactic is the sudden appearance of large, staring, owl-like "eyes," a device used by both the Io and Polyphemus moths. These "eyes" normally remain hidden on the under-wings, but when the moths sense danger, they suddenly spread their wings and startle the would-be attacker.

As a design for survival, camouflage in all its varied forms is extremely effective. And it is a self-correcting adaptation because those creatures with the best camouflage live to breed and pass on successful characteristics. Those whose coloration has not adapted to a changing environment fail to escape notice and are less likely to survive.

Suggested References:

Cott, H. B. *Adaptive Coloration in Animals.* London: Metheun, 1966.

Fogden, Michael, and Patricia Fogden. *Animals and Their Colors.* New York: Crown, 1974.

Heran, Ivan. *Animal Coloration: The Nature and Purpose of Colours in Vertebrates.* New York: Hamlyn, 1976.

McClung, Robert. *How Animals Hide.* Washington, DC: National Geographic Society, 1973.

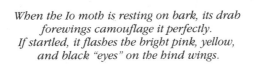

When the Io moth is resting on bark, its drab forewings camouflage it perfectly. If startled, it flashes the bright pink, yellow, and black "eyes" on the hind wings.

Camouflage

FOCUS: Many creatures are shaped or colored to blend into their surroundings, and these different designs of camouflage are critical to the survival of those animals.

OPENING QUESTION: *Why is it helpful for some animals to blend in with their surroundings?*

HARD TO SEE

Objective: To identify some of the different types of camouflage found in nature.

What is camouflage? Have the children draft a definition. Then, show pictures of animals in their normal habitats exhibiting different types of camouflage: matching color or shape, disruptive coloration, countershading, masking, mimicry, as well as animals with warning colors. Discuss the differences, and ask the children to name other animals that display these various forms of protective coloration. Do the children want to revise their definition of camouflage?

Materials:
- pictures or slides of animals illustrating different types of camouflage and protective coloration

SECRET SHAPES

Objective: To use camouflage to hide objects, experiencing how tricky it can be to see the outline of camouflaged objects.

Divide the children into an even number of small groups, 2-3 children in each. Have each group choose a background piece of wallpaper. From different wallpaper scraps, children should design and cut out secret shapes (at least as big as a thumbnail) and glue them onto their background sheet. Pair up the small groups; then have one group place its wallpaper on the floor while the other group stands around it and tries to locate all the hidden shapes. Switch roles and repeat. Afterward, with all the groups' wallpaper samples hanging on a wall, take one simple shape cut out of wallpaper and display it briefly on each of the different wallpaper samples. Is the shape well camouflaged in each of the different backgrounds? Besides shape, what design components help animals to blend in with their surroundings?

Materials:
- wallpaper samples – 8" x 11" pieces and scraps
- glue sticks
- scissors

A BIRD'S EYE VIEW

Objective: To see how matching color can be an effective camouflage.

Ahead of time, scatter colored pieces of yarn on a small area of grass using ten pieces of each color. Tell children that they are hungry birds looking for caterpillars to eat (pieces of yarn represent caterpillars). When the leader says "go," the children pick up as many caterpillars as possible. Allow a few seconds, and then call "stop." Count how many of each color were collected. Which colors were easiest, hardest to find? Give the children a second chance, 5 more seconds, to look for remaining caterpillars, and then count again. Now have the children look for any remaining pieces of yarn; they may be difficult to find, demonstrating how effective camouflage can be. Older children may tally the results for the different trials, and then discuss them using percentages.

Materials:
- two-inch pieces of colored yarn (green, brown, red, yellow, and blue), 10 pieces of each color

HIDDEN IN PLAIN SIGHT

Objective: To experience how difficult it is to see unexpected objects outdoors if their shape, color, or texture blends in with the environment.

Before the activity, place 10 objects on a 30' section of trail. Keep the objects secret. With the children in a line spread apart somewhat, have them walk silently along the trail trying to spot and count (but not pick up or point at) as many objects as they can. A leader at the end of the trail should ask each child to whisper how many objects were seen. If time permits, children can be told how many objects **are** on the trail and given a chance to go back to look again. Ask the children which objects were easiest to see and why. Can the children think of any animals that are colored or shaped similarly? (Don't forget to collect the objects before leaving the trail.)

Materials:
- 10 synthetic objects, some should stand out brightly and some should blend in with their surroundings, e.g., coat hanger, bungi cord, extension cord, yarn, toothpick, magazine, wool sock

CAMOUFLAGED CRITTERS

Objective: To construct a creature that will be well camouflaged for a specific habitat.

Assign a micro-habitat (a shrub, a small area of grass, a tree, a stone wall) to each small group of children. Have the children each make one or two creatures to fit into the assigned habitats and then go place their critters in that habitat. (No fair hiding them under anything.) All groups should then visit each micro-habitat and try to find the critters. Critter creators should point out their creatures if they are not found within a reasonable amount of time, but they should keep their hands behind their backs until then. Afterward, discuss which creatures were easiest to find. Which were most difficult? Why?

Option: This can be done indoors with children selecting different areas of the room in which to place their camouflaged critters.

Materials:
- sticks
- leather and cloth scraps
- pipe cleaners
- tissue paper
- glue
- tape
- scissors

PUPPET SHOW

Objective: To review the concept of camouflage.

Perform, or have the children perform, the puppet show. Afterward, ask the children how the different characters in the puppet show used camouflage.

Materials:
- puppets
- props
- script

ACTIVITY STATIONS:
1) A Bird's Eye View
2) Hidden In Plain Sight

EXTENSIONS

Camo Seek: One child is the seeker and closes his or her eyes while the others have until the count of 20 to camouflage themselves in a designated area of woods. Those hiding must be able to view the seeker at all times. Seeker calls out as she spots other children until she can see no one else. Seeker closes eyes again and counts to 10 while hiders find a closer spot in which to camouflage themselves. Repeat, counting to five. Who got the closest without being seen?

Be a Designer: Challenge the children to design camouflage outfits for the following occupations: skydiver, underwater swimmer, cowboy, lion tamer, school teacher, gardener, librarian, lumberjack, skier, coal miner, mountain climber, football player. They should include the appropriate backgrounds.

Invisible for a Day: Ask the children to pretend they are so well camouflaged for one day that they can move around without being seen. Write a story about what might happen.

Color Me a Survivor: Have the children draw a picture of different animals and color their pictures as accurately as possible, being mindful of how each animal uses its coloration to survive. Then draw in the appropriate habitat.

Camouflage

Characters: Marsha Mouse, Sammy Skunk, Herbert Hare, Freddie Fox, Ellie the Red Eft
Prop: a big birthday cake

Marsha Mouse

Oh, Herbie Hare, I'm so excited about Ellie Eft's party. I know that she's going to be so surprised.

Herbert Hare

Me, too, Marsha. Ellie does love surprises. I'm almost done making the cake for her. Then we can hide near those bushes and jump out to yell "Happy Birthday" when Ellie goes by on her morning walk.

Mouse

But won't Ellie see us when she comes down the path?

Hare

No way! Our coats blend in perfectly with our surroundings, so we'll be well hidden. But that orange skin of hers will warn us when she's getting closer. We'll see **her** before she sees **us**.

Mouse

[jumping up and down] Great! We really will surprise her. I can't wait! I can't wait!

Hare

Take it easy, Marsha. She'll definitely spot us if you keep jumping up and down. It isn't enough to be the same color as your surroundings; you need to stay perfectly still, too, in order to be truly camouflaged.

Mouse

Camo...what?

Hare

Camouflaged. That means that it isn't easy for others to see you because you blend in with what's around you.

Mouse

Oh, I see. Since I'm brown, I'm camouflaged when I'm near things that are also colored brown, like fallen leaves, soil, and those brown bushes over there.

Hare

Right! Those are just the kinds of places where I'm well camouflaged, too. When I'm wearing my **brown** coat, that is.

Mouse

You mean you have more than one coat, Herbie?

Hare

Sure, Marsha. I wear my furry white coat all winter so I can travel on top of the white snow without being seen. Your brown coat would be too easy for Freddie Fox to spot on the snow. It's safer for you to hide under that white blanket.

Mouse

Speaking of the color white, here comes our black and white friend, Sammy Skunk.

Hare

Why don't you tell him about our plan to hide and surprise Ellie while I finish frosting her birthday cake.

Mouse

OK. See you in a few minutes. *[Hare leaves; Skunk enters]* Hi, Sammy. I'm glad you're here for Ellie's party. Herbie wants us all to hide near those bushes so we can surprise her when she comes by.

Sammy Skunk

But Ellie will see me, Marsha. My black and white coat makes me stick out like a sore thumb.

Mouse

Oh, I see what you mean. You aren't very well camouflaged, are you?

Skunk

No, I can't blend in very well, but that's OK. When other animals see me, they run the other way. You know, Ellie isn't camouflaged either, Marsha. Her orange coloring isn't bad taste, it tells others that she tastes bad. But now listen, I don't want to ruin the surprise party. I'd better leave.

Mouse

No, stay here, Sammy. I'll find Herbie, and we'll think of something. *[Mouse exits right; reappears, much agitated]* Run, Sammy, run! Here comes Freddie Fox now! *[Mouse exits left; Freddie Fox enters from right]*

Skunk

Are you sure you want to come this way, Freddie Fox? *[Sammy Skunk turns his tail toward Freddie]*

Freddie Fox

Uh oh. I just changed my mind. Actually I'm heading this way. *[exits right]*.

Mouse

[pops up] Thanks, Sammy. You're a lifesaver!

Skunk

Well, I guess my black and white coat came in handy after all.

Mouse

You bet it did! Gosh, I'm glad you were here, Sammy. I sure am ready to find a safe hiding place! Herbie told me we should go hide near those brown bushes. He said there's a large rock, too, for you to hide behind.

Skunk

I'm ready when you are. *[Mouse and Skunk exit; Hare enters]*

Hare

[calling] Marsha?! Sammy?! The cake's done...Now where'd they go? They must be really well hidden. *[loud whisper]* OK, you two, here comes Ellie now! I'll hide here and when she comes, let's all yell **surprise**. *[exits, counts from back stage]* One, two, three...[enter Ellie the Red Eft; Hare, Mouse, and Skunk pop up from backstage with cake]*

Everyone

Surprise!!!

Ellie Eft

A birthday cake? For me? What a great surprise! And you were all so well hidden, I didn't know there was anyone around here.

Hare

Come on, Ellie, we've got a birthday party to go to. See you later, everybody...if you can spot us, that is. *[all exit]*

Honeybees

HIVES AND HONEY

Honeybees are amazing insects. Imagine making 60,000 flights to gather nectar for one teaspoon of honey or constructing such perfectly designed cells that architects the world over marvel at their strength and their economy. Bees' lives, like their hives, appear complex at first. But close observation of the bustling activity of a beehive reveals an orderly division of labor and remarkable form of communication among the bees in the hive.

Three different types of bees perform all the jobs within a hive: the queen, female workers, and male **drones**. The queen is a special female from early in her life. While all larvae receive royal jelly (a glandular secretion produced by worker bees) for the first three days of their development, a **larva** chosen by the worker bees to be a queen gets royal jelly throughout her 16-day developmental period. Having been reared in an enlarged cell and fed a special diet, the queen grows to be much larger than the workers and longer and slimmer than the drones.

A newly emerged queen will attack other new queens or rip open and destroy unhatched queen cells to try to attain supremacy within the hive. Once her supremacy is established, she moves around over the combs for about a week, then ventures out on her mating flights where she will mate with as many as 10 to 18 drones, usually from other hives. A few days later, the queen starts laying eggs. In theory, she can lay 3,000 eggs per day, but she does not keep this pace up at all times. She lays from early spring to late autumn for as many as five years. Female workers develop from fertilized eggs. Unfertilized eggs develop into male drones.

After hatching, drone bees remain in the larval stage for six days and are fed royal jelly diluted with **pollen**. They then pupate and emerge two weeks later as adults. Drones lack many of the specialized structures present in worker bees. They cannot even feed themselves and must be fed by worker bees. Male bees serve only one function: to fertilize a queen. During a queen bee's mating flight, drones from several colonies will try to mate with her, and those that are successful die shortly afterward. In autumn, the remaining drones are forced out of the hive by the worker bees to face certain death. More drones will be produced again in the spring.

It takes three days for a worker bee larva to hatch from the egg stage. The new larva is fed royal jelly for three days and bee bread (pollen and nectar) and honey for two more days. The larva then spins a cocoon and pupates for 12 days, finally emerging as a sterile female. Worker bees live an average of four to six weeks, though the last brood of the summer lives through the winter.

The life of a worker changes with age. She is involved in hive duties for the first three weeks of her adult life:

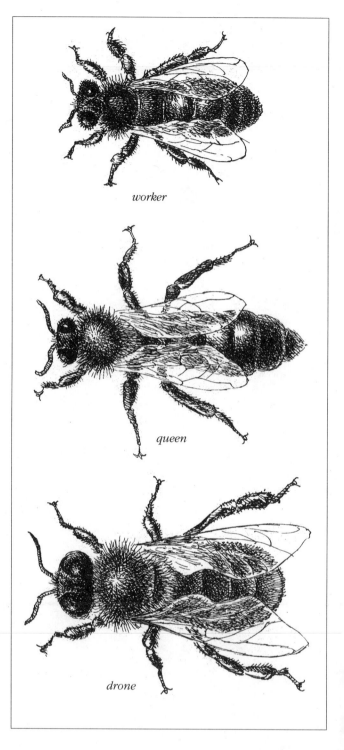

worker

queen

drone

cleaning house, producing royal jelly, feeding the larvae, secreting wax and repairing the hive, packing pollen and honey in cells, and guarding the entrance of the hive. Then she starts taking short scouting trips out of the hive, and spends her final weeks as a field bee.

In order to perform its complicated tasks, a worker bee's body must be very specialized, both internally and externally. Her body parts include the following:

Antennae: contain many smell-sensitive pits, giving the bee a keen sense of smell.

Compound eyes: can differentiate colors except black and red.

Mandibles: gather pollen and mold wax.

Tongue: collects nectar and passes it on to the honey stomach.

Wings: two pairs of delicate wings that lock together with fine hooks, enabling the bee to fly distances up to eight miles.

Stinger: a straight, barbed appendage connected to a gland that secretes stinging fluid; when used, it and other parts of internal organs are pulled out of the bee, causing its death. Drones do not have stingers: queen bees' stingers don't have barbs.

Wax glands: located on underside of the bee's abdomen.

Legs:

First pair – each leg has a comb for removing pollen and other materials from the antennae and a pollen brush to gather pollen from the foreparts of the body;

Second pair – each leg has a pollen brush to remove pollen from the first legs and other body parts, and a spur to pick up wax;

Third pair – each leg has a **pollen basket** for carrying pollen, and a pollen brush and pollen comb for cleaning the bee's body and collecting pollen.

The success of honeybees depends not only on the functional design of their bodies but also on their remarkable dance language. This system of communication allows field bees to pass on their knowledge of good sources of nectar or pollen to other field bees and contributes to the efficiency of the hive. A Round Dance is used to communicate a food source that is less than 100 yards away from the hive. When a scout bee comes back to the hive, it starts dancing in a small circle, going first one way and then the other. Field bees follow this scout bee closely, touching her with their antennae to pick up the scent. Soon thereafter they leave the hive to find the flowers.

When the nectar is farther away than 100 yards, the scout bee does a Wagging Dance. This is a flattened figure eight, with the bee wagging on the center junction of the 8; the more rapid the wags, the closer (or better) the nectar. A bee uses the angle of the sun in relation to her flight path to indicate the direction of the feeding sources to the other bees. On a vertical comb, it is as if the sun were straight up. The bee wags directly up if the food source is toward the sun or, if not, to the right or left at an angle corresponding to the angle between the correct flight path and the sun. By following her and touching her, the field bees learn the direction, the distance, and the specific scent of the nectar. Worker bees in the hive take the nectar collected by the field bees and pass it back and forth until enough of the

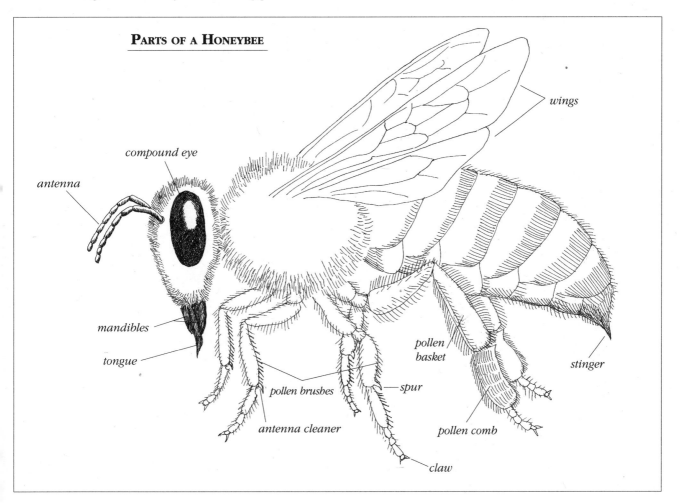

PARTS OF A HONEYBEE

wings

antenna

compound eye

mandibles

tongue

pollen brushes

antenna cleaner

spur

pollen basket

pollen comb

stinger

claw

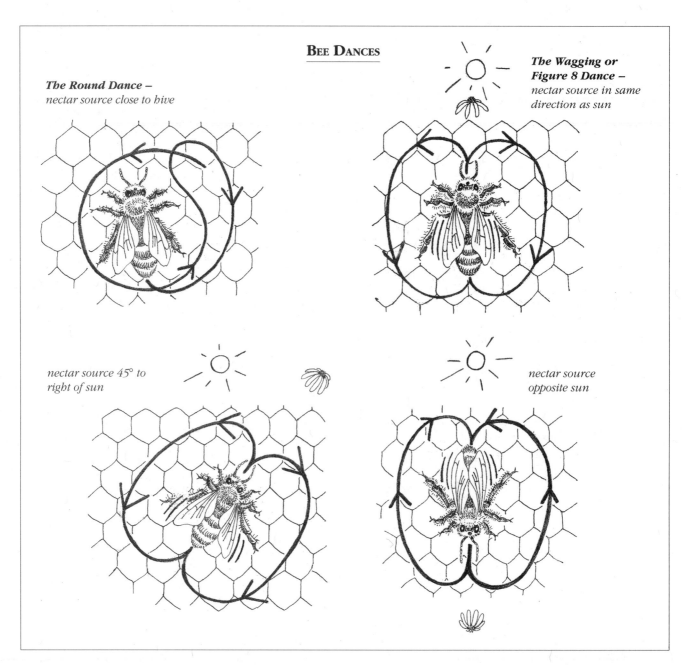

BEE DANCES

The Round Dance –
nectar source close to hive

The Wagging or Figure 8 Dance –
nectar source in same direction as sun

nectar source 45° to right of sun

nectar source opposite sun

excess moisture evaporates for it to be deposited into storage cells. Evaporation continues until the honey is the right consistency, and then the bees seal the cell.

Honeybees did not always build their hives in the wooden boxes we see today; many still don't. Their original dwelling places were the hollow trunks of dead trees. People now use wooden hives with frames to facilitate harvest of the honey. For the combs, the bees themselves produce flakes of wax from abdominal glands. They then mold the wax with their **mandibles** to prepare it for use as a building material.

The honeycomb cells are used for raising young as well as for storage of honey and pollen. The structure of the cells is remarkably efficient. Each cell tilts in such a way that the honey does not trickle out and the larva is easier to feed. The shape is hexagonal, with cells sharing mutual walls, which provides the most efficient use of building materials and space. Square or triangular cells would waste space in the corners and be harder to clean; round cells would not share common walls, leaving gaps between cells.

Whether learning about the design of their hives and bodies or studying their importance as pollinators, wax manufacturers, and honey makers, those who get to know honeybees agree that they are truly amazing insects.

Suggested References:

Frisch, Karl von. *Animal Architecture*. New York: Harcourt Brace & World, 1974.

Frisch, Karl von. *The Dancing Bees*. New York: Harcourt Brace & World, 1953.

Hubbell, Sue. *A Book of Bees*. New York: Ballantine Books, 1988.

Honeybees

FOCUS: Honeybees are fascinating insects whose uniquely designed physical and social structures contribute to their survival and success.

OPENING QUESTION: *What is special about honeybees?*

PUPPET SHOW

Objective: To understand the life cycle of worker bees plus the roles of queen and drone bees in a hive.

Present, or have the children present, the puppet show. Afterward, use illustrations of queen, worker, and drone bees to compare the bodies and roles of the different bees in a colony. What tasks do worker bees perform in a hive?

Materials:
- puppets
- script
- illustrations comparing queen, worker, drone bees

A CLOSER LOOK

Objective: To examine the structural designs of honeybee bodies and beehive cells.

Divide the children into small groups and give each group hand lenses, some dead honeybees, and some honeycomb or a beehive frame. Ask the children to describe what they notice about the eyes, antennae, mouthparts, legs, and bodies of the bees. Can they see the stinger? Use a diagram of a honeybee to point out the important design features of a bee's anatomy. Next have the children use their hand lenses to carefully examine the honeycomb cells. What do the children observe about the shape, size, and texture of the cells?

Materials:
- dead honeybees (beekeepers could supply these)
- beehive frames or a piece of honeycomb
- hand lenses
- diagram of the anatomy of a honeybee

COMB SHAPES

Objective: To compare different possible shapes for the cells of a beehive to determine which is most efficient in terms of space and strength.

Divide the children into groups of 3 or 4 and give each group 2 pieces of paper, 12 circles, and 12 hexagons. Ask the children to determine which shape will be the best design for the cells of a beehive. Tell them to fit as many circles as possible within the edges of one piece of paper and as many hexagons as possible within the other. Without overlapping any pieces, which shape allows the greatest number of cells? Which shape leaves the least amount of open space on the paper? Which would use the least wax for the most cells? What other advantages might there be to hexagonal comb shapes?

Materials:
- 2 sheets of paper, 8 1/2" x 11"
- 12 circles 3" diameter
- 12 hexagons 3 1/8" between opposite points (not between opposite straight edges)

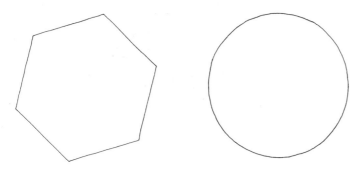

reduced 50 percent of recommended size

DANCE OF THE BEES

Objective: To become familiar with the dance language that bees use to communicate the location of pollen and nectar.

Explain that bees communicate the whereabouts of nectar and pollen by "dancing" on the wall of honeycomb in the hive. Have the children perform both types of dances by following the leader. First, the leader imitates a scout bee that has found food within 100 yards of the hive by doing the round dance, moving in a small circle first in one direction and then the other. After all have performed the round dance, the leader demonstrates the wagging dance used when a food source is farther from the hive. Here the leader dances in the shape of a flattened figure 8 (wagging along the centerline of the 8) with the centerline of the dance pointing toward a prehidden snack. Note that the direction the bee faces as she wags lets the other bees know in which direction they'll find the flowers, and the closer the flowers are, the faster she wags. Children follow the leader in the wagging dance to learn the direction of the hidden treasure or snack, and then take off on their own to find it.

Materials:
- honey and crackers

SNIFF AND SCOUT

Objective: To experience how bees use their sense of smell to find a food source.

Ahead of time, put about 15 cotton balls in each of three plastic containers. Each container of cotton balls should be scented with a different extract. Seal and mark the containers and lids. Just before presenting this activity, arrange cotton ball "flowers" in several different patches on the ground. Each patch will include three clusters of cotton balls, one cluster of each scent. The clusters should be a foot or so apart; the patches should be a good distance apart (25-50' depending on space). For younger children, put only one scent at each patch.

Have the children form groups of three to five, with all groups gathered in the center of the area as though they were bees in a hive. One child from each small group will be a scout bee; the others are field bees. The field bees in each small group should face each other, close their eyes and buzz, while their scout bee goes out from "the hive," gets one cotton ball, and brings it back for her group to sniff. Then, the scout bee does a wagging figure 8 dance (see Dance of the Bees activity) to tell the others which way to go to find that same patch of cotton ball flowers. Once the field bees locate the correct patch, they use their sense of smell to try to identify the cluster of cotton balls matching the scent of the scout bee's. The scout bee follows them to see if they find the correct patch and cluster. Return the cotton balls to the correct clusters, and repeat the activity with a new scout from each group.

Materials:
- clear flavoring extracts such as anise, mint, almond, orange (at least 3)
- plastic containers (same number as extracts)
- cotton balls

HONEYBEE HAUNTS

Objective: To discover if honeybees are visiting nearby flowers and to observe them at work.

In pairs, the children should visit distinctly different types and colors of flowers (e.g., clovers, goldenrods, daisies) to see if honeybees are seen at each kind. Have them record their findings. Afterward, gather pairs together and ask the children to share their observations. Why might different flowers attract different types of insects?

Materials:
- Honeybee Haunt cards
- clipboards
- pencils

HONEYBEE HAUNT CARD

Sit next to one flower and observe it quietly for a few minutes. Record your observations below. Afterward, if you have time, move to a different type of flower and repeat.

Flower #1   Flower #2

Describe the flower.

Is there a bee visiting this flower?

Does the bee have pollen on its legs?

Is the bee collecting nectar?

If the bee is visiting several flowers, are they all the same kind?

Do insects other than bees visit the flower? What kind? Other flying insects? Jumping or crawling insects? Other?

SHARING CIRCLE

Objective: To review and share thoughts about honeybees and their lives.

Ask the children to think of one new thing they've learned about honeybees. Have the children write their ideas on hexagonal pieces of paper. Now have the children, one at a time, share these ideas while taping their hexagons in the shape of a honeycomb to a bulletin board or poster.

Materials:
• hexagonal pieces of paper
• pencils

ACTIVITY STATIONS:
1) A Closer Look
2) Comb Shapes

EXTENSIONS

Beekeeper Visit: Invite a beekeeper to visit with your group and bring in the equipment used to keep bees: smoker, hive tool, veil, and gloves.

Bee Products: Collect a variety of items made from bee products, e.g., beeswax candles, boot or floor wax, honey, lipstick, royal jelly, and even fruits pollinated by bees. Place these on a table with other common household items. Have the children, working in pairs, make a list of those items that contain bee products. Compare lists.

Demonstration Beehive: Visit a demonstration beehive, one with glass walls so the children can find the queen, worker, and drone bees, watch the dances, and notice the cells with honey and with baby bees.

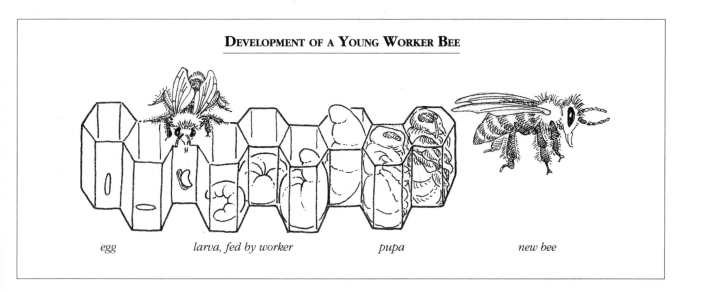

DEVELOPMENT OF A YOUNG WORKER BEE

egg   *larva, fed by worker*   *pupa*   *new bee*

Honeybees

PUPPET SHOW

Characters: Father, Polly the Girl, Magic Fairy, Polly Bee, Queen Bee

Father

There you are, Polly! I've been looking for you everywhere. Want to help me make some strawberry jam today?

Polly the Girl

Sure! But I'd rather have honey than jam. I love honey. Can we make honey, instead?

Father

We can't make honey, Polly. Honey is made by bees!

Polly

Bees make honey?! How?

Father

From nectar. They gather nectar from flowers and change it into honey in their hives. Don't ask me how – I only know about making jam, and I'm going to go get started. Meet me in the kitchen! *[Father exits]*

Polly

Wow! Bees make honey! I wish I were a bee, making and eating honey all day. *[Magic fairy appears]* Well! Who are you?

Magic Fairy

I'm a magic fairy. I make wishes come true.

Polly

So you must have come to turn me into a bee!

Fairy

That's right. But being a honeybee is a lot of work, and I'm not sure you'll like it. If you should change your mind once you're a honeybee, just call out "Zzub, Zzub" and I'll change you back to a person.

Polly

Zzub?? What kind of a magic word is that?

Fairy

Zzub is just "buzz" spelled backward. Now, I've never turned anyone into a honeybee before, so I could use some help. Just say these magic words after me, kids!

Abra Cadabra	*(Abra Cadabra)*
Fiddle-de-fee	*(Fiddle-de-fee)*
Make Polly become	*(Make Polly become)*
A honeybee!	*(A honeybee!)*

[Polly the Girl exits; Polly the Honeybee enters]

Polly Bee

It worked! I'm a bee! I'm a bee! This is so exciting! What do I do first?

Fairy

You'd better go to your new beehive home. It's over in that hole in the big oak tree. The queen bee will put you to work.

Polly Bee

Queen bee? There's a queen bee?

Fairy

Oh, yes. The queen bee is like the boss of all the other bees in the hive. She's rather pampered. She doesn't even do any of the housekeeping...I mean hivekeeping.

Polly Bee

Well, then, what does she do?

Fairy

The queen bee mates with the drones, the male bees, and then she spends her time laying eggs. Later those eggs hatch and out come baby bees.

Polly Bee

You mean the boy bees and the other girl bees do all the work while the queen bee just lays eggs?

Fairy

Well, that's almost right, except the drones don't do any work either.

Polly Bee

So the girl bees are the only ones that do any work? They must stay awfully buzzy, I mean busy.

Fairy

They do, indeed. That's why girl bees are called worker bees – they do all the work.

Polly Bee

Oh, no! Since I was a girl person, does that mean I'm a girl bee, too?

Fairy

That's right, Polly Bee. Now you'd better just buzz over to that beehive and meet Queen Bee-a-trice. Tell her I sent you. She'll explain what you have to do. Farewell! *[Fairy exits]*

240 / Designs of Nature

Polly Bee

[flying across stage] Gee, I hope I like being a honeybee. [calling] Oh, Queen Bee-a-trice. Queen Bee-a-trice!

Queen Bee-a-trice

Yes, yes, were you calling me? Make it quick. I don't want to fall beehind in my royal responsibilities.

Polly Bee

Well, the Magic Fairy turned me into a worker bee. I'm ready to make honey!

Queen

Not so fast! That's what you'll do when you get a little older. Young worker bees start off cleaning cells in the beehive and feeding the baby bees.

Polly Bee

Cleaning the hive and feeding the baby bees?

Queen

Indeed. Then you'll make wax and help with building the combs.

Polly Bee

Gosh, there's lots to do in a beehive – cleaning, feeding, building...

Queen

Yes, there is, and we're not through yet! Next you'll have four days of guard duty, protecting the hive entrance, and then you'll become a field bee.

Polly Bee

A field bee? You mean, going out to get nectar?

Queen

That's right. You'll collect nectar and pollen from flowers and bring it back to the hive.

Polly Bee

Gee, I know we need nectar to make honey. But what do we want pollen for? And how am I supposed to collect it and bring it back?

Queen

We use the pollen to make food for the babies. And don't worry, you'll have no problem carrying it back on your hairy legs.

Polly Bee

Wow, there's a lot I didn't know about being a bee. So, after I clean the cells, feed the baby bees, repair the hive, guard the entrance, and collect pollen and nectar, then can I make honey and eat it?

Queen

Certainly. But you do realize that most of the honey you make won't be for you to eat. All of us bees will eat it.

Polly Bee

All of us? Even the bees that didn't collect any nectar?

Queen

That's right. All the honey is shared. Now, it's time for you to get working and for me to get laying. [Queen exits]

Polly Bee

Gosh, this sounds like an awful lot of work for a little bit of honey. I think I could eat more honey when I was a person. I want to be a girl again. If only I could remember the magic words that fairy told me to say. Glub, glub? No. Fuzz, fuzz? No, that's not it. Zip? Zap? Zoob? Zeeb? Oh, what was it? [pause hoping audience shouts "zzub"] Oh yeah, Zzub, Zzub. How could I have forgotten that?! Oh, Magic Fairy, Magic Fairy, Zzub, Zzub!!

[Polly Bee exits; Polly the Girl appears]

Polly the Girl

It worked! Thank goodness. Gosh, I'd better go find Dad before he wonders what's been keeping me. I'll just have to explain that I was awfully bizzzy. Bye.

Earth and Sky

What is the connection between plants and the air we breathe, between honeybees and the earth's magnetic field, between ancient shellfish and limestone rocks? In examining the physical characteristics of the world around us, children discover that the living and the nonliving are interwoven in an intricate tapestry.

Life does not exist apart from the physical world. Plants produce the oxygen that forms a portion of the earth's atmosphere. Honeybees depend on cues from the sun's position and the earth's magnetic field in order to find their way. Over the millennia, the remains of shellfish accumulate on the ocean floor where they are pressed and cemented into stone.

Learning about the physical environment involves examining the properties of rocks and air, sunlight and sound, and wind and water. Through the activities in this theme, children will explore many different aspects of the physical world as well as the forces that affect and shape the earth over time. These explorations will help children to understand the connections between living organisms and the nonliving systems that support them.

Finding Your Way

CLUES THAT GIVE US DIRECTION

Have you ever gone out for a walk or a drive with no destination in mind? You turn down new paths or roads, exploring any that look interesting, happy just to be wandering and seeing new sights. You turn a corner and suddenly know precisely where you are! Some combination of landmarks makes the place immediately recognizable, and you once again know in what direction you must head to return home.

People and other animals rely on cues from the environment to guide them as they travel in search of food, water, shelter, or a safe place to raise their young. Animals are able to find their way between summer and winter ranges that are often separated by thousands of miles. They can locate their own nests among hundreds of others in crowded colonies. They can navigate successfully in the dark, under water, and in areas where they have never been before. The important ability to orient oneself in the natural world and navigate using sensory cues is vitally important to all creatures.

Many cues, though often not consciously noticed, tell us where we are and give us our sense of direction. Do you know where the sun rises and sets at your house? What smells and sounds tell you when you are near a familiar place? Can you picture the route you take to and from school or work? Observations like these lead us to form a spatial image or mental map of our home areas. As we gain more familiarity with an area, we add more information to our map. This allows us to move about within these areas without getting lost.

Just as people create mental maps, other animals also have a spatial memory of the area in which they live. Learning the features of this familiar area or territory is important to survival as it allows an animal better access to resources such as protective cover, food sources, or nesting sites. The digger wasp relies on the location of landmarks to find its nest hole in a sand bank crowded with other similar nests. It can be tricked into returning to the wrong hole when objects near its nest, such as pine cones and twigs, are repositioned while it is away.

Though many animals seem to have a mental map of their surroundings, not all rely on sight. Bats and dolphins emit high-pitched sound pulses that bounce off objects in their surroundings. Their mental maps must be an acoustical picture of the world that allows them to navigate and to hunt for **prey** in their dark or watery worlds. Salmon, on the other hand, recognize their natal stream by chemical cues. They know when they're home because it smells right.

But what about finding our way when we move beyond familiar territory? Sign posts, trails, roads, and maps serve as guides helping people to find their way in unfamiliar places. Do animals communicate direction to each other as well? Certainly some animals leave directional cues for their fellow creatures. Ants and termites leave scent trails to and from foraging sites. Female silkworm moths release a chemical scent that leads courting males to their perch. Honeybees perform a symbolic dance within the hive that tells the other field bees which direction to fly, in relation to the sun, to reach a patch of flowers far away from the nest. (They can even navigate this way on cloudy days because their eyes can see the polarized sunlight that penetrates the clouds.) Many of these visual and scent cues help animals find their way within a certain range around their home area, but what about longer trips well beyond this range?

Many animals leave their homes and travel over extremely long distances during their annual migrations. Caribou migrate over much of Canada, song birds and hawks migrate across two continents, arctic terns travel from pole to pole, and monarch butterflies travel thousands of miles across North America to winter in Mexico. How is it that these animals can find their way over such great distances without getting lost?

the right way to hold a compass

Most of these animals rely not just on one strategy for finding their way, but on a combination of many different sensory cues. Day-migrating birds, such as hawks, can tell direction by the sun and also follow major landscape features such as mountain ranges and rivers as they travel to and from their wintering areas. Night-migrating birds may use the rays of the setting sun as they depart. In the deep of the night, some birds can steer by the stars. Others are able to hear the extremely low-pitched sounds made by waves crashing on a shore, helping them to follow coastlines. And some birds have a magnetic "sense," using the alignment of the earth's magnetic field to find their way.

To travel great distances, people have also relied on the sun, the stars, prevailing winds, and the earth's magnetic field for navigation. Over 3,000 years ago, the Phoenicians were already using the North Star for sailing at night. They taught the art of stellar navigation to the Greeks. This skill allowed these ancient civilizations to establish trade routes throughout the Mediterranean.

Compasses have helped people to find their way for at least a thousand years. The earth has a magnetic field that resembles that of a huge bar magnet with a north and a south pole. This is due to the interaction of motion and electrical currents deep in the earth's core. When we use a compass, we make use of two magnets, the needle of the compass and the earth itself. The compass needle always points north and south because its magnetic poles are attracted to the opposite poles of the earth's magnetic field.

In its earliest form, a compass was simply a dish of water in which a rod of magnetic ore was floated on a small piece of wood or reed. Compasses were used by mariners since the time of the Vikings, for a compass could indicate direction even when the skies were cloudy. Used in conjunction with maps, compasses can help us find our way in many circumstances.

There are many mysteries yet to unravel about how animals, including humans, navigate. What is it that shapes this sense of direction? And can we hone this ability to find our way? When we pay attention to the landscape, noting its details with all of our senses, we gain confidence to further explore our environment. Such explorations, whether a drive in the country, a cab ride uptown, or a wilderness hike, expand our spatial maps and add to our understanding of the different places we call home.

Suggested References:

Dingle, Hugh. *Migration: The Biology of Life on the Move.* New York: Oxford University Press, 1996.

Grier, James W. *Biology of Animal Behavior.* Missouri: Times Mirror/Mosby College, 1984.

Levy, Sharon. "Navigating With a Built-in Compass," *National Wildlife,* Oct./Nov. 1999.

McFarland, David. *Animal Behavior.* California: Benjamin Cummings, 1985.

Ridley, M. *Animal Behavoir: A Concise Introduction.* California: Blackwell Scientific, 1986.

Tinbergen, N. *Curoius Naturalists.* New York: American Museum of Natural History, Doubleday, 1968.

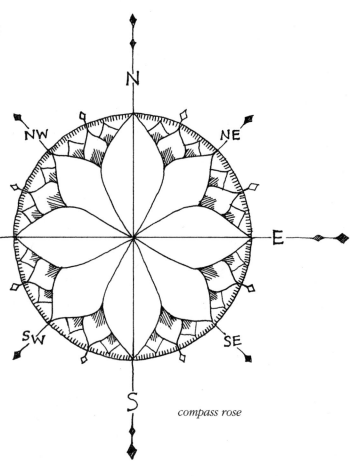

compass rose

Finding Your Way

FOCUS: Animals, including humans, use various cues to recognize their surroundings and orient themselves in the natural world.

OPENING QUESTION: *What cues do different animals use to find their way?*

FINDING YOUR WAY PUPPET SHOW

Objective: To learn some different methods animals use to find their way.

Perform, or have the children perform, the puppet show. Afterward, review the different methods of orientation and navigation used by the animals in the puppet show. Do people use any of these methods?

Materials:
- script
- puppets

MENTAL MAPS

Objective: To discover that prior observations and memory help us locate familiar places and objects.

Have the children close their eyes and imagine that they are standing in the doorway of their kitchen. Smell the brownies cooking in the oven? Ask them to point at the clock, the refrigerator, the sink, the garbage can, the cookie jar, and other kitchen landmarks. Try this activity again, asking them to point to familiar places within their school, such as the school office, the cafeteria, the gym, the playground. How were they able to locate these objects and places with their eyes closed? Discuss the role a mental map plays in finding one's way.

SNEAK A PEEK

Objective: To experience how observations and memory help in making mental maps.

Work in small groups. While the rest of the group members sit with their eyes closed, have one child place from three to six assorted objects (e.g., acorn, cone, pebble, leaf, seed pods, small toys) in the cups of an egg carton and close the lid. (For younger children, use half egg cartons and three to four items.) Now direct the rest of the group members to open their eyes and study the arrangement for a few seconds. Provide each group with a second carton and a duplicate set of objects, and ask them to recreate the pattern from memory. Repeat until each child has had a turn arranging the objects. Ask the groups to share their strategies for remembering the placement of the objects.

Materials:
- egg cartons
- an assortment of small objects in duplicate

SIMON SAYS *(grades K-2)*

Objective: To learn the cardinal directions.

Introduce the cardinal directions: north, east, south, and west. Children may use the simple rhyme "Never Eat Shredded Wheat" to remember the order. Point out that if you know where one direction is located then you can figure out all the rest. If you face north and hold your arms out, your right hand points east, your back is to the south and your left hand points west. Using familiar local landmark clues (such as where the sun rises or sets, a nearby city, lake, or mountain range) ask the children to assign a direction to each of the 4 walls of the classroom and label them with signs or pictures.

Materials:
- signs or pictures to indicate north, south, east, and west

Now play "Simon Says" to reinforce these cardinal directions. Before you begin, review the rules of the game: you can only move if the leader uses the words "Simon says." When a child moves without hearing this, she or he must sit down. Try some of the following instructions or make up your own: take two baby steps to the south; wave to the east; turn and snap your fingers to the west; turn and clap to the north; fly to the south; stick your tongue out to the north. If your group becomes too loud during the activity, use the command, "Simon says not to talk until the activity is over."

Direction Drill

Objective: To learn the parts of a compass and how to use a compass to locate the cardinal directions.

Briefly explain that the earth's magnetic field acts like a giant bar magnet. The floating steel needle in a compass is attracted to the magnetic north pole of the earth so it always points in that direction.

Pass out compasses to pairs of children. Hold up an enlarged model of a compass and review the various parts. Bring the children outside and have them stand in a line. Practice using the compasses following these simple directions:

1. Put the compass string around your neck and hold the compass level with the ground, with the **base plate** touching your chest. The **direction of travel arrow** should be pointing away from you. This arrow shows you the way to head. Pretend that you and the compass are permanently attached like this, and always turn together as a unit. Practice turning your body, and watch the **magnetic needle** swing back to a northerly direction.

2. Turn the **compass housing** until the letter "E" (for East), marked along the circumference, is lined up with the **direction of travel arrow** on the base plate.

3. Now turn as a unit with the compass until the red end of the magnetic needle is inside the **orienting arrow.** Many people refer to this orienting arrow as the "bed."

4. Explain to the children that they are now facing east. To actually travel east, find an object that is lined up with the direction of travel arrow and then proceed toward it, stopping occasionally to check the compass.

5. Now have the children try to find headings for south, west, and north.

6. With older children, try having them find a number bearing such as 330°.

Share this simple rhyme to help the children remember the steps: "Put the red (red end of the magnetic needle) in bed (inside the orienting arrow) then head (in the direction indicated by the direction of travel arrow on the base plate)."

Materials:
• compasses
• enlarged model of a compass

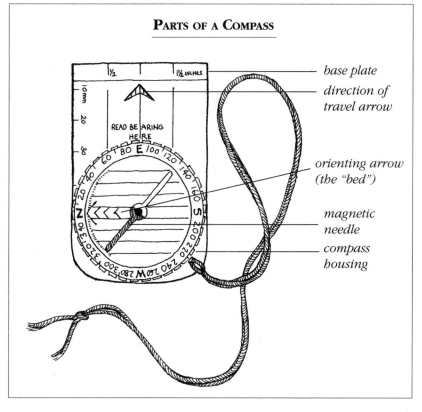

Parts of a Compass

base plate

direction of travel arrow

orienting arrow (the "bed")

magnetic needle

compass housing

TREASURE TREK

Materials:
- cards or sheet of directions for each group
- compasses
- (optional) items to collect at each station, special snack, or "treasure"

Objective: To practice using directions to follow a course or locate features of the landscape.

Before meeting with the children, set up a compass course appropriate to their age and skills. The course should begin and end at the same point and include a series of from five to eight stations. Write a list of instructions describing how to get from station to station. Instructions should include such things as cardinal directions, up- or downhill, number of steps, reference to familiar landmarks. At each station on the course, have the children complete a task, answer a question, or collect something as in the examples below.

Note: For younger children, instead of following a compass course, stand in one spot outside and have the children find the direction to different landmarks in the area. For example, ask them what direction it is to the church steeple or what direction it is to the school slide.

RHYMING CLUES:
- Fly to the west like a bird on the wing, then take a seat on the playground _____ (fill in the blank).
- Head to the east if you are able, and sit right down at the picnic _____.

CLUES WITH A TASK:
- Standing on home plate, in what direction is the white building? ____ The bridge?____
- Face the gazebo. What direction are you facing?_____ How many steps does it take you to get to the gazebo?____
- Walk due north to a fence. Find a leaf of Virginia Creeper and make a leaf-rubbing.

FOR OLDER CHILDREN:
- Stand on the northwest corner of the basketball court. Head 320° for 20 paces and look under a rock for the next clue.

TREASURE HUNT:
- Have the groups collect a token at each station and then turn them in at the end for a final clue that will lead them to a hidden surprise or snack.

Arrange the children in small groups accompanied by an adult, and have each group start at a different station on the course, or set up several different courses, one for each group. Give each group a list of directions and have them follow the course to its conclusion, stopping to do tasks along the way. Have the children take turns being the leader as they move from station to station.

SHARING CIRCLE

Objective: To share new awareness of how we and other animals find our way.

Form a sharing circle and ask each child to complete the following sentence: "One thing that could help me find my way is _____."

EXTENSIONS

Map and Compass: Bring in copies of detailed maps (topographical or road maps) of your town. Have the children find their houses, the school, roads, and familiar landmarks like ponds or mountains. Older children can try to find the bearing to their houses from the school.

Make a Compass: Magnetize a steel sewing needle by stroking it from eye to point with a strong bar magnet at least 50 times. Cut a groove in the top of a circular piece of cork and place the needle in the groove. Now float the cork and needle in a bowl of water. When the cork stops moving, the needle should be pointing north, just like the magnetic needle in the compass. Have the children compare the direction the floating needle points with the direction the compass needle points.

Finding Your Way

Characters: Benjy Bear, Dora Digger Wasp, Hannah Honeybee, Sammy Salmon, Thelma Thrush

Benjy Bear

[singing] Oh, there's a rumble in my tummy, and some honey would be yummy. *[now mumbling]* I guess it's time to find my way to some lunch. *[buzzing noise offstage]* Hey, that sounds like a bee. Maybe lunch is close at hand!

Dora Digger Wasp

[coming up from below Benjy] Excuse me, buster, do you mind?

Bear

Oh, it's you, Dora Digger Wasp. I thought you were a bee. What's that you're doing?

Wasp

What do you think? I'm digging a hole for my eggs.

Bear

Are you going to put some honey in there?

Wasp

No, Benjy. I'm putting in some insects for my babies to eat when they hatch. Now, I don't want anyone else to find this hole, so I'll just push some sand over the entrance, and off I go. *[exits]*

Bear

Hey, that's great. I can't even see where the hole was! I wonder if she'll be able to find it when she comes back. *[buzzing noise; Wasp reappears]* Why, look at that. She went right to her nest hole! How did you do that?

Wasp

It's easy. You see, my nest hole is right between that pine cone over there and these pebbles. I have a mental map right here in my head. I remember the landmarks close to home. Now I must get more food, so off I go. *[exits]*

Bear

Well, if my mental map is right, I'm heading in the right direction for some honey. *[buzzing noise and Bee appears]* Why, hello there, Hannah Honeybee! I was just thinking about you. Is your hive nearby?

Hannah Honeybee

No, Benjy, my hive's far away. I came here for this tasty clover.

Bear

Really? How did a little bee like you find your way this far from home?

Honeybee

Oh, no problem. We bees use the sun to guide us.

Bear

Hey, that's pretty neat. But what if it's cloudy out?

Honeybee

Oh, cloudy days do make me a bit grouchy. But we have special eyes so we can still see the sun's rays despite the clouds. Now, I'm so full of nectar, I'd better beee going! *[exits]*

Bear

Bye, Hannah. I guess the sun is pretty useful for finding your way. It always rises over there *[points east]* and sets over there *[points west]*. Now the sun is right overhead, so that means...it's lunch time! *[Salmon appears]*

Sammy Salmon

Uh oh. *[nervous voice]* I think he means me. Don't you remember me, Benjy?

Bear

Is that you, Sammy Salmon?

Salmon

Yes, it is, and please don't come any closer.

Bear

Why, Sammy, you sure have grown. When I last saw you, you were just a small fry.

Salmon

True, but I've been out at sea for three years now, so I've grown up a bit.

Bear

Out at sea? But that's **miles** from here! Why did you go out there?

Salmon

Oh, we salmon migrate out to sea when we're young and come back home when we're ready to mate and lay eggs.

Bear

I've heard of birds migrating, but I didn't know some fish did, too. How did you ever find your way back?

Salmon

Well, I used my eyes to navigate back to my home river. But, when I got closer, I just used my nose.

Bear

Your nose?

Salmon

Sure. To me, the stream where I was born smells different from other streams. Once I picked up the scent, I just followed my nose until I got back here again.

Bear

Well, my nose tells me you'd be good to eat.

Salmon

Nobody knows that as well as I do! So long, Benjy! *[exits]*

Bear

Shucks, I wouldn't eat an old friend. I'm just going to follow my nose right over to that patch of blueberries and have some lunch. *[munching sounds]* MMMmmm! *[Thrush enters]*

Thelma Thrush

Hey, I'm hungry too, so would you mind moving your big paw so I can get that bug under there.

Bear

Oh sure, sorry, Thelma Thrush.

Thrush

And there's another one, and another one. Got to fatten up for the trip.

Bear

You must be getting ready to migrate south. I hear it's quite a long trip. How do you find your way?

Thrush

Well, for one thing, I migrate at night and use the stars to navigate.

Bear

Now that's a stellar idea!

Thrush

And we also use familiar landmarks to help us find our way.

Bear

Oh, you mean pine cones and pebbles? Like Dora Digger Wasp?

Thrush

No, Benjy. Remember, we fly high up in the sky. We follow big things like mountain ranges and rivers. And birds that follow the coastline listen for the sound of the waves crashing on the shore.

Bear

Surf's up!

Thrush

And then, of course, there's always a compass.

Bear

You carry a compass?

Thrush

No, but some birds can sense the earth's magnetic field, and that helps us to find the right direction. We're lucky that we can use lots of different cues to help us find our way.

Bear

Yup, we animals sure do have a pretty good sense of direction, and I'm going to direct myself right home for a snooze. So long, Thelma.

Thrush

So long, Benjy. *[both exit, then Benjy returns]*

Bear

Um, can anyone direct me to my den? I think I'm lost!

Erosion

SHAPING THE LANDSCAPE

Picture in your mind the Grand Canyon, a spectacular mile-deep river gorge cut into the rocks of the Colorado Plateau. Now picture the little gullies that are carved into muddy road banks by heavy rains. These two features of the landscape may seem unrelated. However, both are the result of erosion, the process by which rocks and soil are worn away and carried off to be deposited elsewhere on the earth. The most important agents of erosion are water, glaciers, wind, the pull of gravity, and weathering by chemical and mechanical processes.

In its many forms, water is one of the most powerful agents of erosion. From the splash of a single raindrop, to the raging force of a storm-swollen river, water has a tremendous effect on the earth around us. Rain causes erosion in two ways. First, the impact of raindrops when they hit the ground causes loose particles of soil to be scattered in all directions. Although the force of a single raindrop is little, the collective effect of a rainfall can move surprising amounts of soil. Second, in heavy rainfalls, when the ground is saturated, water flows over the surface and carries away exposed soil. As the water carries material downhill, gullies and channels are created in places where the ground is softer than its surroundings. These may eventually become stream channels. They are further eroded as the rushing water washes rocks and soil from the banks and these loose rock fragments scrape and scour the streambed.

Ocean waters are also agents of erosion as waves pound against the shore with tremendous force. Rocks are broken into fragments that are carried off and further eroded by the battering action of waves and currents. Over the millennia, sediments on the ocean floor are compressed into new kinds of rock. The beautiful smooth and rounded rocks found on beaches or river bottoms attest to the polishing action of crashing waves and rushing streams.

Just beneath the earth's surface, another form of erosion is taking place. **Ground water** containing carbonic acid leaches minerals as it percolates through underground rocks. This water slowly eats away at underground limestone formations to form caverns. At the same time, it creates beautiful stalactites and stalagmites, drip by drip, from the dissolved minerals that it carries.

Water in its frozen form can also cause erosion. Glaciers, formed of huge masses of snow compressed into ice, have repeatedly advanced and retreated over the surface of the earth. Although they move slowly, they are tremendously powerful because of their great weight. Rocks carried along on the bottom of the ice scour the earth's surface, carving out valleys and lakes, basins and rivers. Erosion by glaciers even carved out the Great Lakes. When glaciers retreat, they leave large amounts of rock fragments, ranging from boulders to pebbles, from sand to rich soil, in their wake. This deposition created Cape Cod and Long Island as well as coastal islands like Martha's Vineyard and Nantucket. The ice sheets that cover Antarctica and Greenland, and the glaciers still present on the highest mountain ranges, continue to wear away and transport rocks and soil, shaping the landscape in a slow but irrevocable way.

Wind erosion is most important in areas where there is little rain and little or no vegetative covering, such as the open sands of a beach, desert, or dry prairie. Wind can cause erosion simply by carrying away loose material. These wind-borne particles can then cause further erosion through abrasion, or "sandblasting," as they scour rock and earth surfaces. Unusual rock and land formations in desert areas, such as

steep cliffs and sandy beaches

arches and table rocks, were created by abrasion. If you have ever felt the sting of blowing sand on your skin, you are well aware of the force behind wind-blown sand particles.

Other agents of erosion that are less obvious are gravity and **weathering.** The pull of gravity causes downward movement of materials. This can happen very slowly as in soil creep, the gradual downward movement of soil on slopes, or it can be a dramatic and visible event, such as a sudden rock slide or avalanche. Weathering is the breakdown of rocks on the surface of the earth through chemical or mechanical processes. When exposed to alternate heating and cooling, rocks expand and contract until they finally break apart. Pockets of water in the rocks expand when they freeze, causing fragments and shards to separate. Plant roots and fungal hairs have a similar effect as they work into cracks and openings in rocks. Chemicals in the water or air also erode and change the composition of exposed rocks. All of these processes break rocks into smaller and smaller fragments, eventually creating the mineral grains that form the basis of soil.

Along with these natural causes of erosion, human activity has come to play a significant role as well. Urban and industrial development as well as poor agricultural and forestry practices create many opportunities for wind and water to carry off valuable topsoil. Whenever vegetation is removed, soil becomes vulnerable to the action of erosion. However, erosion control methods can help to curtail the loss of soil. Plantings of vegetation and placement of rocks can be used to stabilize steep slopes and stream banks. Contour plowing, terrace farming, and crop rotation help to hold the soil in cultivated fields. Care and maintenance of roadside ditches can prevent road washouts and the runoff of soil into lakes and streams. Wise use of the land and attention to erosion is essential in order to conserve our precious topsoil.

Although erosion may rob one area of soil, sand, and rock, this material will be deposited elsewhere. Erosion of soil and rock from mountainsides brings soil, minerals, and nutrients to low-lying areas. Sand carried by the wind is deposited to form sand dunes. Beaches and barrier islands constantly change as material is eroded from one place and dropped off in another. Each time a river floods and then subsides, it leaves behind mineral and nutrient-rich deposits transported from other areas.

If erosion were the only force affecting the landscape, the world would eventually be devoid of mountains, hills, lakes, and ponds as material that was removed from the high, exposed features of the landscape was deposited in the low areas. Why do mountains still exist after millions of years of erosion – including the glacial ages? This is because there are powerful forces deep inside the earth that bring molten material upward from the interior. There is a balance between the forces that uplift the earth's crust into mountains and the agents of erosion that wear it down again.

The process of erosion both wears away and builds anew. It creates soil by wearing rocks down to their mineral fragments and then sweeping them away to be deposited elsewhere. It gouges out valleys and lakes and then fills them in with sediments. It erodes honeycomb caverns and then builds new rock formations within them: It wears gigantic boulders down to sand grains and then deposits them in layers so deep that the pressure and weight of so much material turns them into stone again. The force of erosion is at work all around us, continually shaping and reshaping the contours of the earth.

Suggested References:

Beiser, Arthur. *The Earth*. New York: Time-Life, 1963.

Skinner, Brian J., and Stephen C. Porter. *The Dynamic Earth*. 3rd edition. New York: John Wiley & Sons, 1995.

Tank, Ronald. *Focus On Environmental Geology*. London: Oxford University Press, 1973.

Tarbuck, Edward J., and Frederick K. Lutgens. *The Earth: An Introduction to Physical Geology*. 3rd edition. Columbus, OH: Merrill, 1990.

eroded stream bank

Erosion

FOCUS: The shape of the landscape changes over time due to the erosive forces of wind, water, gravity, weathering, and ice as well as human activity.

OPENING QUESTION: *What is erosion and how does it shape the landscape?*

EROSION PUPPET SHOW

Objective: To learn some of the causes and effects of erosion.

Perform, or have the children perform, the puppet show. Afterward, review the causes of erosion and its effects on the various puppet characters.

Materials:
- script
- puppets
- props

SHAPING THE LAND

Objective: To recognize examples of erosion and identify their cause.

Show pictures of land changed or shaped by erosion. Ask the children to identify and discuss what might have caused the erosion: wind, water, ice, weathering, gravity, activity by humans or animals.

Materials:
- pictures of different places where erosion has occurred

SPLASH

Objective: To discover the power of raindrops falling on loose soil.

Working with a small group of children, set a large jelly-roll pan or cookie sheet on the floor and place a tablespoon of flour in a mound in the center of the pan. Ask the children to predict what will happen when a drop of water hits the flour. Now have a child hold the baby bottle at waist height over the pan, tip the bottle, and gently squeeze out a drop of water so that it falls on the flour. What happens to the flour? Have the children take turns squeezing out drops of water and watching to see how far the flour scatters. Have the children hold the bottle at head height and try again. Now how far does the flour go? What would be the effect of raindrops falling from the sky onto loose soil? onto pebbles? onto a large rock?

Materials:
- large jelly-roll pan or cookie sheet
- tablespoon
- flour
- hard plastic baby bottle with nipple
- water
- (optional) soil, pebbles, rock

ROCK AND ROLL

Objective: To examine the effects of abrasion on rocks.

In small groups, direct the children to place ten washed stones and two cups of water into a clear plastic container marked "Shake" and place another ten similar stones and an equal amount of water into a second container marked "Don't Shake." Screw the containers' lids on tightly and set aside the container marked "Don't Shake." Now pass around the first container and have each child take a turn shaking it at least 20 times. Keep track of the total number of times it is shaken, with 100 shakes as a minimum. Pour the water from this container into a clear jar and observe. Compare to the water in the container that was not shaken. Why are there fine particles of dirt floating in the water of the first container and where did they come from? Where might a similar process occur in nature? Pass around some smooth river rocks and pieces of beach glass from the ocean for the children to feel and observe. How did they get so smooth?

Materials:
- plastic containers with tight-fitting lids (e.g., plastic peanut butter jars)
- washed stones
- water
- measuring cup
- clear jar
- smooth river rocks and/or beach glass

smooth, rounded pebbles

SWEPT AWAY

Objective: To explore how wind helps shape landscapes.

In small groups, place a few small rocks and a couple of pieces of wood in a shallow baking pan, and then add a handful of uncooked rice. Stretch a piece of plastic wrap tightly over the pan. Ask the children what will happen to the rice when blown with a straw. Have the children take turns blowing through plastic drinking straws toward the rice. What happens? (For fun, you could hide two pennies under the rice for the children to discover.) Encourage them to move the straw around and direct it at different locations, noticing places where the rice builds up or gets hollowed out. Where might this kind of erosion be found? Afterward, share results and discuss how wind affects deserts, mountaintops, sand dunes, and beaches.

Materials:
- uncooked rice
- rocks
- pieces of wood
- shallow baking pans
- straws (one per child)
- (optional) pennies

SLIPPING SOIL

Objective: To experiment with various materials and designs to limit or control erosion.

Have the children work in small groups and give each team a dish basin that contains moistened sand or soil packed so it forms a slope. The top of the slope is at one end of the basin and the bottom is 2-3 inches from the opposite end. The exposed bottom of the pan needs to be clean of soil. Explain that this represents a bare hillside on which all the trees have been cut, with a clear stream running at its base. Ask the children to predict what will happen to this bare slope when it rains? Demonstrate by slowly sprinkling one jar of water on the top of the slope. Discuss what happened to the bare slope as well as to the stream below it. Now, explain that each team must develop and carry out a plan to reduce the amount of soil runoff, using the following questions to help them formulate their erosion control plan:

Materials:
- plastic dish basins
- soil or sand
- sprinkling watering can or quart jar with holes punched in lid
- water
- rocks, hay, twigs, and other natural materials

How can the surface soil be protected?

Is there any way to hold the soil in place?

What could be added to absorb the water rather than let it run off into the stream?

Is there any way to divert the water flow?

Provide time for the children to gather materials from outdoors to use in their construction. Once groups have implemented their erosion control plans, ask each group to present its plan, the materials used, and the reasoning behind their ideas. Then test each group's construction by sprinkling one quart of water on the top of their slope. Observe the water as it moves down the slope, noting any runoff and sediment that reaches the stream. Which designs worked best? Have you seen any of these methods used along highways, stream banks, or on hiking trails?

sand dunes anchored by plants

EROSION HUNT

Objective: To look for signs of erosion in the school yard.

Divide the children into small groups and assign a different part of the school yard for each group to study. Provide clipboards and maps of the assigned areas. Instruct the groups to hunt for any signs of erosion in their designated area and to mark those signs on their map, noting how the erosion could have been caused. Have the groups come together to share findings and discuss ways that erosion could be affecting the various animals and plants that live nearby. What can be done to slow down or prevent the erosion identified in their school yard?

IF I WERE A FISH...

Objective: To consider the impact of erosion on animals' lives.

Ask each child to think of an animal. (To save time with younger children, give each a card with the picture of an animal on it.) Taking turns, have each child tell what animal they chose. Can the child or anyone else in the group think of a way that this animal might be affected by erosion?

EXTENSIONS

Sowing Seeds: Plant grass seed on some of the slopes built in Slipping Soil. Let it germinate and grow; then try sprinkling a jar of water over the slope. Observe and discuss the results. Compare these results with the results when water was sprinkled on bare soil. Also compare with the results when the soil was protected by erosion control devices.

Erosion Hot Spots: Using the results from Erosion Hunt, have each group place color-coded dots (yellow for wind erosion, blue for water, and red if caused by people or animal activity) on a master map indicating the identified erosion spots. After all the dots are on the master map, discuss the findings. Are there clusters of dots? What erosion control plans could the children implement in their school yard?

Save the Soil Service Projects: The students could organize a project to control erosion on one or more of the eroded areas identified in their Erosion Hunt. Be sure to get appropriate permission.

Materials:
- clipboards
- pencils
- maps of school yard

ACTIVITY STATIONS:
1) Splash
2) Rock and Roll
3) Swept Away

Erosion

Characters: Billy Boulder, Ricky Rock, Rushing River, Wind (*just a voice*), Tiny Billy Boulder, Tiny Ricky Rock

Prop: water spray bottle

Ricky Rock

Mornin', Billy Boulder.

Billy Boulder

Hey, Ricky Rock.

Rock

What do **you** want to be when you grow up, Billy?

Boulder

What do you mean, "grow up"? I'm a boulder!

Rock

So, don't boulders grow up?

Boulder

No! We boulders grow **down.** You've gotten smaller over the years, too, Ricky. It's all because of erosion.

Rock

What's erosion?

Boulder

It's what wears us rocks down. The powerful agents of erosion are constantly wearing away at us, breaking off pieces and grinding against us. Why, before you know it, we're dirt.

Rock

Agents of erosion? They must be powerful to turn **you** into dirt!

Boulder

They sure are. I bet you've met a few in your day, too.

Rock

Really? Who are these agents of erosion anyway?

Boulder

Well, for one thing, there are glaciers. You know, those huge masses of snow and ice that inch along, scraping out valleys and digging out lakes. They're so strong they can lift and carry big rocks like us for miles and miles. Why, they're what put us here in the first place.

Rock

Wow, you mean I was bulldozed by a glacier! Vroom! Vroom!

Boulder

You sure were! And then there's **gravity.** It's always ready to pull things downhill. In fact, I'm feeling a little shaky myself. I think I'm going to fallllllllllll. *[exits]*

Rock

Gosh, I guess I won't be seeing Billy for a while. Hmm, I wonder if there are other agents of erosion. Hey, it looks like rain. It feels like rain. *[spray water at audience]* Wow, it's really raining now. Look how the rain is splashing the dirt away. It's carving little gullies all around me. Help! I'm being lifted by the water, and I'm slipping down the mountainside. *[move Ricky along the stage]* Ouch! It hurts to bump along. *[Rushing River appears]*

Rushing River

Sorry, Ricky Rock. You're right in my way, so I'm taking you for a ride.

Rock

You sure seem to be in a hurry! I'll bet you're an agent of erosion.

River

I sure am. I'm the Rushing River, and the force of my waters as I move through here is carving out this streambed.

Rock

Well, that's fine, but look how you're tearing apart that stream bank and carving away pieces of that nice field over there.

River

I didn't mean to do that, but someone cut down all the trees on the bank and mowed hay right to the edge. Maybe they didn't realize that the roots of those trees were holding the bank together.

Rock

And now your water is full of sand and silt. It must be hard for fish and other water creatures to breathe when you get so full of dirt.

River

You're right, Ricky. It takes a while after a big storm for me to settle down. Now I'll just put you down here on this beach while I go on my way out to the sea. *[rushing water sound as River exits]*

Rock

Too bad. I was having fun on my river ride. This beach seems kind of boring – so empty, just sand and pebbles. *[spooky howling wind noise]* I wonder what that noise is?

Wind

[a windy voice, no puppet] That's me, the Whistling Wind.

Rock

I can hear you, but I can't see you. Are you an agent of erosion, too?

Wind

Of course, can't you feel me eroding you?

Rock

Yes, now that you mention it, I can feel you battering me from every side. *[shakes from side to side]* Ouch, you're blowing sand on me! That stings!

Wind

I'm sorry Ricky, but when there's loose sand in my way, I can't help blowing it around. And when I blow hard, I'm like a sandblaster. I've carved some beautiful shapes out of rock that way.

Rock

Go away! I don't want you carving any beautiful shapes out of me!

Wind

Lucky for you, I'm on my way out to sea. Catch you later, Ricky.

Rock

Phew, that was close. I wish Billy were here so I could tell him about all the agents of erosion that I've met. Let's see, there was the pouring rain, the rushing river, the whistling wind, and yikes! *[rumbling noise, puppet shakes]* Hey, it feels as if those waves are getting closer. Oh no, here they come! Guess I'm sunk! So long, everyone. *[exits]*

[Tiny Billy Boulder and Tiny Ricky Rock (now pebble-sized puppets) appear. High voices]

Tiny Billy

Hi.

Tiny Ricky

Hi. Who are you?

Billy

I'm Billy Boulder.

Ricky

Billy Boulder? *[laughs]* You? You're just a tiny pebble!

Billy

[offended] Well, I used to be a big boulder, but I got eroded.

Ricky

You mean you're really Billy? Why, I used to know you when you were a big boulder. Remember me? I'm Ricky Rock!

Billy

Ricky Rock? Look at you. You're no bigger than a peanut!

Ricky

Well, you're no bigger than a pea!

Billy

I guess we've both seen some erosion in our days.

Ricky

Yeah, we sure have. So, Billy, what do you wanna be when you get older?

Billy

I wanna be dirt. For things to grow in.

Ricky

Yeah, me too! And we're almost there!

Billy & Ricky

Hurrah! Yippee! Hurrah!

Pebbles and Rocks

ARCHIVES OF EARTH'S HISTORY

The rocks and pebbles at your feet, under the ground, or at the edge of the road are treasures overlooked by most people. Every pebble and rock represents a piece of the earth's history and may be millions of years old. Rocks are an important part of our daily lives for we build with them, refine them into metals, and burn them as fuel. Even more important, the weathering of rocks gives us soil, essential in growing and sustaining so many living things.

Rocks are formed in one of three ways. They may be **igneous** rocks, formed after lava from volcanoes or molten **magma** deep inside the earth cools. They may be **sedimentary** rocks, made from the compressed and cemented sediments of ancient seas. Or they may be **metamorphic** rocks, formed when igneous or sedimentary rocks are further transformed by exposure to intense heat and pressure.

Igneous rocks form from rock material that is heated so much that it becomes liquid. When it cools, it solidifies into new rock. Igneous rocks are divided into two types, depending upon whether they harden slowly within the earth (intrusive) or are pushed out of the earth where they cool rapidly above the surface (extrusive). Slow cooling of magma (i.e., a few thousand years) produces coarse rocks, such as granite, that contain large crystals. If the magma cools quickly, only small crystals have time to form in the new rock. Extrusive rocks range from shiny, glass-like obsidian, to pumice, a rock made from volcanic froth that is so light it floats on water.

Sedimentary rocks are made from many layers of eroded material. Through the processes of **weathering** and **erosion**, rock fragments are broken down and then carried away by wind and water. These rock particles finally settle out in quiet water or protected areas. Over millions of years, layer upon layer of sediments are compressed under the weight of overlying material and cemented together to form a stratified rock. Various plant and animal remains may be contained in these layers, and some are preserved as fossil imprints in the rock. Some of the most common sedimentary rocks are shale, formed from minute clay and silt-size particles; sandstone, made of sand-sized grains cemented together; and limestone, which is made from the calcium-rich remains of ancient sea creatures.

Metamorphic rocks tell of times when the earth's landscape was changing more dramatically. These stones began as either igneous or sedimentary rocks but were then transformed further by heat and pressure. They may have been squeezed when flat lands were pushed up to become mountains or when the earth's crust buckled as tectonic plates collided. Magma erupting from deep in the earth may have heated surface rocks and melted their crystals. When exposed to intense heat, the crystals of igneous rocks often spread out into bands or fuse into clumps. Grains of sedimentary rocks melt together into a denser stone, and rock layers become twisted. Slate is a metamorphic rock formed from sedimentary shale that has been subjected to enormous pressure. Marble is an even-grained, crystalline metamorphic rock whose parent rock was limestone or dolomite. Quartzite is a metamorphic rock formed from sandstone.

Since the earth's beginning, rocks have continually been formed, changed by heat and pressure, worn away into tiny pieces, and then cemented together again. Rock carried downward in deep ocean trenches melts and becomes liquid magma. Cooled magma recrystallizes into rock. Rocks exposed to weathering are broken down into sediments, and sediments are compressed into stone. This endless cycle of change has been going on since the earth's crust started to form at least 4.5 billion years ago, and it still continues today.

Rocks are made of **minerals**. Each mineral has a specific chemical composition and structure. Identifying rocks depends on recognizing the minerals in them. Some rocks contain a single mineral, such as halite or rock salt, and others contain a variety, such as granite, which contains quartz, feldspar, mica, and other minerals. A few simple tests can be helpful in identifying rocks and minerals. For example, only a few minerals, such as magnetite, are attracted to magnets, and only rocks made

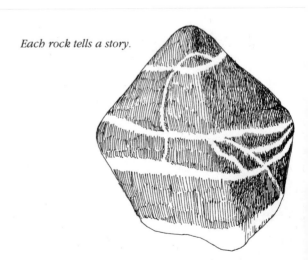

Each rock tells a story.

from calcium-rich shells, such as limestone or chalk, will react with vinegar to produce bubbles of carbon dioxide gas. Geologists identify minerals by comparing their hardness (resistance to scratching) and examining the shape of their crystals. Another characteristic of minerals useful in identification is the color of a mineral's streak, which is made by rubbing it across an unglazed porcelain tile. Other identifying features include cleavage, or how a rock breaks, and luster, the quality of light reflected from its surface.

quartz crystals

Not only are rocks the building blocks of our planet, they were also important in the development of human society and culture. During the Stone Age, people depended on rock to make simple tools and weapons. Later, during the Bronze and Iron ages, people dug metals from the earth's crust to forge more sophisticated implements. Even today we depend on rocks. In fact, it would be almost impossible to get through the day without using some rock or mineral product.

Metals and minerals make everything from jewelry to medicine, from cars to computers. We season our food with salt mined from rock salt. We drink from glasses made of quartz sand or aluminum soda cans made from the mineral bauxite. We brush our teeth with a paste made from magnesium or limestone. Our jewelry is made from precious metals and gemstones, and we may wear a watch that contains a tiny quartz crystal. Photographic film contains silver, and pencils contain the mineral graphite. Ceramic dishes and bricks are made from clay, and statues are carved from rocks like granite or marble. We may use chalk, a rock made from shells, to write on a slate blackboard

or on a sidewalk made of ground limestone, shale, and gypsum. We get most of our energy from mining or pumping fossil fuel from rocks underground, and in addition, petroleum gives us such varied products as plastics, nylon, and polyester. Moreover, almost all the food we eat is derived from the soil beneath our feet, and soil is largely made up of rock fragments produced by weathering and erosion.

Every rock tells part of the story of how the earth was formed, shaped, and changed over time. Learning to decipher the story can be an exciting and rewarding challenge, for rocks are the foundation, both literally and figuratively, of our entire civilization and of all life on Earth.

Suggested References:

Beiser, Arthur. *The Earth*. New York: Time-Life, 1963.

Fenton, Carrol L., and Mildred Adams Fenton. *Rocks and Their Stories*. New York: Doubleday, 1951.

Goldsworthy, Andy. *Andy Goldsworthy: A Collaboration with Nature*. New York: H. N. Abrams, 1990.

Parker, Steve. *Rocks and Minerals*. New York: Dorling Kindersley, 1993.

Ranger Rick's Naturescope. *Geology: The Active Earth*. Washington, D.C., 1987.

Skinner, Brian J., and Stephen C. Porter. *The Dynamic Earth*. 3rd edition. New York: John Wiley & Sons, 1995.

Zim, Herbert S., and Paul Shaffer. *Rocks and Minerals*. New York: Golden Press, 1957.

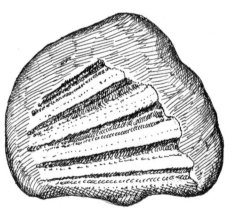

the fossilized print of an ancient sea shell

Pebbles and Rocks

ACTIVITIES

FOCUS: Every pebble and rock tells a story of how the earth was formed, shaped, and changed over time.

OPENING QUESTION: *What are some different characteristics of pebbles and rocks?*

PEBBLE PASS

Objective: To examine rocks using the sense of touch.

Divide the children into groups of ten or fewer, and have each group sit in a circle and listen to the directions. Pass out one small rock to each child, telling the children not to peek but to keep the rocks hidden in their hands. Give the children time to feel and become familiar with the rocks. To begin playing, have the children put their rock in their right hand. Now, at your signal, the children should pass their rock to the neighbor on their right, and receive a rock with their left hand. Have them feel the new rock and then put it in their right hand, ready to pass again. Continue passing and receiving rocks without peeking. (It may help to have them chant, "Pass a rock, get a rock, put it in your right hand" while they pass rocks.) Tell the children to raise their hand as soon as they think they have their original rock back. What characteristics helped them to recognize their own rock?

Materials:
- rocks, acorn- to walnut-size, of varying textures and shapes, one per child

PEBBLES AND ROCKS PUPPET SHOW

Objective: To compare how igneous, sedimentary, and metamorphic rocks are formed.

Perform, or have the children perform, the puppet show. Afterward, review how the three basic rock types are formed. Pass out examples of each kind for the children to examine.

Materials:
- puppets
- script
- props
- samples of igneous, sedimentary, and metamorphic rocks

ROCK RIDDLES

Objective: To examine several different characteristics commonly used in rock identification.

Place rocks and examination tools on a table. Working in small groups, have the children test some properties of the rocks using the equipment provided and find the rocks with the characteristics listed below. Some rocks may fit in more than one category.

Materials:
- rock set that includes hematite (iron), talc, magnetite, limestone, pyrite, halite crystals, quartz crystals
- small jar of vinegar
- unglazed ceramic tile
- magnet
- hand lenses
- Rock Riddles cards

ROCK RIDDLES CARD

Find a rock that leaves a dark streak on a tile.

Find a rock that feels greasy.

Find a rock that fizzes when dropped in a jar of vinegar.

Find a rock that can be picked up by a magnet.

Find a rock that has cube-shaped crystals.

Find a rock that has hexagonal (six-sided) crystals.

Find a rock with a metallic shine.

260 / Earth and Sky

ROCK CYCLE STORIES

Objective: To understand the rock cycle and to become familiar with some examples.

Use a rock cycle diagram to show the children how one kind of rock can be changed into another over time. Now, place the four rock sets – labeled A, B, C, and D – on tables. Working in pairs or small groups (younger children with an adult), give each team the Rock Cycle Stories to read while they examine the rock sets. Have the children decide which story goes with which rock set and write their answers on the story sheet. What clues helped them to reach these decisions?

ROCK CYCLE STORIES

Sand-Sandstone-Quartzite

Rocks are constantly being worn down into smaller and smaller pieces by wind, water, and ice. These small pieces may then be the building blocks for different rocks. In one rock family, grains of *sand* settle in layers and are buried and squashed together. In time, this sand hardens and is cemented together into *sandstone*, a sedimentary rock. If you look at sandstone closely, you can still see grains of sand. When exposed to tremendous heat and pressure, the grains of sand in sandstone melt and form a shiny hard rock called *quartzite*, a metamorphic rock.

Shells-Limestone-Marble

Millions of creatures in the ocean live inside hard shells. When these creatures die, their *shells* eventually settle to the bottom of the ocean. Pieces of shell are squeezed and cemented together into sedimentary rocks called *limestone*. When limestone is exposed to intense heat and pressure deep inside the earth, *marble,* a metamorphic rock, is formed.

Peat-Lignite-Hard Coal

Would you believe there are rocks made from plants?! Long ago in swamps and bogs, *plants* fell to the wet ground and began to rot. Later, this soft, rotting plant material was buried and pressed together to form *peat*. Over millions of years, these plant remains were further squeezed and pressed under more rock layers into a brown sedimentary rock called *lignite*. Continued pressure and heat convert lignite into *hard coal*, a metamorphic rock. Most coal is found deep underground, sandwiched between other layers of rock. For hundred of years, people have mined coal to use as fuel.

Clay-Shale-Slate

Picture your local river or stream after a heavy rain. It is often brown and muddy from rock *particles suspended in the water*. Eventually these particles settle out, forming a fine layer over the bottom of the stream. Many layers of these particles build up into thick *mud* or *clay*. Over time, this clay is pressed and hardened into a flat, layered sedimentary rock called *shale*. When the shale is further heated and squeezed incredibly hard, *slate,* a metamorphic rock that easily splits into sheets, is formed.

Materials:
- rock cycle diagram
- rock sets
- copies of Rock Cycle Stories
- clipboards
- pencils

A Houseful of Rocks

Objective: To discover the variety of rock and mineral products we use everyday.

Place a collection of household items on a table, including some that are made from rocks and minerals and others that are not, such as those listed below:

Rock Products: salt, drinking glass, toothpaste, chalk, pencils, scissors, jewelry, aluminum foil or cans, paper clips, pottery or bricks, photographic film, a quartz watch, cement, blackboards, statues, and petroleum products such as plastics, petroleum jelly, lipstick, nylon stockings, polyester clothes.

Products not from rocks: wool, beeswax, cotton, leather, cork, rubber, paper, cardboard, wood, jute twine, straw baskets, bamboo.

Ask the children to count how many products they can see that are made from rocks or minerals. Afterward, compare the numbers reported and discuss the items, being sure to clarify whether each one is a rock product or not. Were there any surprises? Can the children think of other rock products that are part of their daily lives?

Materials:
- an assortment of rock and mineral products
- an assortment of plant or animal products

Pebble Pickup

Objective: To look for similarities and differences among common pebbles and rocks.

Take the children outside and ask them to collect 10-20 favorite pebbles. Now have the children sort their rocks based on any characteristics that they choose – size, color, shape, texture, designs, how crumbly, etc. Afterward, have the children share their system of classification with the class. Have the children label and display the collections in egg cartons or create designs with their rocks and then photograph them for display. Have each child pick out a favorite rock for the Sharing Circle activity below.

Materials:
- pebbles
- egg cartons
- (optional) camera

Sharing Circle

Objective: To share new discoveries about pebbles and rocks.

Sitting in a circle, have each child hold up a favorite rock selected from the Pebble Pickup and tell the group one characteristic they especially like about their rock.

Materials:
- pebble or rock selected from the Pebble Pickup activity, one per child

ACTIVITY STATIONS:
1) Rock Riddles
2) Rock Cycle Stories
3) A Houseful of Rocks

Extensions

Field Trip: Take the class to a rock outcropping, road cut, quarry, gravel pit, sand pit, river bank, or natural beach to observe and investigate the local geology. Collect samples if allowed, record observations, and make a list of questions for further research.

Interviews: Have the children interview someone who works with rocks and rock products. This could be a state geologist, a professor from a local college or university, a potter, a soil scientist, a quarry or gravel pit worker, a stone mason, or a sculptor.

What's Made of Rock Scavenger Hunt: Conduct a scavenger hunt in the classroom or around the school building and grounds for rock and mineral products. See who can identify the greatest number of rock and mineral products.

Soil Studies: Have the students grind some soft rocks into powder. Fill some tiny pots with a mixture of sand, clay, and these ground rocks. Fill another set of pots with pure organic matter, such as compost and peat moss. Fill a third set of pots with a mixture of the sand, clay, and ground rocks and the organic matter. Have the students plant seeds (the same type) in each of the three soil mixes. Explain that each pot should receive the same amount of light and water. Ask the students to predict what will happen to the seeds in each pot. Have the students observe and record the plant growth for a month. Which soil mix grew the healthiest plants?

ROCK CYCLE DIAGRAM

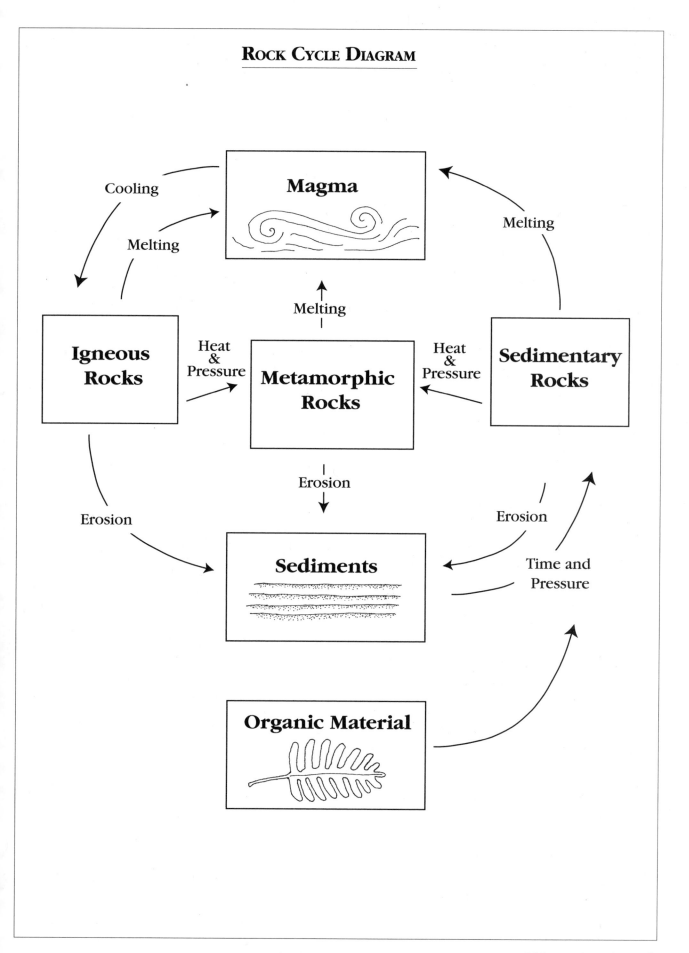

Pebbles and Rocks

Characters: Teacher (real person or hand puppet), Grandma Granite, Sandy Sandstone, Maggie Marble

Props: small box of rocks, slice of white bread, slice of wheat bread, crunchy peanut butter, jelly, spreaders, rolling pin, paper plate

Teacher

I'm so excited. I'm teaching a unit on rocks for the first time. I'll just look through this box of rocks to get some teaching ideas.

Grandma Granite

[enters] May I be of some help?

Teacher

A talking rock? I can't believe it. Is this some kind of joke?

Granite

Certainly not. Let me introduce myself. I'm Grandma Granite. I'm an igneous and proud of it!

Teacher

Oh my, I've never met a talking rock before. You **must** be a genius!

Granite

Not a genius, dear! I said I'm an igneous. An igneous rock!

Teacher

Oh, I am sorry, Grandma. As you can see, I have a lot to learn about rocks. Just what is an igneous rock?

Granite

Igneous means formed by fire! Think of the word "ignite." You know how hot boiling water is, dear? Well, deep inside the earth where I was formed, it was ten times hotter than that! Why, we rocks just melt when it gets that hot. Then we cool slowly over time and become…well, rock hard.

Teacher

I see. So igneous rocks are formed when melted rock cools and hardens. Then does lava from volcanoes form igneous rocks, too?

Granite

That's right! Volcanic rocks are close relatives of mine. Now, it's time for you to meet an old friend of mine, Sandy Sandstone. Here he is now. *[Sandy Sandstone enters; Grandma Granite exits]*

Teacher

Nice to meet you, Sandy. I'm an elementary school teacher.

Sandy Sandstone

Well, you might be **ele**mentary, but I'm **sedi**mentary! Sedimentary rock. I'm made up of sediments – small bits of rocks, minerals, or shells.

Teacher

How can you be made up of tiny rock pieces? You look very solid to me.

Sandstone

Well, it's all just part of the rock cycle. Over millions of years, rocks get broken down into tiny grains of sand. Then the sand grains settle in layers and get pressed and glued together into rocks like me.

Teacher

Hmm. All those layers sound a bit complicated.

Sandstone

Not really. It's as simple as making a peanut butter and jelly sandwich!

Teacher

Hey, I'm planning to have a PB&J for lunch.

Sandstone

Good! You can listen to my story while you make your lunch. A long time ago a river was flowing out to sea carrying white sand. The sand sank to the bottom and was cemented into a layer of white sandstone.

Teacher

[holds up slice of white bread] This can be the white sandstone.

Sandstone

Years later there's a big flood. Tons of mud and rocks are carried out to sea, sinking to the bottom and covering the white sandstone.

Teacher

[spreads peanut butter on the bread] Here's a layer of mud and rocks.

Sandstone

For thousands of years, colorful shellfish live and multiply on the bottom of the sea, and their crushed shells form another layer.

Teacher

[spreads on jelly] Maybe they were purple shells!

Sandstone

Over time, the river cuts a different channel. Now it flows more slowly, carrying tiny particles of brown clay that settle out to form a layer of darker rock.

Teacher

[places slice of wheat bread on top] Here's the brown rock to top it off!

Sandstone

And there you have it: a sedimentary rock – sandwich!

Teacher

Now peanut butter and jelly geology is something I can really sink my teeth into! Are there different kinds of sedimentary rocks?

Sandstone

There sure are. I was made from sand, but some sedimentary rocks are made from clay and others from shells. Why, I even know some rocks that were made from plants! You've heard of coal, haven't you?

Teacher

Of course. People burn coal for heat. Amazing, a rock made of plants!

Sandstone

We sure are amazing, and people use us rocks every day! *[Maggie Marble enters; Sandy Sandstone exits.]*

Maggie Marble

Like me, for instance! I'm Maggie Marble. People like to make statues, buildings, and monuments out of marble. I'm positively metamorphic!

Teacher

Did you say metamorphosis? Are you made of caterpillars or something?

Marble

No, but like caterpillars, I've changed from one form into another. That's what metamorphic means, changing form. You see, I used to be sedimentary, just a chunk of limestone. Then pressure and heat deep inside the earth changed me into marble.

Teacher

I never imagined heat and pressure could make such a change!

Marble

Oh, yes. Why, imagine baking your sandwich in a pizza oven. It'd burn to a crisp! And if you want to see how pressure can change things, try using a rolling pin over your sandwich.

Teacher

OK. *[flattens sandwich with rolling pin]* Oh my, this sandwich sure has changed form!

Marble

Nice marbling. Bye! *[exits]*

Teacher

Now I'm not hungry! What **will** I learn next? *[Grandma Granite enters]*

Granite

Well, dear, there are hundreds of different kinds of rocks, but we're all made in one of three ways! Can you remember them all?

Teacher

Of course. There are igneous rocks, like you, Grandma, formed from fire. Then there are sedimentary rocks made of sediments pressed and glued together in layers. And lastly, there are metamorphic rocks, changed by heat and pressure!

Granite

Now you sound like an expert. Glad we could give you some support.

Teacher

Thanks, Grandma Granite. You rocks support us in many ways, and I'm sorry I've taken you for granted all these years!

Granite

That's OK, dear. After all, I **am** granite! Goodbye, now, and have fun learning more about rocks!

Breath of Life

EARTH'S INVISIBLE BLANKET

Air is everywhere. It blankets our planet, filling every nook and cranny, protecting and nourishing Earth's inhabitants. Because we cannot see, feel, smell, or taste air, it often goes unnoticed, and yet we could not live without it. A person can live for weeks without food and days without water, but only minutes without air. Air is truly the breath of life.

But what is air? It is the invisible mixture of gases that surrounds the earth. This protective cover shields our planet from harmful radiation while at the same time trapping some heat from the sun's rays so that life can exist. Air contains many materials essential to living things, such as nitrogen, oxygen, carbon dioxide, and hydrogen. It transmits sound waves and carries smells. It contains the water vapor and fine particles that form clouds and produce life-giving rain and snow.

Like other materials, air has properties. Air expands when heated and contracts when cooled. This causes the circulation and mixing of Earth's atmosphere that brings us wind and weather. Air has volume and takes up space. An air-filled bag cannot be squeezed flat because of the air it contains. Air has weight and exerts pressure. Air pressure measures the amount of push exerted by an air mass; the more molecules of air within a given volume, the greater the push. At sea level, air exerts a pressure of about 15 pounds for every square inch. When air is forced into a small space, or compressed, as in a bicycle or automobile tire, it has great power. Tires filled with air to a pressure of 30 pounds per square inch can support the weight of a truck. Understanding the properties of air can help us to understand how birds fly, how fish dive, how smoke rises from a chimney, or why the wind blows.

The principal gases in air are nitrogen (78%) and oxygen (21%). Tiny quantities of other gases, including carbon dioxide (.035%), make up the remaining one percent of dry air. We use oxygen to produce the energy needed to carry out all of our body functions. Just as a burning candle uses oxygen and fuel to produce energy in the form of heat and light, the human body uses oxygen and food to produce energy for our cells.

Plants, on the other hand, take in carbon dioxide from the air and water from the soil and convert these compounds into sugars and starches through the process of **photosynthesis.** Oxygen, the byproduct of this process, is released into the atmosphere. In fact, much of the oxygen animals depend upon for energy comes from plant photosynthesis. With the exception of certain bacteria that use sulfur to obtain metabolic energy, air is necessary for all life on Earth.

goldenrod

Animals get oxygen from the air in a number of different ways. As humans breathe, the rib cage expands outward and the diaphragm downward, allowing air to fill the lungs. The lungs contain tiny air pockets lined with a rich supply of blood vessels. The oxygen in the air pockets diffuses through the moist lung membranes into the blood supply, reaching all parts of the body through a system of arteries and vessels. Carbon dioxide is removed from the body in a reverse process.

Other animals obtain oxygen through a variety of mechanisms. Fish, many aquatic insects, and some immature **amphibians** have gills that remove oxygen from the water. Terrestrial insects have holes called **spiracles** in the sides of their bodies that connect to a

Atlantic salmon

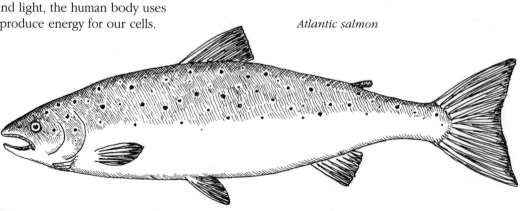

series of interior air ducts through which oxygen travels to all parts of their bodies. Frogs, worms, and other creatures with moist skin can receive oxygen from the air by diffusion directly through their skin.

Not only is air the source of oxygen and carbon dioxide, but it is important in other ways as well. Air can act as an insulator. When trapped in tiny spaces such as those between fur or feathers, air helps to prevent animals from losing body heat. Air is also important for animal locomotion. Air provides lift for flying birds, and rising air thermals allow **birds of prey** to float effortlessly above the land. Air trapped under the fleshy capes of flying squirrels allows them to glide from tree to tree just as a parachute breaks the fall of a skydiver. Air can also be used for flotation. Fish have swim bladders, air chambers that enable them to float and that are filled or emptied as the fish rise or sink. Air is important for animal communication, too. It transmits sound waves and chemical odors that animals use to locate prey; to learn of approaching danger; or to communicate with mates, rivals, or young. It can even be used in defense. Animals like puffer fish and horned lizards can use air to inflate their bodies in order to frighten away enemies.

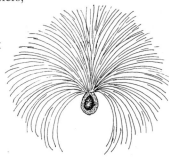

milkweed seed

Many plants depend on air for carrying **pollen** or **spores** or for giving their insect pollinators a lift. Numerous plants have evolved aerodynamic seeds designed to be carried by the wind, such as the helicopter seeds of maples and ashes and the parachute seeds of milkweed and dandelions.

Because both plants and animals are so dependent on air, the contamination of air by pollutants, both natural and manmade, raises concerns. Air has always carried natural pollutants, such as dust, pollen, and spores, that can be irritating to respiratory tissues. Of greater concern are manmade pollutants, especially from the burning of fossil fuels in power plants and automobiles. Polluted air makes breathing more difficult and can have devastating effects on living creatures. Similarly, cutting large tracts of forests affects the amount of oxygen present in the atmosphere because of the loss of photosynthesizing trees. These are just a few of the ways that human activities affect air quality.

Life on Earth is nourished by the invisible blanket of air that surrounds it. Although we cannot see the air, we can see swaying branches and birds on the wing. Although we cannot smell the air, we can enjoy the fragrances that it carries. Although we cannot touch the air, we can feel the wind in our faces. How fortunate to be able to take a deep breath of air and relish this unique and vital treasure.

Suggested References:

Bailey, Jill, ed. *The Way Nature Works*. New York: Macmillan, 1992.

Barry, Roger G., and Richard J. Chorley. *Atmosphere, Weather and Climate*. 7th edition. New York: Routledge, 1998.

Beiser, Arthur. *The Earth*. New York: Time-Life, 1963.

Skinner, Brian J., and Stephen C. Porter. *The Dynamic Earth*. 3rd edition. New York: John Wiley & Sons, 1995.

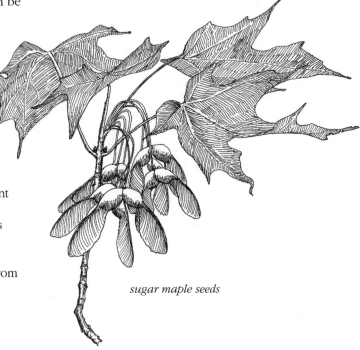

sugar maple seeds

Breath of Life

FOCUS: We cannot see, smell, feel, or taste the air around us, and yet it is vitally important to us and to all other life on Earth.

OPENING QUESTION: *Why do animals and plants need air?*

PUPPET SHOW

Objective: To learn that air is important to both plants and animals.

Present, or have the children present, the puppet show. Afterward, review the various ways that different animals obtain oxygen. Why do animals need air? How is air important to plants?

Materials:
- puppets
- script
- props

AIR CHARADES

Objective: To discover some other ways that air is important to living things besides providing oxygen to breathe.

Divide the children into pairs or small groups. Give each group a card describing one way in which air is used by living things. Have the groups rehearse, and then have each perform a short pantomime showing how air is important to their plant or animal. The other children try to guess the animal or plant and the use of air being demonstrated.

Materials:
- (optional) props such as furry clothing, combs and rulers, paper cutouts of feathers, paper helicopters, cloth capes or towels, balloons
- Uses of Air cards

USES OF AIR

~ Furry animals like bears keep warm on a cold and snowy day because their fur traps warm air close to their bodies.

~ Crickets rub the combs on their wings together to create sound waves that travel through the air and tell other crickets where to find them.

~ Birds use wings to fly about on the air. Soaring birds use rising warm air currents to support them while they glide.

~ Maple seeds spin around and are carried by the wind to land away from the shade of the parent tree, in a place where they can germinate and grow.

~ Flying squirrels use flaps of skin to glide through the air so that they can "fly" from tree to tree in search of food or to escape predators.

~ Puffer fish fill their bodies with air to make themselves look big and scary so they can frighten off bigger fish that might want to eat them.

AIR POCKET

Objective: To observe that air takes up space.

Working in small groups, have one child push a crumpled tissue into the bottom of a glass. Test to be sure the tissue stays in place when the glass is upside down. Now have a child hold the glass upside down and push it straight down into an aquarium filled with water until the glass is submerged. Does the water go up into the glass? Now pull the glass straight up and observe the tissue. Why is the tissue still dry? What kept the water from filling the glass? Fill a plastic zip-closing bag with air and then close it. Have the children feel the bag and try to squeeze it flat. What's in the bag?

Materials:
- clear drinking glass
- aquarium
- tissue
- plastic zip-closing bag

NEWSBREAK

Objective: To observe that air exerts pressure in all directions.

In small groups, lay several sheets of newspaper on a smooth, flat table or desk. Arrange them so that the longest edge is along the edge of the table. Place a wooden paint stick under the paper so that one third of it hangs over the edge of the table. Carefully smooth and flatten the paper against the table. Have a child hit the paint stick with a single, quick blow. What keeps the paper from flying up into the air? (air pressing down on it with great pressure) Why is it easier to lift the paper if you press on the paint stick slowly? (Air exerts pressure in all directions, so air that flows in underneath the paper helps to lift it.)

Materials:
- several sheets of full-size newspaper
- wooden paint sticks

POP THE TOP

Objective: To see that when air is compressed it pushes outward with great strength.

Using a bicycle pump with a ball needle attached, insert the needle into an empty plastic milk jug, one or two inches from the base of the jug. Place the cap securely on the jug. While one child holds the jug so that the cap is pointed away from the rest of the children, have a second child pump air into the jug until the cap is forced off. Ask the children to explain what is causing the cap to fly off.

Materials:
- clean, empty half-gallon plastic milk jug with plastic flip-off cap
- bicycle pump with ball needle

SCRAMBLED EGGS

Objective: To see that a flame lowers the air pressure inside a bottle by using up oxygen.

Note: *This activity involves fire and should be done as a demonstration by an adult.*

Light a candle and place it under a glass milk bottle. Light a second candle and place it outside the bottle. Ask the children to predict which will burn longer. Why does the candle in the bottle stop burning after a while? Now, with the bottle upright, light a 2" square piece of paper and drop it in the bottle. Quickly place a peeled hardboiled egg in the bottle's mouth. Why does the egg suddenly slip into the bottle? (burning paper uses up oxygen, lowering the air pressure in the bottle, so now air pressure outside the bottle is greater than that inside) How can you get the egg back out? (Fit a flattened straw into the neck of the bottle, hold the bottle upside down, and blow into the straw. Get ready to catch the egg.)

Materials:
- candles
- matches
- glass milk bottle
- peeled hardboiled egg
- 2" square of paper
- drinking straw

OUTDOOR AIR SEARCH

Objective: To find examples of air in action in the world around us.

Outside, have the children work in small groups to find the items listed below and record their examples.

AIR SEARCH CARDS
Find:
~ Signs of moving air
~ Something air has changed
~ Something that you can do that lets you feel the air
~ Air that is dirty
~ Something using air under pressure
~ Something using air to fly
~ Something floating in the air
~ Something besides a person that breathes air
~ Something that was moved by air
~ An odor carried by the air

Afterward, gather children in a group and ask each team to report its favorite example of air in action.

Materials:
- Air Search cards
- clipboards
- pencils

PARACHUTE PLUNGE

Objective: To explore how air trapped in a parachute slows the speed of a falling object.

In pairs, have the children make a simple parachute by tying the handles of a plastic grocery bag to a small pine cone. Launch the parachutes from a window or from an elevated platform outside and watch how they fill with air and float to the ground. Now, drop cones with and without a parachute at the same time. Why do the parachutes slow the cones down? (air trapped under the parachute is compressed so it pushes upward, counteracting the pull of gravity on the cone) Have the children experiment with other materials for making parachutes. Do some materials and designs work better than others?

Materials:
- plastic grocery bags
- small pine or spruce cones
- pipe cleaners or yarn (to attach bags to cones)
- (optional) assorted construction materials, such as paper plates, paper cups, cellophane, scrap fabric, gauze

SHARING CIRCLE

Objective: To share new awareness of air and its importance to life on Earth.

Gather the children in a circle and have each child complete this sentence: "I can't see the air, but I know it's there because _____."

ACTIVITY STATIONS:
1) Air Pocket
2) Newsbreak
3) Pop the Top

EXTENSIONS

Breath of Life: Fill a dish basin with a few inches of water. Then fill a gallon milk jug to the top with water. Cover the top of the jug with your hand, quickly turn the jug over, and place it in the dish basin so that the mouth of the jug is under water. It should remain filled with water. Keeping the mouth of the jug under the surface, fit one end of a flexible tube inside the jug and have the other end come out above the water. Push a straw into the exposed end of the tubing. Now, one at a time, have the children take a deep breath and blow into the straw until they don't have any breath left. How much air is in our lungs?

Exercise Energy: Have the students count their breaths for one minute while sitting quietly. Then have them run in place for one minute and count breaths again. Why do you breathe more when you exercise? Use a diagram of lungs and the circulatory system to trace the path a breath takes through our bodies.

Air Pollution: Discuss how air pollution and smoking can damage the lungs. Invite a health care provider to your class to discuss the importance of healthy lungs and clean air. (Demonstration lung sets are available from the American Lung Association.)

Rising Waters: Light a candle and place it in a pan containing several inches of colored water. Lower a glass milk bottle carefully over the candle until the bottle's mouth is under water. Why does the water rise up into the bottle?

Breath of Life

Characters: Henry, Treasure Chest, Herbie Hare, Sammy Salmon, Wendy Worm, Greta Grasshopper, Goldie Goldenrod

Henry

Well, I'm all done! Cleaning out my room was a big job. Hey, wait a minute. What's this old box doing here? Better check what's inside before I throw it away. *[looks inside]* Empty! Well, off you go to the recycling bin.

Treasure Chest

Empty box! What are you talking about? I am not a box. I am a treasure chest! And, I am **not** empty. I'm full to the brim with a most valuable treasure – something very important for you and every living thing!

Henry

Huh? I don't see anything inside you! Is it something very small?

Chest

No, but it's very invisible!

Henry

Well, I don't think there's anything inside you besides air.

Chest

That's it exactly! I knew you'd get it!

Henry

You mean, that's it? Air! I thought you meant a real treasure. Who needs a lot of air?

Chest

Why, all animals need air and the oxygen in it. Your body uses oxygen to give you energy to run and play and…clean your room.

Henry

Some treasure! Well, I may need oxygen to breathe, but air isn't much of a treasure for all the animals that don't need oxygen.

Chest

And who might these animals be, may I ask?

Henry

Well, you know, fish, worms, insects…

Chest

Perhaps you had better ask the experts themselves whether they need air. Since I am a magic treasure chest, I will now produce one for you. Abracadabra! *[Hare appears from behind box]*

Henry

Are you an expert on air?

Herbie Hare

Oh, sorry! I thought he said "expert hare"! Oh well. Bye. *[exits]*

Chest

Hmm, I guess my magic was off by a hair. I will try this again. Abracadabra.

Henry

Experts! Ha! I wonder what kind of expert he's going to come up with this time! *[Salmon enters]*

Sammy Salmon

Me for instance – an expert on fish! And let me tell you, fish most certainly **do** need oxygen. Why, if there's not enough oxygen in the water, we fish will die.

Henry

Gosh! I didn't know there was oxygen in water!

Salmon

There sure is, just like there's oxygen in the air.

Henry

So you breathe water? If I did that, I'd drown.

Salmon

That's because you have lungs. I have gills, and they take the oxygen right out of the water. Speaking of which, I'd better go. I'm beginning to feel like a fish out of water. *[exits]*

Henry

Well, maybe people and fish need air, but I don't think earthworms need air. *[Worm enters]* Um, are you another expert on air?

Wendy Worm

Oh no, I don't have any hair. I'm nice and slippery. Hair! Oooo, gives me the creeps!

Henry

I didn't say hair. I said air. The stuff we animals breathe in our lungs, or gills if you have them.

Worm

Well, I'm an animal too, but I don't have lungs **or** gills. I breathe air through my skin.

Henry

Through your skin? Really? Now that's a different way to breathe.

Worm

Well, if you're an earthworm, it's the **only** way to breathe!...But it doesn't work if I get dried out, so I'd better worm my way back into the damp soil. So long. *[exits]*

Henry

So long. Well, I guess I was wrong about fish and earthworms not needing air. But I'm sure insects don't need it. *[Grasshopper appears]*

Greta Grasshopper

Wrong again, Henry! All animals need oxygen, and insects are animals too.

Henry

Right...So, are you going to tell me you insects have lungs? Or maybe you breathe through your skin, too.

Grasshopper

No lungs or skin. I get air through holes in my sides. From there it goes all through my body in little tubes.

Henry

Why the little tubes?

Grasshopper

The tubes carry the oxygen to every part of my body. Why, air travels through your body in little tubes, too – your arteries and veins.

Henry

Wait a minute. You can't fool me. My arteries and veins are full of blood.

Grasshopper

Sure! Because it's your blood that **carries** the oxygen you need. Oxygen's what keeps us hopping. Speaking of which, here I go! *[hops off stage]*

Henry

Well, I guess all animals need air, but I'm sure plants don't need it, and they're living things, too. *[Goldenrod appears]* Who are you?

Goldie Goldenrod

Goldie Goldenrod, expert in botany, at your service.

Henry

Oh no, not another expert!

Goldenrod

Oh yes, and this is my golden opportunity to tell you that we plants use air and water to make our food. Why, we even **make** the oxygen that you animals need to breathe.

Henry

Wow! So plants need air to live, too?

Goldenrod

Quite so. And without air, we'd be...why we simply wouldn't be! The very thought makes me wither! *[falls over and exits]*

Chest

And now, my friend, what did you learn about air?

Henry

I guess I learned that air is pretty important to living things.

Chest

Very good! And what do you think of my precious treasure now? Am I just an **empty box?!**

Henry

No, you're not an empty box at all. You are a real treasure chest, full of precious air. And I'm going to put you somewhere safe. But, you know what you should do first?

Chest

What should I do?

Henry

You should take a bow. You're a great magician!

Chest

Oh, it was nothing. Nothing at all. But if you insist...*[bows forward and back]*

Henry

Bravo, bravo! *[wait for applause]* Come on now, off we go. *[lifts box and exits]*

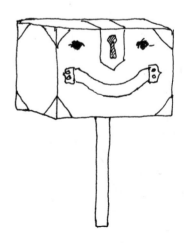

Sound Symphony

GOOD VIBRATIONS

In the natural world, we are surrounded by a wonderful variety of sounds, from the symphony of birdsong on an early spring morning to the eerie yipping of a lone coyote on a moonlit night. Although we can appreciate the beauty and variety of sounds that we hear, we should also be aware of their importance. The ability to produce sounds, to hear sounds, and to distinguish sounds is crucially important to survival in the animal world.

Sound is created when an object vibrates and causes the air around it to vibrate as well. As the object pushes forward, it compresses the air next to it. Air particles push against neighboring particles. As the object vibrates back in the reverse direction, it allows the air to expand, and air particles move back to fill the emptied space. The vibrations of compressed and expanded air are called sound waves; these travel outward in all directions from the vibrating object.

Sound waves can also travel through denser substances like liquids and solids. In fact, they may travel faster in these substances because the molecules are closer together. Sound travels through air at a speed of 1,100 feet per second, through water at 4,700 feet per second, and through steel at 16,400 feet per second.

Two rocks banged together underwater sound louder than in the air. In the early days of railroads, people discovered that by putting an ear to the track, they could hear an approaching train long before they could hear the airborne sound of its approach.

When a sound wave encounters an obstacle in its path, the sound wave may go around the obstacle, it may travel through the object, it may be absorbed, or it may be reflected. We can hear sounds around corners, over hills, and through holes. We can hear sounds through some walls because the sound waves cause the walls to vibrate, and thus the waves are transmitted through to the other side. We can block sounds from passing through walls by using certain materials like cork or acoustical tiles, which absorb sound waves in tiny holes and prevent them from being transmitted. Just as carpeting and draperies are good at absorbing sounds in a home, lush foliage and dense leaf litter can dampen the movement of sound waves outdoors.

When sound waves in the air come up against large dense surfaces that do not vibrate easily, such as high brick walls or rocky mountainsides, they are reflected rather than passing through them, and we hear these returning sound waves as echoes. When we stand in front of a wall and clap our hands, we first hear the sound by the shortest path – directly through the air from our hands to our ears. We may then hear a second

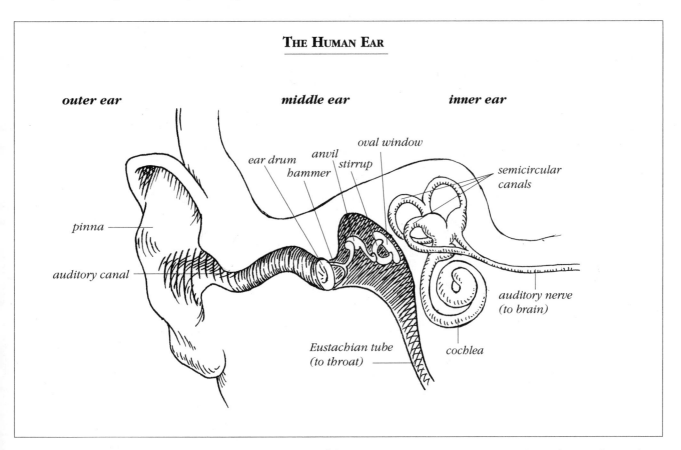

THE HUMAN EAR

outer ear

middle ear

inner ear

oval window

ear drum
anvil
hammer
stirrup

semicircular
canals

pinna

auditory canal

auditory nerve
(to brain)

Eustachian tube
(to throat)

cochlea

clap when the sound waves that bounce off the wall reach our ears. If we stand closer than 30 feet from the wall, then we cannot distinguish the two sounds because both waves reach our ears at almost the same time. As we move farther and farther away from the wall, we can hear a greater time lapse between the first clap and the echo. In valleys and canyons, echoes may repeat as the sound waves bounce from wall to wall.

Sound is an effective way to communicate, especially in the natural world. Sound can travel between animals in the dark, under water, under the snow or leaf litter, or even in a dense forest, where visual signals might be futile. Sound can be used for communicating over long distances, through barriers or screens, or around corners. Animals use the sounds they make to attract mates or to warn others of their species about danger. Sound also can tell animals much about their surroundings. Many predators locate their **prey** by sound, and prey animals may be warned of the presence of enemies by the telltale rustling of leaves or breaking of twigs. Humans depend on this ability as well. Automobile sounds warn us of approaching traffic when we cross a road; a rising wind may signal bad weather on its way; the crying of children tells us they need help. The ability to distinguish among the many sounds in our environment helps animals, including humans, to survive.

Human ears are designed to receive sound waves through the air and then translate them into nerve impulses that are sent to the brain. The outer ear funnels sound waves down the auditory canal to the eardrum, a tight piece of skin stretched across the canal. The airborne vibrations cause the eardrum to vibrate, which makes three tiny bones in the middle ear vibrate as well. The sound waves traveling along these bones are transmitted to the liquid within the inner ear. From there they stimulate sensory nerves and are carried to the brain, where they are interpreted.

Animal ears come in many forms; not all are membranes, and not all are located on the head, but all are designed to receive and interpret vibrations. Mosquitoes have plumes on their antennae that receive sound waves, and cockroaches have sound-sensitive hairs on projections at the ends of their abdomens. Many grasshoppers have a **tympanum** (like an eardrum) on the front legs, while **nocturnal** moths have them on the sides of their bodies. Rattlesnakes have no external ears on their bodies. Instead, they receive vibrations through their jaws, which they rest on the ground in order to hear. Dolphins, which use **echolocation** to find their prey, also receive sound vibrations through their jaw bones.

Animals not only receive sounds, they also produce them, and they do this in many different ways. Some animals produce sounds using vocal cords. Fruit flies send messages through the air by vibrating their wings, and elephants rumble their stomachs. Whales produce underwater sounds that can be heard by other whales hundreds of miles away. Mole rats and soldier termites send signals through the ground by hammering their heads against the ceiling of their subterranean tunnels.

These sounds – traveling through the different mediums of air, water, and soil – vary not only in volume but also in pitch. The pitch of a sound, how high or low it sounds, depends on the frequency of the vibrations: the faster the vibrations, the higher the pitch. The average human ear can hear sounds with frequencies as low as 20 vibrations per second and as high as 16,000 vibrations per second. Sounds of higher frequencies are considered ultrasonic – beyond the hearing range of the human ear – but they are not beyond the hearing range of many animals. Dogs can hear whistles pitched above the human range of hearing. Bats hunt and navigate by emitting ultrasonic cries that bounce off, or echo from, insects they are hunting or obstacles they are trying to avoid. Some moths can hear these ultrasonic sounds emitted by bats and thus can better avoid their predators.

The honking of geese flying overhead, the chirping of crickets on a summer evening, and the springtime trilling of tree frogs are all sounds that are vitally important to the animals that produce them. Sound serves as an important means of communication and provides valuable information about the world around us. Additionally, listening to the sounds of the natural world gives us much pleasure and enriches our lives.

Suggested References:
Bailey, Jill, ed. *The Way Nature Works*. New York: Macmillan, 1992.
McFarland, David. *Animal Behavior: Psychobiology, Ethology, and Evolution*. California: Benjamin Cummings, 1985.
Physics Today. The World Book Encyclopedia of Science, Vol. II. Chicago: World Book, 1989.
Ridley, M. *Animal Behavior: A Concise Introduction*. California: Blackwell Scientific, 1986.
"Special Issue: The Mystery of Sense," *Discover: The World of Science*, vol. 14, n. 6, June 1993.

a spring peeper singing

Sound Symphony

FOCUS: Sound waves, produced by vibrating objects, play important roles in the lives of animals, from warning of danger, to attracting mates, to finding food.

OPENING QUESTION: *What can we learn from sounds in our environment?*

SOUND PANTOMIME

Objective: To experience the challenge of identifying common sounds using only one's hearing.

Have the children sit with their eyes closed and listen while you make from five to ten common sounds behind a screen. After each sound, ask the children to try to identify what they heard. Now play a tape of some animal sounds and have the children try to identify these as well. What are some reasons why animals might use sounds to communicate?

Materials:
- common objects that produce sounds, such as ball to bounce, keys to jingle, comb to play, paper to tear or crumple, straw and water for bubbling noises, carrot to chew, twigs to snap, Velcro, zipper, stapler, pen to click
- tape of animal sounds
- tape player

WHAT IS SOUND?

Objective: To discover how sound is produced when objects vibrate.

Give each child or pair of children a rubber band or a paint stick. Ask the children to experiment with these objects to produce sounds. What motion did they observe in the objects when sound was produced? Discuss how vibrating objects create sound waves by causing the air to vibrate.

Materials:
- rubber bands
- paint sticks

SOUND WAVES AND EARDRUMS

Objective: To understand how sound waves move through the air and can cause membranes, like eardrums, to vibrate.

Divide the children into small groups and give each group a coffee can over which a piece of thin plastic wrap has been tightly secured with rubber bands. Sprinkle colored decorating sugar on the plastic wrap. Now have a child clap hands above the coffee can. What happens to the sugar? Next, have a child tap a tuning fork to make it ring and then hold it pointing toward, but not touching, the plastic wrap. What makes the sugar jump around? (the sound waves in the air make the plastic membrane vibrate) Now place some sugar on the table and repeat. What happens? How do we hear sounds? (our eardrums are thin membranes, like the plastic, that vibrate in response to sound waves) Show pictures of some animal ears (e.g., human, owl, grasshopper, frog, moth) and discuss some of the different ways that animals hear sounds.

Materials:
- plastic wrap
- coffee cans
- rubber bands
- colored sugar
- tuning forks
- pictures of human and other animal ears

SOUND MATCH

Objective: To investigate the human ear's ability to distinguish among similar sounds.

Set up a collection of film canisters filled with a variety of everyday materials, two of each type. Have pairs of children shake and match the canisters by sound. (Provide several sets so that many children can play at one time.) Ask them to guess what is inside before opening the canisters. How might an animal's ability to tell sounds apart help it to survive?

Materials:
- sets of ten paired film canisters filled with a variety of materials such as beans, rice, pebbles, cornstarch, salt, tiny bells, paper clips, metal screws

HANGER BANGER

Objective: To compare how well sound travels through solid materials versus air.

For each pair of children, provide a "hanger banger" made by tying the middle of a 36" piece of string to the hook of a metal clothes hanger. One child of each pair should wind the ends of the string around his or her index fingers and hold the hanger in midair by this string. The partner then taps the hanger with a metal spoon. What does it sound like? Now have the children holding the hangers touch their fingers to their ears, while the partners tap the hangers again. What was the difference in the sound? Be sure that every child has a chance to be the listener. Have some other supplies on hand for experimentation. (Metal cake racks can be fun for playing a tune!) Why is the sound louder when you hold the string to your ears? (sound travels better through string than through air because the molecules in solids are closer together than in a gas)

two ways to make a hanger banger

Materials:
- metal coat hangers
- string
- scissors
- metal spoons
- (optional) metal cake racks

STOP THE BELL

Objective: To compare the ability of different materials to absorb sound waves.

In a small group, give the children an alarm clock and three different containers, such as a cookie tin, an empty plastic lunch box, and a suitcase stuffed with clothes. Have the children put the alarm clock in each container and listen for its ring. Which container was best at muffling the sound? Which made the noise seem louder?

Materials:
- alarm clock or egg timer
- cookie tin
- plastic lunch box
- suitcase filled with clothing or a pillow, or a box lined with cork or soundproofing material

SOUND SEARCH

Objective: To use our sense of hearing to experience the variety of sounds outside.

Outside, have the children work in small groups to find the items listed below and record their examples. Walk to several different locations and compare the sounds heard at each.

Materials:
- Sound Search cards
- clipboards
- pencils

SOUND SEARCH CARD	Location #1	Location #2
The loudest sound you can hear		
The softest sound you can hear		
A high sound		
A low sound		
A sound from the natural world		
A sound from something manmade		
A sound from far away		
A nearby sound		

Afterward, have each group share its most unusual example.
What were some of the differences in sound at the two locations?

CLAP FOR ECHOES

Objective: To discover that sound waves bounce off walls, and we hear these as echoes.

Have a group of children face the outside wall of a building, standing at least 30 feet away from it. Have one person clap hands to produce a sharp "crack" of sound. Listen for the echoing sound that comes from the direction of the wall, a split second after the hands are clapped. Why do we hear a second clap? (some sound waves reach our ears directly, and others travel to the wall, bounce off it, and then reach our ears) Move closer to the wall and try again. Move farther away and note the time that elapses between clap and echo. Why did the echo reach our ears later when we stood farther away from the wall? (it had farther to travel) Look for other nearby places where echoes might be produced, and try them out. What features of the landscape might produce echoes?

SHARING CIRCLE

Objective: To share favorite sounds or new discoveries about sound.

Sitting in a large circle, have the children complete this sentence: "My favorite sound is _____." Older children might share their favorite new discovery about the nature of sound.

After sharing, pass around a noisy snack like celery, carrots, pretzels, crackers, or corn chips. Have the children try covering their ears while chewing.

Materials:
• noisy snack food

ACTIVITY STATIONS:
1) Sound Match
2) Hanger Banger
3) Stop the Bell

EXTENSIONS

Sound Sayings: Have the children make a list of common sound sayings and their meanings, such as "in one ear and out the other."

Sound Charades: Act out common sounds without making any noise or using any words (e.g., a balloon popping, a bee buzzing).

Why Two Ears?: Have a blindfolded child sit in the center of a circle of classmates. When the leader signals one child in the circle to make noises, the blindfolded child must point to the noisemaker. Now have the blindfolded child cover one ear and try again. Compare how easy or hard it is to locate sounds with one ear or two.

Hose Horns: Have one child hum into a garden hose while another child listens. Then add a funnel to the listener's end of the hose and listen again. Discuss how the funnel makes the sound louder by concentrating the sound waves so they don't spread out as much.

Bat and Moth: Play a game to learn about how bats use echoes to hunt for moths. Two children, one blindfolded and one not, stand within a circle formed by other students. Each time the blindfolded child calls out the word "bat," the other child must answer "moth." The blindfolded child must try to find and tag the prey by listening for its reply.

Water, Water Everywhere

JOURNEY THROUGH THE WATER CYCLE

The water on the earth and in our atmosphere today is the same water that was here billions of years ago. It is thought to have formed from chemicals present in the mass of rocks and gases that condensed to become our planet. The earth's water is not stationary, but rather it is constantly in motion. The sun's heat energy and the force of gravity keep it circulating endlessly from the earth's surface to the atmosphere and back again in a process known as the water cycle.

Water from the atmosphere reaches the earth by precipitation as rain, snow, sleet, or hail, or by condensation as fog, frost, and dew. About three-quarters of this precipitation falls directly into the oceans, and the rest falls on the land. It may evaporate from the surface of the ocean, or it may follow a variety of routes before returning to the atmosphere. It may become part of a lake or a mountaintop snowfield, or be frozen in a glacier for thousands of years. It may seep into the soil providing moisture for plants to grow or bubble up in a spring to be consumed by animals and stored for a time in their bodies. It may percolate deep into the earth to become

ground water and make its way slowly through natural outlets into streams, rivers, or underground channels, eventually finding its way to the sea.

The sun's heat causes water to evaporate and turn into water vapor. This occurs wherever water is exposed to the sun's radiation, such as the surface of oceans, lakes, ponds, or glaciers. Water can also evaporate from the soil, from vegetation, or even from our own skin. Warm air containing water vapor rises up in the atmosphere until it meets cooler air. Clouds form when the cooled water vapor condenses around tiny particles in the air, becoming water droplets, snowflakes, or ice crystals, depending on the air temperature. When these droplets or crystals become heavy enough, the force of gravity pulls them to earth as precipitation to continue through the cycle once again.

Water is absolutely essential to all forms of life, from microscopic one-celled organisms to the largest trees and mammals. Living tissues consist mostly of water. Our bodies are more than two-thirds water, and some fruits may be 95 percent water. The vital materials of life are transported by water within the bodies of animals and plants. It is generally believed that life evolved in water, and vast numbers of plants and animals still live in watery environments.

THE WATER CYCLE

Animals depend on water for survival, ingesting it in the liquids they drink and the food they eat. A loss of only 15 percent of a person's body water, or two to three days without water, can be fatal. Animals return water to the water cycle by excretion, perspiration, decomposition, and just by breathing. When you "see your breath" on a cold day, you're seeing the water vapor that is given off as a byproduct of respiration. It is visible because it condenses into water droplets as soon as it meets the cold outside air.

Plants are as dependent on water as animals are, and plants are an important part of the water cycle. They hold moisture in the ground and gradually return water to the atmosphere. A large healthy tree may draw in hundreds of gallons of water in one day through the tiny hairs at the tips of its roots. This water, which contains dissolved nutrients from the soil, moves throughout the tree, supporting its life processes. Water is a necessary component of **photosynthesis,** which takes place in the tree's leaves. The sugars and starches produced in photosynthesis are transported, dissolved in water, to every part of the tree. Excess water is released from the tree as water vapor, exiting through tiny openings in the leaves in a process called **transpiration**. As much as 90 percent of the water taken in by plants is transpired through the leaves. In this way, plants return large amounts of water to the atmosphere, where it continues to travel through the water cycle.

One unusual property of water important to the water cycle is its tendency to be drawn into very narrow spaces in some materials, a process known as **capillary action**. It results from the attraction between molecules of water and other materials, and among the water molecules themselves. Water is drawn into the fibers of a paper towel, into the spaces in the soil, and into the root hairs of plants by this process. Capillary action helps water and sap circulate within plants, and it helps move blood inside our bodies. Capillary action in the soil draws water upward from the water table, providing moisture for living things. Similarly, it helps water seep downward through the spaces in the soil, refilling the ground water reservoirs that are so important for our supply of fresh water.

Another important property of water is that ice, its solid form, is lighter than liquid water. In most substances, solids are heavier and more dense than liquids. And, indeed, water does become heavier as it is cooled, but only until it reaches a temperature of 39°F. When it is cooled below this temperature, water becomes less dense, and so ice floats. This property is important for life on Earth. If ice were heavier than liquid water, it would sink to the bottom of the ocean where the sun's rays could not reach it to melt it. Eventually, the ice would build up until only a thin layer of water would persist above the surface of the ice. The water cycle would stall, and life on Earth would perish.

Although water covers three-quarters of the earth's surface, 97 percent of it is salt water. Fresh water makes up less than three percent of all the water on the earth, and more than two-thirds of this is frozen in glaciers and polar ice caps. The remaining water, less than one percent of the earth's total supply, is fresh water in lakes and streams, in the soil, in ground water, in the atmosphere, and in plants and animals. Most of the water that we use comes from ground water – water that is present in porous rock beneath the soil – and from lakes and streams. Unfortunately, many of our activities result in water pollution. Additionally, the rate at which we use fresh water exceeds the rate at which it is returned to ground water reserves. Fresh water is a limited and valuable resource on our planet.

The earth's water is nearly as old as the planet itself. When enjoying a cool, refreshing glass of water, we are tasting one of Earth's most ancient and precious resources. The water we sip may once have wetted a dinosaur's back, risen in a cloud above an ancient volcano, or carried an Egyptian pharaoh's funeral barge to its pyramid tomb. Through the ages, a drop of water may travel an infinite number of different paths as it circulates within the earth's water cycle.

Suggested References:

Beiser, Arthur. *The Earth*. New York: Time-Life, 1963.

Press, Frank, and Raymond Siever. *Earth*. San Francisco: W. H. Freeman, 1978.

Skinner, Brian J., and Stephen C. Porter. *The Dynamic Earth*. 3rd edition. New York: John Wiley & Sons, 1995.

Tarbuck, Edward J., and Frederick K. Lutgens. *The Earth: An Introduction to Physical Geology*. 3rd edition. Columbus, OH: Merrill, 1990.

Water, Water Everywhere

FOCUS: Earth's water is a finite resource, never increasing or decreasing, yet continually changing form as it circulates from the land or oceans to the air and back again in a never-ending cycle.

OPENING QUESTION: *What is the water cycle?*

PUPPET SHOW

Objective: To visualize a water drop's journey through the water cycle.

Perform, or have the children perform, the puppet show. Review the steps in Wally Water Drop's travels through the water cycle.

Materials:
- puppets
- script
- props

WATER, WATER EVERYWHERE MURAL

Objective: To identify some of the places water is found and to discuss some of the many pathways within the water cycle.

Show the children a rough outline of a landscape. Give each pair of children a picture representing a place in the water cycle where a water drop might be found. Have each pair come forward and attach their picture to an appropriate spot in the landscape. Ask them to describe one way water might have gotten to this place and one place it might go next.

Materials:
- large landscape outline
- pictures of a pond, stream, glacier, soil, clouds, a water well, animals, trees and other plants, rain, snow, ocean
- tape

MOVING WATERS

Objective: To observe how water travels through some substances by capillary action.

Put about a quart of water in a dishpan and add some food coloring. Have each child cut a strip of paper or cloth, about one inch wide by six inches long, from a selection of materials. Next, have each child place one end of his or her strip into the basin so that it hangs about one inch into the water, and place the other end into an empty cup next to the basin. Why does the water move through the strips? (water molecules are attracted to other materials and have a tendency to move into tiny spaces, a process called capillary action) Where might this process occur in the natural world? (water moving through the soil or traveling up the stems and trunks of plants) Recheck the cups after an hour to see which material transported the most water.

Materials:
- dishpan
- food coloring
- water
- cups
- a variety of materials such as construction paper, newsprint, paper towels, pieces of fabric (cotton t-shirt, denim, fleece, cloth diaper)
- scissors

EVAPO-RACE

Objective: To observe how wind and warmth cause water to evaporate.

Work in small groups. Give each team of children three small plates with several smooth stones on each. Use a spray bottle to wet all the stones thoroughly. Have each team place one of its plates of rocks in a spot protected from wind and heat. Place a second plate under a heat lamp. Have teams fan their third plate with paper fans. Which pebbles dry the fastest? Where does the water go? Explain that air picks up moisture and carries it away as invisible water vapor. Why does moving air (wind) speed evaporation? (fresh, dry air that has more room to hold moisture is moving over the wet surface) How does heat speed evaporation? (warm air holds more moisture than cooler air)

Materials:
• plates
• smooth stones
• spray bottle
• water
• heat lamp
• folded paper fans

WATER IN THE AIR?

Objective: To view the condensation of water vapor into water droplets.

Give each pair or small group of children a clean, dry, empty glass jar. Pour a small amount (1/2 cup) of very hot water into each jar. Have the children put a metal can containing a few ice cubes on top of the jar. Wait a few minutes and then have the children look for water droplets on the bottom of the can. Where did the water droplets come from? (steam from the hot water condenses into water droplets when it contacts cooler surfaces) What are clouds made of? (water droplets that have condensed around tiny particles floating in the air)

Materials:
• quart-size glass canning jars
• thermos of very hot water
• small metal cans
• ice cubes

WHERE'S THE WATER SCAVENGER HUNT

Objective: To find evidence of water in the world around us.

Outside, have the children work in small groups to find the items listed below and record their examples. No two items may have the same answer. Challenge the children to find examples the other groups may not have considered.

Materials:
• Water Scavenger Hunt cards
• pencils

WATER SCAVENGER HUNT CARD
Find:

Moving water

Two things that store water

Something that is made of water droplets

Somewhere water goes

Something that water has changed

Water you can hear

Water in a very high place

Three different things that drink water

Water that is dirty

Something that water helps to grow

Afterward, have each group share its most unusual example. Was it hard to find signs of water?

GOING IN CYCLES

Objective: To understand the varied routes that a water droplet can take as it travels through the water cycle.

Materials:
- signs (or labeled buckets) for station markers
- bags
- labeled (or color-coded) game chips
- (optional) travel logs and pencils

Ahead of time, create signs and matching game chips to represent the following six components of the water cycle: oceans, lakes and rivers, clouds, plants and animals, ground water, and glaciers. Place the signs around a large area outside and put a bag of labeled game chips, in the proportions shown below, by each sign. Explain to the children that they will each follow the path of a single water drop through the water cycle. Have the children keep a travel log of their journey. To begin, divide the children among the stations so that there are at least two standing in line at each sign. At the word "go," the children at the head of each line must choose a game chip from the bag to find out their next move. Have them record their next station in their trip log, return the game chip to the bag, and then proceed to the station indicated. Now the next players in line pick game chips and move to their next positions. If the game chip indicates that a player stays at the same location, the child should simply go to the end of the line at that station. Children continue playing until the leader asks them to stop or they have completed ten journeys. Did anyone stay a long time in one place? Where were the most water drops at any one time and why? Point out the wide variety of possible trips any one water droplet could take.

Station:	Numbers and types of game chips at each station:
Oceans	6 oceans, 3 clouds, 1 plants & animals
Lakes & Rivers	4 oceans, 2 clouds, 1 lakes & rivers, 1 plants & animals, 1 ground water
Clouds	3 oceans, 2 clouds, 2 glaciers, 1 lakes & rivers, 1 plants & animals, 1 ground water
Plants & Animals	4 oceans, 2 clouds, 1 glaciers, 1 lakes & rivers, 1 plants & animals, 1 ground water
Ground Water	4 oceans, 2 lakes & rivers, 2 plants & animals, 2 ground water
Glaciers	3 oceans, 2 clouds, 3 glaciers, 1 lakes & rivers, 1 ground water

oceans

clouds

lakes & rivers

plants & animals

glaciers

ground water

BUT NE'ER A DROP TO DRINK

Materials:
- gallon jug of water
- measuring cup
- tablespoon
- teaspoon

Objective: To visualize how much of Earth's water is fresh water, and how much of that is available for our use.

Fill a gallon jug with water. With the children standing in a circle, explain that this gallon represents all of Earth's water. Now pour out 1/2 cup of water into a measuring cup and explain that this represents Earth's fresh water and the water left in the gallon jug represents the salt water in our oceans. Remove five tablespoons of water from the cup, representing the water frozen in glaciers and ice caps. (Add this water to that in the gallon jug.) Pointing to the small quantity of water left in the cup, ask the children to think of all the ways we use fresh water. With each idea they share, remove a teaspoon of water from the cup. Discuss with the children the reasons why we should conserve and protect fresh water.

SHARING CIRCLE

ACTIVITY STATIONS:
1) Moving Waters
2) Evapo-Race
3) Water in the Air?

Objective: To share and listen to each other's views on water.

In a circle, have the children take turns completing the following sentence: "One thing I really like about water is _____."

EXTENSIONS

Percolation Race: Pierce several small holes in the bottom of 3 clear plastic drinking cups. Fill each cup with one of the following: gravel, sand, soil, clay, or peatmoss. Have the children hold the cups over clear glass jars and add 1/2 cup of water to each cup. Watch to see the water percolate down through the material in the cups and begin to drip out the bottom. Through which material did water move the fastest? Which material allowed the least water to come through?

Water Cycle in a Bag: Fill a plastic bag with some water and seal well. Tape it to a bright, sunny window at an angle so that the water is pooled in one corner. Have the children watch what happens over time.

Color Changers: Place celery stalks or white carnations in water containing a few drops of red or blue food coloring. Why do the plants change color after a long while? Now cover the top of each celery stalk with a plastic bag. Why does moisture appear inside the bags?

Sink That Ice!: Have the children place an ice cube in the bottom of an empty jar and then pour in water to try to cover the ice. Why doesn't the ice stay at the bottom? What would happen to the water cycle if ice were heavier than water and would sink to the bottom of the ocean?

Water, Water Everywhere

PUPPET SHOW

Characters: Wally Water Drop, Winnie Water Drop, Marsha Mouse, Willie Worm, Sammy Salmon, Tree

Props: metal sheet (for thunder sounds), water mister

Winnie Water Drop

[bumping into Wally] Oh, excuse me, Wally Water Drop. Did I bump you?

Wally Water Drop

That's OK, Winnie Water Drop. I'm used to it. Other water drops are always bumping into me. They're like that.

Winnie

It's gotten crowded up here in this cloud. I wonder if we're going to fall soon.

Wally

Oh, probably. We usually do. We'll be going around in cycles until our heads spin.

Winnie

Don't you mean going around in **circles?**

Wally

No, I mean cycles – the water cycle. All water drops move around in the water cycle from here in the sky, down to the earth and back, over and over again.

Winnie

Water cycle – sounds like a ride at the fair. I can't wait. I'd love to be rain or maybe get frozen into a snowflake. I think it'd be great fun to be snow!

Wally

Oh, yeah, snowing is fun. Last time that happened to me, I ended up stuck in a glacier for 10,000 years.

Winnie

I guess sometimes a trip through the water cycle must be kind of slow. *[spray water at audience]* Oh goody! We're starting to rain.

Wally

[thunder noise] Oh great, a thunderstorm. Wouldn't you know. Instead of gliding to earth in a nice rainfall, we're going to get tossed about in high winds and dumped out in a downpour.

Winnie

Sounds like fun to me! Here comes the wind, Wally. I'm off for a ride through the water cycle. See you when we get back to the clouds! *[exits]*

Wally

The wind's taking me along, too. With my luck, I'll probably end up in some dirty old gutter! *[exits]*

Marsha Mouse

[enters singing] I'm singing in the rain, da da da da da da. Nothing like a nice shower for freshening up. And now a little blow dry for my fur. *[shakes]* There, all done. *[Wally appears on mouse's back]* Oh, no! I'm all wet again. I thought the rain was over. What are you doing here, Wally?

Wally

I was in that storm, but the other raindrops went rushing off and left me behind. Now I'm stuck in your fur, and I'll never catch up.

Mouse

Where did all the other droplets go?

Wally

Oh, they're probably in puddles or trickling off into streams, or they might be seeping down into the soil – just to get away from me. They're like that.

Mouse

I can give you a jump start if that'll help you catch up. Here goes. *[shakes; Wally flies off]*

Wally

Whoa! *[voice fading as he exits]*

Mouse

So long, Wally! *[exits]*

Wally

[reappears] It sure is dark down here in the soil. I can't see anything, but I can feel myself slipping downward. Ouch! Those roots have sharp elbows. Oof! That must have been a rock. It's just like them to leave a rock in my way.

Willie Worm

Hi there, Wally. You must be the last of the raindrops from the storm.

Wally

I'm always last, and now I'm lost! I just want to keep moving through the water cycle.

Worm

I can lead you, Wally. I know all the tunnels in the soil. In fact, my friends and I made most of them. Just follow me, and I'll get you to an underground spring.

Wally

I hear it's very cold in underground springs. I'll probably freeze.

Worm

This one wells up into a nice warm lake. Here it is, Wally. Time for you to spring into action! *[exits]*

Wally

Very funny – worm jokes. Well, here I go into the spring. With my luck I'll end up in the ocean. *[Wally exits; Salmon enters]*

Sammy Salmon

Why, fan my fins and tickle my tail! This cold spring feels great. *[Wally appears]*

Wally

I knew I'd end up in the ocean. And now I'm going to get swallowed by a shark!

Salmon

A shark! Where's a shark? Hide me! *[hides behind Wally]*

Wally

Aren't you a shark?

Salmon

Me? Oh, you meant me! I guess I am the terror of the lake – if you're an insect. But I'm not a shark – just a lake salmon.

Wally

So I guess this isn't the ocean then.

Salmon

Ocean?! No, it's just a lake. And this is my favorite spot because it's right over a cold, bubbly spring – feels great!

Wally

You might feel great, but I feel seasick! All your fin-flapping and tail-wagging is sweeping me away toward the bank.

Salmon

Just waving goodbye, Wally. Hey, remember to look out for trees! *[Salmon exits]*

Wally

Trees? Why would I want to look out for trees? *[Tree enters]*

Tree

Because we take in water through our roots.

Wally

Oh, great. Now I'm going to get stuck in a tree!

Tree

I'm an important part of the water cycle, Wally. I pull water up from my roots to give my leaves a drink. They let the water out again through tiny holes. You won't be stuck in me for long.

Wally

That's what they all say. Uh, oh! Here I go. *[exits, then reappears above Tree]* Hey, what do you know? I did make it out of the tree! And I'm floating in the air – heading right up to the clouds. *[Tree exits; Winnie enters]*

Winnie

[bumps into Wally] Oh, excuse me, Wally. Did I bump you? It's gotten crowded up here in this cloud. Do you think we're going to rain soon?

Wally

Why do I feel as if I've heard all this before?

Winnie

That's because we're going around in cycles – the water cycle!!

Wally

Yeah, we just keep going around and around. But I'm not one to complain. Let's go. *[spray water at audience; Waterdrops exit]*

Wind and Clouds

READING THE SKY

All winds, from the gentlest breezes to the strongest gales, are created when warm air rises and expands while cooler, denser air flows in to replace it. Winds stir and mix the atmosphere, picking up moisture as they blow over the earth, forming clouds and carrying them away. Wind and clouds bring us weather of all kinds as they swirl around the globe.

Although many factors are involved in understanding and predicting the weather, much can be learned from studying wind and clouds. Sailors, farmers, and others whose lives depended on it recognized the connection between wind, clouds, and weather, and their observations have been passed down through the ages as folklore. Many old sayings, especially those pertaining to wind and clouds, are confirmed by what we know today from scientific study.

Winds are the result of differences in temperature on the earth, differences caused primarily by unequal heating of the earth's surface by the sun. As air is heated, its molecules become more energetic and spread out, so it is less dense and therefore exerts less "pressure" on the earth. The molecules in colder air have less energy, and they tend to compact and become more dense, causing more "pressure." Of course, the atmosphere is free to flow, so air will move from areas of higher pressure toward areas of lower pressure and, in moving, becomes wind. This happens on many different scales, from air moving near a woodstove to air moving around a planet.

The winds that make the weather patterns over the earth are the consequence of a lack of heating at the poles, while the equator receives an excess of heat. The hot air in the tropics becomes lighter and less dense, and rises. The colder, heavier air at the poles sinks. As the air in the Tropics rises, the colder air from the poles flows in to replace it, while the rising, warm air compensates by traveling toward the poles. This simple pattern is complicated by the fact that Earth is spinning. This deflects the air flow and creates bands of moving air called "prevailing winds" that flow from the east or west at different latitudes. The prevailing "westerlies" are winds that dominate the air flow over most of North America, and they cause most of our weather to travel from west to east.

A wind vane on a barn is a familiar sight in North America, for farmers have long recognized the importance of wind direction in predicting the weather. Wind vanes are simple tools that swing on an axis and are balanced so they point into the wind. Winds are named according to the direction from which they blow. Thus, a wind blowing from the north is a north wind and an easterly wind is one that blows from the east.

A few simple guidelines, though not hard and fast rules, can be used to interpret local weather based on wind direction. In general, in the Northern Hemisphere, north winds tend to bring cold weather, and south winds are apt to bring warmer temperatures. Winds from the northwest, west, and southwest generally bring good weather. Winds from the northeast, east, and south often bring stormy weather.

When the wind direction changes, a change in the weather can be expected. Wind that shifts in a counter-clockwise direction usually brings stormy weather. These directions are all reversed in the Southern Hemisphere. It is important to observe the direction of upper winds since surface winds can be affected by features of the landscape such as mountains and valleys.

Observing the movement of twigs and branches, or the flapping and waving of flags, even the flow of smoke from chimneys, can help us know how hard the wind is blowing. Knowing the speed of the wind was especially important for old-time sailors, for it helped them to know how much canvas they could spread and when to prepare for a storm. In 1805, Admiral Beaufort of the British Royal Navy developed a scale to evaluate wind speed based on his observations of the wind's effect on the sea, from creating ripples to large waves, and its effect on his ship. This scale is still being used today, and it has been adapted for use on land as well. It provides a simple visual guide for evaluating wind force.

Clouds, with their billowy shapes and muted colors, are a constant source of enjoyment to the observer, but they are also useful as visual guides to the weather. Scientists classify clouds according to their shape and altitude in the sky. There are two basic shapes of clouds that are easy to distinguish from one another. **Cumulus** clouds look tall and puffy, like giant cotton balls. **Stratus** clouds, the second type, generally cover a large portion of the sky in low, gray, fog-like layers.

Clouds are then grouped by altitude into low, middle, and high clouds. Cumulus and stratus clouds below 10,000 feet are considered low clouds. When they are found at middle altitudes, the phrase "alto" is added, such as altocumulus or altostratus clouds. The highest clouds are referred to as **cirrus** clouds, and they can be flat, as in cirrostratus, or more rounded puffs or ripples known as cirrocumulus. However, some cirrus clouds show little form or organization, often appearing as wispy, transparent brush-strokes or streamers called "mares' tails." These formless clouds are known simply as cirrus. Rain-producing clouds include the phrase "nimbus" in their name. **Nimbostratus** clouds are heavy, gray clouds low in the sky that are producing rain or snow. **Cumulonimbus** clouds, the true thunderheads, are similar to cumulus clouds in appearance but much larger. They can reach towering heights of 40,000 to 75,000 feet above the earth and often produce violent thunderstorms.

The type and shape of clouds are determined by the amount of moisture present in the air and by the winds to which they are exposed. Air carries water as invisible water vapor. When the air is cooled, the water vapor in it condenses into tiny water droplets that are visible to us because they reflect light. Thus, clouds are made of tiny water droplets or ice crystals. Wind affects the shape of clouds by evaporating them in some places or adding moisture to them in others. Cumulus clouds are formed when there are updrafts from rising warm air, and moisture is carried upward to ever-increasing heights where it condenses into water droplets. Towering cumulonimbus form when these updrafts become extreme. Stratus clouds form where there is little vertical air movement, and moisture condenses into clouds in a large, relatively still, horizontal layer of air. Cirrus clouds form high in the atmosphere where it is so cold that water vapor changes directly to ice crystals. These clouds are blown by the wind into long wisps and hair-like strands.

Clouds tell us much about the coming weather. Cumulus clouds are called "fair weather clouds" because they are generally a sign of clear skies and sunny weather. Stratus clouds rarely produce more than a drizzle because the air is still, so water droplets have little chance to collide with each other and combine into raindrops. Nimbostratus clouds produce rain, and cumulonimbus clouds often produce thunderstorms and sometimes hail. Cirrus clouds are the precursors of a warm front, which they precede by hundreds of miles. Thus, precipitation usually follows mare's tails by two to three days. A mackerel sky is a term used by sailors to describe high, rippled, cirrus-type clouds called cirrocumulus. These clouds warn of an approaching warm front and very unstable air, and they are nearly always followed by a storm. This is reflected in the old-time sayings, "A mackerel sky, Not 24 hours dry," and "Mackerel scales, Furl your sails."

For ages, people have tried to read the sky in order to prepare for the weather. Even with sophisticated tools, weather forecasting is a difficult job. However, studying wind and clouds can often provide the basic information that helps us dress for today and plan for tomorrow.

Suggested References:

Ahrens, C. Donald. *Essentials of Meteorology*. St. Paul, MN: West, 1993.

Forrester, Frank H. *1001 Questions Answered about the Weather*. New York: Dover, 1981.

Lee, Albert. *Weather Wisdom*. New York: Doubleday, 1976.

Lehr, Paul E., R. Will Burnett, and Herbert S. Zim. *Weather*. New York: Golden Press, 1987.

Mandell, Muriel. *Simple Weather Experiments with Everyday Materials*. New York: Sterling, 1990.

Wagner, Ronald L., and Bill Adler, Jr. *The Weather Sourcebook*. Old Saybrook, CT: Globe Pequot Press, 1994.

CLOUD TYPE CHART

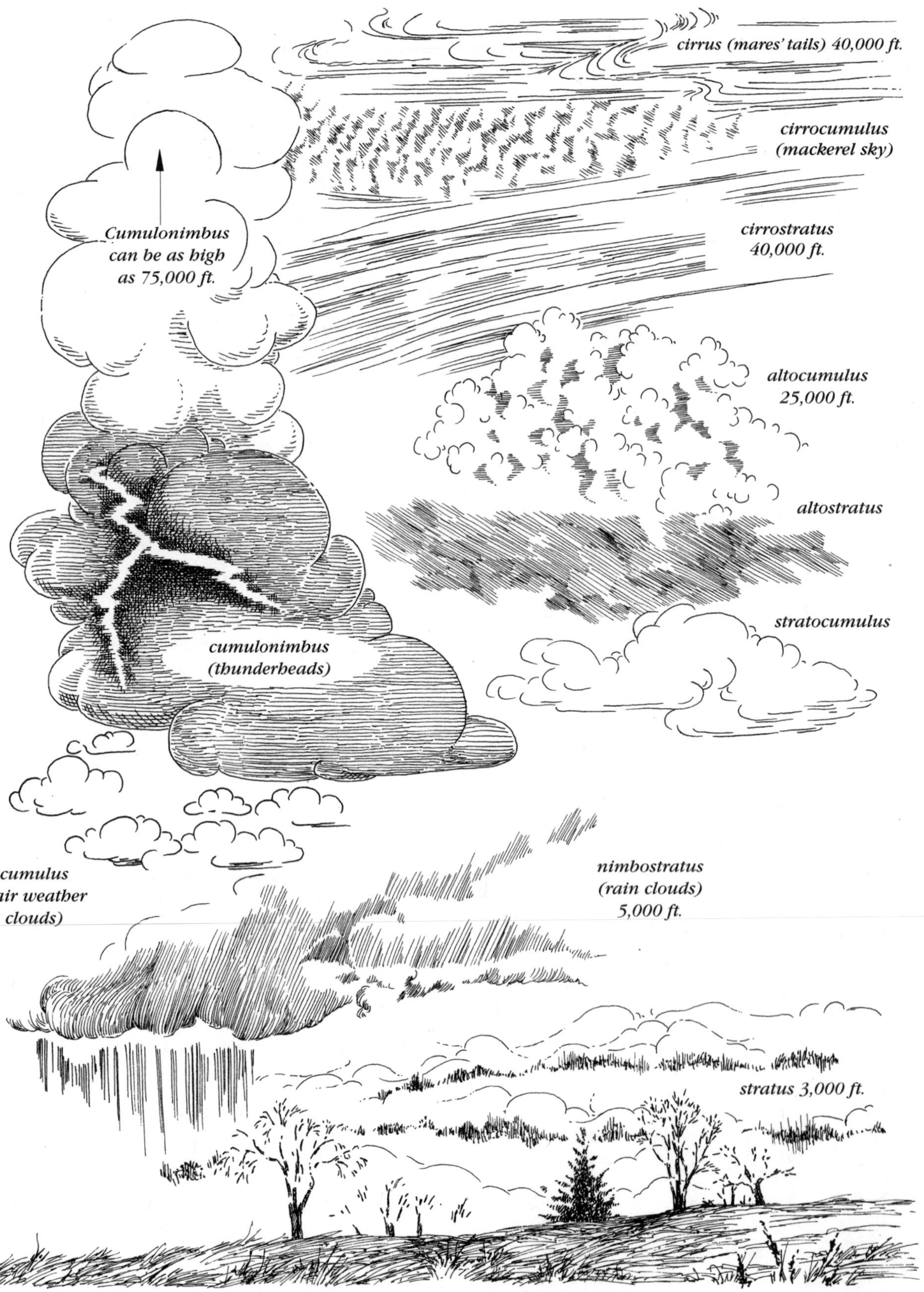

cirrus (mares' tails) 40,000 ft.

cirrocumulus
(mackerel sky)

cirrostratus
40,000 ft.

Cumulonimbus
can be as high
as 75,000 ft.

altocumulus
25,000 ft.

altostratus

stratocumulus

cumulonimbus
(thunderheads)

cumulus
(fair weather
clouds)

nimbostratus
(rain clouds)
5,000 ft.

stratus 3,000 ft.

Wind and Clouds

FOCUS: Winds carrying moisture-laden clouds bring weather, both fair and foul, to all parts of the planet.

OPENING QUESTION: *What can wind and clouds tell us about the weather?*

PUPPET SHOW

Objective: To learn how wind and clouds can give us clues about the weather.

Present, or have the children present, the puppet show. Afterward, discuss how wind and clouds can give us some clues about the weather.

Materials:
- puppets
- script
- props

HOT AND COLD

Objective: To see that warm air rises and expands, while cold air sinks and shrinks.

In small groups, have one child fit a balloon over the neck of an empty bottle. Is there any air in the balloon? Now place the bottle in a bowl of hot water. Why does the balloon stand up after a while? Now place the bottle in a bowl of cold water. Why does the balloon deflate? What happens to the air in our atmosphere when it is heated or cooled? How does this create wind? (as warm air rises, cooler air flows in to replace it)

Materials:
- small balloons
- 20-ounce plastic soda bottles
- bowls
- thermos of very hot water
- ice water

WIND VANES

Objective: To experiment with reading wind vanes and to learn about the connection between wind direction and weather.

To make the base of a simple wind vane, hold a straw horizontally, insert a straight pin down through the midpoint of the straw and vertically into the eraser on the end of a pencil. When holding the pencil upright, the straw should swing freely in a circle.

Give each small group of children one of these simple wind vane bases. Have the children tape paper of various sizes and shapes (triangles, squares) to each end of the straw. In each group, the children should take turns holding the wind vane in front of an electric fan to see how it orients with respect to wind direction. Which end points into the wind? Have the children come up with a rule to interpret wind vanes such as: "The smaller end always points into the wind."

Using a compass rose to illustrate, explain that winds are named by the direction from which they blow (e.g., a wind blowing <u>from</u> the north is called a north wind). What kind of weather usually comes with a west wind? (good) An east wind? (stormy) Ask the children if they would expect a north wind to bring colder or warmer weather.

Materials:
- drinking straws
- pencils with erasers
- straight pins
- construction paper
- tape
- electric fan
- compass rose

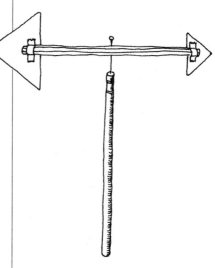

CLOUDSCAPES

Objective: To observe different types of clouds and understand how they are affected by wind.

In small groups, have the children sort a collection of sky pictures by grouping clouds with similarities, and then discuss the characteristics they used in sorting. Use a cloud chart to review the cloud types mentioned in the puppet show and the weather that is associated with each. Find a picture that shows a cloud made by vertical air currents. (cumulus or cumulonimbus) Find a kind of cloud that forms when there is little vertical air movement. (stratus) Find a picture that shows a cloud blown into long wisps by the wind. (cirrus) Do any of the clouds in the pictures look like combinations of different types?

Materials:
- pictures or photos of clouds
- cloud chart

WEATHER WATCHERS

Objective: To evaluate wind speed, wind direction, and cloud type outdoors and use this information to make a prediction about the weather.

Ahead of time, set up stations outside, marking each with a numbered stake. At station 1, place the wind vane and a jar of bubble mix. At station 2, provide a copy of the Wind Force Scale and a flagpole or small flag on a dowel. At station 3, provide a simple cloud identification chart and a plastic tarpaulin if the ground is wet. Take the children outdoors, and divide them into groups with a leader. Give each group a clipboard with a Weather Watchers card, a pencil, and a compass. Groups should try to answer the following questions:

Materials:
- Weather Watchers cards
- clipboards
- pencils
- wind vane
- bubble mix with wand
- compasses
- Wind Force Scale
- flag
- plastic tarpaulin (optional)
- cloud chart

WEATHER WATCHERS CARD

Wind Direction (station 1):
According to the wind vane, what is the main direction of the wind?
Can you see another wind vane or a flag? Does it show the same wind direction?
Blow some bubbles and note in which direction they float.
What features of the landscape could affect wind currents in this area?
What place might be protected from the wind today?

Wind Speed (station 2)
Observe the movement of branches, leaves, trees, grasses, smoke, and flags, and compare to the Wind Force Scale.
What is the speed of the wind according to the Wind Force Scale?
What signs of wind in action help to determine this?

Clouds (station 3):
Lie on your back, and look up at the clouds.
Describe the clouds.
What kinds of clouds can be seen?
Are the clouds moving? Which way? (Face the direction from which the clouds are coming, and use a compass to determine this direction.)
Is this the same as the wind direction found at station 1?
What type of weather is associated with these clouds?

Final Question: What can you forecast about tomorrow's weather from your observations about wind and clouds?

Afterward, have the groups reassemble to compare their findings about wind and clouds and discuss their predictions about the weather.

SHARING CIRCLE

Objective: To reflect on and share new information about wind and clouds.

Sitting in a large circle, ask each child to complete the following sentence: "One new thing I learned about wind or clouds today is _____."

ACTIVITY STATIONS:
1) Hot and Cold
2) Wind Vanes
3) Cloudscapes

EXTENSIONS

Heat Sink: Have the children fill a small jar with sand and another small jar with water. Place a thermometer in each jar and have the children record the starting temperatures. Now place both jars in a sunny window and compare the temperatures after 15 minutes. Which jar was hotter? Ask children to imagine a seashore on a sunny day. Would land and sea heat up equally fast? How would this affect air temperature? Discuss how uneven heating of the air causes wind.

Kite-flying: Ahead of time, make predictions about how well kites will fly and in what direction they will face based on observations of wind speed and direction. Then have the children fly some kites to test their predictions.

Cloud Concentration: Play a game of concentration with cloud types. Younger children can match cloud pictures. Older children can try matching the cloud picture to its name.

Bubbles and Breezes: Blow soap bubbles outside and follow them, noting the direction in which they move. Compare to a nearby wind vane. How are the bubbles affected by walls and buildings?

Classic Wind Vanes: Give the children stencils of several traditional wind vane designs and have them guess which direction they will point. These can then be cut out, mounted, and tested in front of a fan.

WIND FORCE SCALE

0  **Calm**
less than 1 mph

Smoke rises straight up. Nothing in motion.

1  **Light Air**
1-3 mph

Smoke drifts. Weather vanes still. Trees barely move.

2  **Slight Breeze**
4-7 mph

Leaves rustle. Weather vanes turn. Wind felt on face. Flags stir.

3  **Gentle Breeze**
8-12 mph

Leaves and small twigs in motion. Dust, loose paper raised from ground. Flags wave out 1/3 from pole.

4  **Moderate Breeze**
13-18 mph

Small branches move. Dust, dry leaves, loose paper driven along. Flags flap.

5 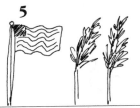 **Fresh Breeze**
19-24 mph

Small trees sway. Large branches in motion. Dust clouds raised. Flags ripple.

6  **Strong Breeze**
25-31 mph

Large branches in constant motion. Wind begins to whistle. Umbrellas hard to hold. Flags beat.

7  **Moderate Gale**
32-38 mph

Whole trees in motion. Walking against wind difficult. Flags extended.

8  **Fresh Gale**
39-46 mph

Twigs break off trees. Walking is very hard.

9  **Strong Gale**
47-54 mph

Twigs break off trees. Walking is very hard.

10  **Whole Gale**
55-63 mph

Trees uprooted. Much damage.

11  **Storm**
64-72 mph

Widespread damage.

12  **Hurricane**
Anything over 73 mph is a hurricane.

Adapted from the Beaufort Scale and Massachusetts Audubon Society.

Wind and Clouds

Characters: Henry, Mother, Casey Cumulus, Stella Stratus, Celia Cirrus, Wind
Props: flag, thunderhead cloud

Henry

Hey, Mom, have you seen Grandpa? I need to talk to him about my field trip!

Mother

He's not home, Henry. And you already told him about your field trip tomorrow.

Henry

Yes, but I needed to ask him about the weather. What if we get a tornado or a hurricane or a blizzard? The whole trip will be called off!

Mother

Now calm down, Henry. We can listen to the weather report tonight.

Henry

Tonight?! I can't wait that long!

Mother

Well, maybe you can make a prediction yourself. Think of some of Grandpa's weather sayings, like "Keep your eye on the sky." There's a lot of truth in some of that folklore.

Henry

OK. I remember a lot of Grandpa's sayings. I can look at the sky on the way to school. Bye, Mom.

Mother

Bye, Henry. Have a good day. *[exits]*

Henry

[walking along] OK, I've got my eye on the sky, but what am I looking for? All I see are a bunch of clouds. *[Cumulus enters]*

Casey Cumulus

Wait a minute! I'm not just a bunch of clouds. I'm Casey Cumulus, and I can tell you a thing or two about the weather.

Henry

Really? That's great! You see, I have this field trip tomorrow, and I need to know what the weather will be. So, what can clouds tell me?

Cumulus

Well, first of all, you have to learn who's who in the sky. Fluffy clouds like me are called cumulus clouds.

Henry

Oh, you're easy to recognize. Grandpa calls you fair weather clouds.

Cumulus

That's right! I'm usually around on sunny days.

Henry

You change shapes a lot, too. Sometimes you look like dragons or dinosaurs. Right now, you look like a big cotton ball, and I'd like to jump right into you!

Cumulus

You would? Well, come along then. I'll take you for a ride to meet some other kinds of clouds.

Henry

OK! Here goes. *[jumps behind Casey so only his head shows above the cloud]* Hey, this cloud isn't fluffy at all. It feels like fog! I guess clouds are just a bunch of water drops. *[Stratus appears]* Oh look, there's another cloud.

Stella Stratus

Hello, Henry. I'm Stella Stratus Cloud.

Henry

Stratus? Why are you called that?

Stratus

Stratus means layer. We stratus clouds form a thick layer all over the sky.

Henry

Oh, I've seen you before. You make the whole sky look as if it's covered with a gray blanket. What kind of weather do you bring?

Stratus

Oh, anything from just overcast to drizzle, rain, or snow. We're not fair weather clouds like some clouds I know! Bye, Henry. *[exits]*

Henry

So long, Stella Stratus. *[tilt upward]* Looks as if we're going up higher – and here's another kind of cloud.

Celia Cirrus

Hello, I'm Celia Cirrus Cloud.

Henry

A serious cloud?

Cirrus

[giggling] Not serious. I'm a cirrus. That's Latin for curl.

Henry

You do look like strands of hair. Are you made of water drops like other clouds?

Cirrus

No, it's so cold up here, we're made of ice crystals. The wind blows us into these long wisps. Some people call us mares' tails.

Henry

That reminds me of one of my Grandpa's sayings: "Mackerel skies and mares' tails make tall ships carry low sails." I can see you're a mare's tail, but what's a mackerel sky?

Cirrus

Well, that's another kind of cirrus cloud that looks like the stripes on a mackerel fish.

Henry

Oh, I get it. So, when I see the sky looking fishy, I can expect bad weather?

Cirrus

Yes, in a day or two – **cirrus** trouble! Bye! *[exits giggling]*

Henry

Well, I'll be in trouble if I don't get to school soon. You'd better let me off, Casey. Thanks for the ride. *[jumps out]*

Cumulus

My pleasure.

Henry

I guess I won't have to worry as long as I only see clouds like you around.

Cumulus

Unless you see a cloud like me growing taller and darker *[Thunderhead appears in front of Cumulus]*…like this!

Henry

A thunderhead! Help! *[Thunderhead exits]*

Cumulus

Ho, ho, ho, just kidding. Can't resist changing shapes, you know. Those thunderheads can bring pretty bad storms, but don't worry, I'm not traveling with any of them today.

Henry

Phew, I'm glad about that!

Cumulus

But I do need to travel on; the wind's carrying me off. Goodbye, Henry. *[exits]*

Henry

Bye, Casey, thanks for the ride! That's funny; I don't see any wind…whoa! *[shakes as though blown by wind; Wind enters]* But I can feel it! Put me down!

Wind

Oh, sorry. Sometimes I get a bit carried away.

Henry

I was the one getting carried away! I just wanted to know about the weather.

Wind

Well, I **bring** you the weather. If I'm blowing from the north, I'll bring you some cold weather. But if I'm blowing from the south, I might warm things up a bit.

Henry

That's pretty easy to remember.

Wind

But one thing I never do is stay around for very long! Goodbye, Henry! *[exits]*

Henry

Bye, Wind…Hey, wait! Come back! I forgot to ask about tomorrow's weather! Oh, well. I just remembered another one of Grandpa's sayings: "When the wind is in the west, then the weather's at its best; when the wind is in the east, neither fit for man nor beast!" I'll just check which way the school flag is blowing. *[wave flag]* Yippee! The wind is in the west, and the clouds are like fluffy cotton balls! I'm going to go tell my class that we're definitely going on a field trip tomorrow!

Sun Power

ENERGY FOR LIFE

Warming and illuminating the earth from 93 million miles away, the sun provides the energy for all living things on our planet. This glowing star produces vast amounts of energy, most of which is radiated out into space. However, a tiny fraction – less than one-billionth of the total – reaches Earth. It is this energy that heats and lights the planet, creates our wind and weather, and powers the water cycle, making it possible for life to exist.

The sun's energy reaches us as heat, light, and other forms of radiation. The heat energy, or **infrared** radiation, received from the sun maintains a temperature range on the earth in which life can survive. **Reptiles, amphibians,** and many **invertebrates** depend on the sun's heat to warm their bodies because they produce little heat of their own. Even mammals and birds, which produce heat energy from the food they eat, can survive only within a relatively narrow range of temperatures.

Energy from the sun is also the moving force behind the water cycle. It heats the earth's surface, causing water to evaporate and warm air to rise and circulate. This produces the wind and weather that eventually return the water to the earth as precipitation. The water cycle continually renews the supply of fresh water available to, and essential for, terrestrial life.

Equally essential to life is food, and sunlight is the basis of the food chain. Plants convert the sun's energy into food in the process of **photosynthesis**. All green plants have **chlorophyll**, which enables them to absorb sunlight and use the energy to build carbohydrates for growth and other life processes. In turn, animals eat plants (or other animals that have eaten plants) and obtain the energy they need to live and grow.

In addition to food and warmth, the sun's radiation provides the visible light that we need in order to see. For many animals, sight is an important sensory link to the environment and, thus, an essential adaptation for survival. Animals use vision for finding their way, for recognizing mates, for spotting danger, and for finding food. Without light, animals would be unable to see, and the world would be colorless.

Energy from the sun is also important as a source of fuel for human activities. The fossil fuels we use today were made from ancient plants and animals that were converted over the ages into coal, oil, and natural gas. We harness power from winds that are generated by the sun's unequal heating of the earth. Solar power comes from sunlight collected by solar panels and converted to electrical or thermal energy. Even hydro-electric power is fueled by the sun's energy since the water of rushing streams is a part of the water cycle.

The sun's rays do not reach the earth equally at all times. Because the earth rotates on its **axis** (an imaginary line through the poles), the sun's rays reach only the part of the earth

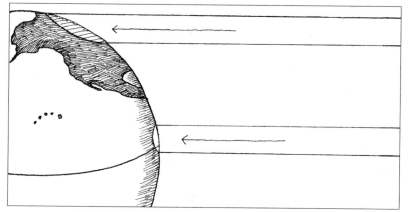

Because of the earth's curved shape, the sun's rays are more concentrated at the equator, less so at the poles.

that is facing the sun at any given time. This causes us to experience night and day. Earth takes 24 hours to complete one full **rotation**. This makes the sun appear to rise in the east, move across the sky, and set in the west, even though it is actually the earth and not the sun that is moving. While spinning on its axis, the earth is also traveling around the sun. It takes one year for Earth to complete one full **revolution** around the sun.

Because the earth is a sphere, some of the sun's rays shine directly on the earth while others reach the earth at an angle. Direct rays deliver their energy to a smaller area and are therefore more concentrated, creating more heat than slanted rays that reach a larger area of the earth's surface. Because the sun shines directly on the earth between the Tropics but reaches the poles at an angle, the climate is much hotter at the equator than at the poles.

If the earth's axis were perpendicular to the sun's rays, day and night would always be of equal length, and there would be no seasons. However, the earth's axis is actually tilted about 23° with respect to the sun, which results in variations throughout the year in the amount of the sun's energy striking a given point on the earth's surface. When the North Pole is tilted toward the sun, the Northern Hemisphere experiences summer because the rays of the sun strike the earth more directly above the equator. When the north end of the axis is tilted away from the sun, at the opposite end of the earth's orbit, the sun's rays strike the earth more directly below the equator, and it is winter in the Northern Hemisphere. Because of the earth's tilt, there are fewer hours of sunlight in winter, and this contributes to the overall colder temperatures. In the summer, days are longer and there are more hours of sunlight. In fall and spring, when the sun is shining directly over the equator, moderate weather and more equal days and nights occur.

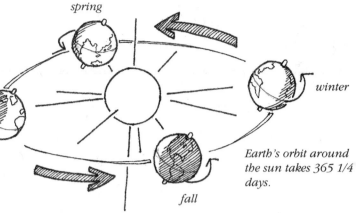

SEASONS IN THE NORTHERN HEMISPHERE

spring

summer

winter

fall

Earth's orbit around the sun takes 365 1/4 days.

Along with the effects of sunlight on the earth, one must also consider the effects of shadows. Solid objects block some of the rays of light that strike them, producing a darker area or "shadow" opposite the source of light. Over the course of a day, shadows outside gradually change their size and position. In the morning when the sun first appears above the eastern horizon, shadows form to the west of trees, houses, mountains, or other features of the landscape. As the earth turns toward the east, the shadows gradually move until they lie to the east of objects in the evening when the sun is shining from the west. They are longest at dawn and dusk when the sun's rays are at the most acute angle, and shortest at midday when the sun is most nearly overhead.

Shadows have an important effect on the ecology of an area. In the north temperate zone, the north slopes of hillsides are in shadow most of the day, and the plant and animal communities that live in these areas differ greatly from those that are found on hotter, drier, south-facing slopes. On the north-facing slopes we find shade-loving plants and animals that require cooler, moister habitats. Streams that are shaded by trees have a lower water temperature and are home to cold-loving species like trout, which avoid the higher temperatures of open, sunny portions of streams. When a tree falls or a forest is burned or cleared, a patch of forest floor is exposed to the sunlight, and a whole succession of plants occurs. Sun-loving, pioneer species such as poplar, birch, and white pine spring up first, and as they grow they provide needed shade for the later growth of shade-loving trees like sugar maples and hemlocks. Shading and shadows create variations in habitats that contribute much to the diversity of plant and animal communities.

Similarly, plants and animals respond, sometimes dramatically, to the seasonal changes of day length and the sun's intensity. During the shorter and cooler

days of fall, some trees show their beautiful oranges, reds, and yellows before losing their leaves, while a snowshoe hare's brown summer coat slowly turns white. The warmer and longer days of spring stimulate sap to flow and buds to grow, birds to build nests, and other animals to stir after their winter's rest. Humans, too, respond to changing seasons with different clothes and activities. All of these seasonal variations are due to the amount of heat, light, and other energy the earth receives from the sun.

At the center of our solar system, the sun's powerful gravity holds our planet in orbit and keeps it from spinning off into space. The sun is also central to our lives and to all life on Earth, for it provides us with the warmth and light, the wind and weather, and the food and fuel necessary to our existence.

Suggested References:

Barry, Roger G., and Richard J. Chorley. *Atmosphere, Weather and Climate.* 7th edition. New York: Routledge, 1998.

Linacre, Edward, and Bart Geerts. *Climates and Weather Explained.* New York: Routledge, 1997.

Neiburger, Morris, James G. Edinger, and William D. Bonner. *Understanding Our Atmospheric Environment.* San Francisco: W. H. Freeman, 1982.

Smith, Robert Leo. *Elements of Ecology.* 3rd edition. New York: Harper Collins, 1992.

painted turtles basking in the sun

Sun Power

FOCUS: The sun provides warmth and light, generates wind and weather, and powers the water cycle, making it possible for life to exist on Earth.

OPENING QUESTION: *Why is sunlight important to living things?*

PUPPET SHOW

Objective: To learn some of the ways that life on Earth is affected by the sun.

Perform, or have the children perform, the puppet show. Afterward, review how the sun affects the lives of the various plant and animal characters.

Materials:
- puppets
- script
- props

WHY DAY, WHY NIGHT?

Objective: To observe how day and night result from the earth's rotation on its axis.

Using an orange or apple to represent Earth, thread a skewer vertically through the center to represent the axis, put a rubber band around the middle to represent the equator, and mark the approximate location of your neighborhood with a tack. Have two children stand in a central location a few feet apart. One child will hold the earth model (the orange) while the second child shines a flashlight on it. Dim the room lights if necessary. Now have the child holding the orange rotate it slowly on its axis. Ask the other children to call out when it is daytime in your neighborhood and when it is night. Explain that it takes 24 hours, or one day, for Earth to make one full rotation on its axis.

Materials:
- orange
- bamboo skewer
- rubber band
- flashlight
- thumb tack

WHY THE SEASONS? *(grades 3-6)*

Objective: To understand how the orbiting of the earth around the sun and the tilt of its axis give us the different seasons.

Use a large melon or squash to represent the sun and place it on a coffee can or similar container on top of a table. Insert four skewers partway into the melon, positioning the skewers equally spaced in a flat plane around what would be the equator of the melon. These skewers represent some rays of the sun. Use an orange (prepared as described in the activity Why Day, Why Night?) to represent the earth. Hold the orange so that its equator is level with the skewers in the melon, and its axis is vertical. Now move the orange around the melon to demonstrate how the earth orbits around the sun.

Explain that the earth's axis is actually tilted at about a 23° angle. Tilt the orange about 23° toward the sun, and ask the children where the sun's rays are directed now. (above the equator – the Northern Hemisphere) Tilt the orange away from the sun. Where do the direct rays point now? (below the equator – the Southern Hemisphere) Explain how it is summer in the Northern Hemisphere when the North Pole is tilting toward the sun, and winter when the North Pole is tilting away from the sun. Now move the orange around the melon, keeping its axis tilted. Show how the Northern Hemisphere is tilted toward the sun at one end of its orbit and away from the sun at the opposite end of its orbit. As you move the orange around the melon, ask the children when it is summer in your neighborhood and when it is winter. When is it spring and fall?

Materials:
- coffee can
- large melon or squash
- bamboo skewers
- orange
- thumb tack
- rubber band

STRAIGHT OR SLANTED (grades 3-6)

Objective: To compare the energy of rays striking a surface directly or at an angle.

Have the children work in pairs, and give each team a flashlight, pencil, and piece of paper. Have one child shine the flashlight straight down onto the sheet of paper while the other child traces the circle of concentrated light, noting its brightness. Now have them shine the flashlight from a similar distance but at an angle and trace the patch of light again. Which angle produced the larger spot of light? Which produced the brighter (hotter) spot of light? Use a diagram to show how the sun's rays shine directly on the earth close to the equator, but reach the poles at an angle because the earth is curved. Where would it be hotter?

Materials:
- flashlights
- pencils
- paper
- diagram showing direct and indirect rays of sun striking Earth

SHADOWS AND SHAPES

Objective: To investigate how shadows are produced and how they change depending on the position of the light source.

Have the children work in pairs, and give each team a piece of white paper, a flashlight, and a small animal figurine. Have the children stand the figurine in the middle of the sheet of paper and shine the flashlight on it. What causes a shadow to appear on the paper? What happens to the shadow when the flashlight is moved from side to side? In what ways do shadows outside change during the day? Ask the children to think of some ways in which shadows and shade are important to living things.

Materials:
- flashlights
- white paper
- small animal figurines

MELT DOWN

Objective: To compare how fast ice cubes melt in full sun and in shadow.

Challenge the children to an ice cube melting contest outside. Give each child two clear plastic drinking cups containing two ice cubes each (more if it's windy so the cups won't blow over), one labeled "sun" and one labeled "shade." Now have each child place one cup in a shady spot and the other cup in full sun. After a few minutes (10-15 minutes – when the ice cubes are only slightly melted), have the children carefully remove the ice cubes and notice how much water they have in each cup. Combine the results for the whole group by having the children pour all the water collected from the "sun" cups into one measuring cup, and all the water from the "shade" cups into another measuring cup. Ask the children to explain the results.

Materials:
- clear plastic cups labeled "sun" and "shade"
- ice cubes
- two clear measuring cups

SHADOW PLAY

Objective: To observe and measure shadows, and to note their movement in relation to the sun.

On a paved area outside, have the children stand facing away from the sun. Have them look for their shadows. Have the children guess whether their shadow is longer or shorter than their height. Partners can trace and measure each other's shadows (use yarn if pavement is not available), being sure to outline the feet and mark the drawing with that child's initials. Have the children revisit their shadow outlines after 15-30 minutes and note any changes. Why have their shadows moved outside of the outlines?

Materials:
- chalk
- measuring tapes or string
- (optional) yarn

Sun Seekers Scavenger Hunt

Objective: To notice outdoor signs of the sun and the seasons.

Divide the children into small groups for outdoor exploration. Give each group a Scavenger Hunt card to follow.

Sun Scavenger Hunt Card

Locate the sun with eyes closed.

Find something as long as your shadow.

Find two shadows that touch. Are the objects touching?

Find something that doesn't cast a shadow.

Find something that reflects sunlight.

Find something warmed by the sun.

Find something that looks a different color in sun vs. shade.

Find the temperature in the sun.

Find the temperature in the shade.

Find an animal hiding from sunlight.

Find a leaf that is hidden from the sunlight. (Look under rocks or logs.) What color is it?

Find something bleached by the sun.

Find three signs of the season.

(On an overcast day, try to determine where the sun is. Look for shadows. Can you find any? Why not?)

Materials:
- Sun Scavenger Hunt cards
- thermometer in the shade
- thermometer in the sun

Sharing Circle

Objective: To review some ways in which the sun affects the lives of plants and animals.

Ask the children to think of ways the sun affects the lives of plants and animals, including their own. Write, or have the children write, their ideas on strips of yellow or orange paper, and then tape these strips around a large yellow paper circle, mounted on the wall, to form the rays of the sun.

Materials:
- large yellow paper circle
- yellow and orange strips of paper
- tape

Extensions

Solar Cooker: Have the children try to make their own solar cookers. Provide them with a variety of supplies such as shoe boxes, aluminum foil, plastic wrap, black paper, and other heat and light absorbing and reflecting materials. Explain that they will try to use the sun's energy to melt grated cheese on tortilla chips for a snack.

Sun Dial: To make a sundial outside, place a broomstick on the ground and use a compass to orient it so that it lies in a north/south direction. Lift the north end of the broomstick, and hammer the south end into the ground so that the broomstick makes an angle with the ground that is equal to your latitude. Place some pebbles on the ground in a large circle, with the south end of the broomstick as the center. Every hour, check the position at which the broomstick's shadow crosses the circle of pebbles and drive a stake into the ground at this point to mark the hours of the day.

Sun Tea: Make sun tea by placing some herbal tea bags in a large clear container of water and placing it in a sunny location for at least 30-45 minutes.

Sun Prints: Make sun prints by placing solid objects on colored construction paper and placing these on sunny windowsills. After a few days, remove the object and its darker impression will be left on the faded paper. You can also purchase special blueprint or sun print paper for even more dramatic results.

Sunflowers: Plant sunflowers.

Sun Power

Characters: Benjy Bear, Polly the Girl, Rachel Raspberry, Hannah Honeybee, Terry Turtle

Prop: sunglasses for Benjy

Benjy Bear

Grrrrr. I hate this hot weather. Why does it have to be so hot? Why can't it rain for a change?

Polly

What's the matter, Benjy? You don't seem to be in a very sunny mood today.

Bear

Sunny! Ugh! This sun is just un**bear**able! It's so hot and bright and uncomfortable! I wish the sun would just go away!

Polly

Go away?! Do you know what you're saying? The sun is so important, we couldn't live without it!

Bear

What's so important about it?

Polly

Well, for one thing, the sun gives us daylight and all the seasons. How would you like it if it were dark all the time, or if we always had the same season? I wouldn't! I like going swimming in the summer and skiing and skating in the winter!

Bear

Well, I wish it were winter right now. You could be skiing all you wanted and I would be sound asleep, so I wouldn't care if it was sunny and hot.

Polly

Oh, Benjy, why don't you just stop thinking about the sun and come pick some raspberries with me? *[exits]*

Bear

[follows slowly, still grumbling] I like that idea of being in the dark all the time. I could be happy in the dark all right. *[Raspberry pops up]*

Rachel Raspberry

Without food?

Bear

Who said that?

Raspberry

I did. Rachel Raspberry. You couldn't live very long without anything to eat, you know! Without sun, you wouldn't have any food.

Bear

I wouldn't? Why not?

Raspberry

Well, we plants need the sun's energy to make our food. Without the sun, plants couldn't grow leaves or seeds or juicy berries for animals like you to eat!

Bear

You mean you plants couldn't live without sunshine?

Raspberry

That's right, Benjy. We plants are solar powered!

Bear

Hmm...But even if there weren't any plants, I could eat other things – like honey, for instance! Bees could still be making honey in their hives. They don't need sun.

Raspberry

I don't know about that, but I see Hannah Honeybee over there. Maybe you'd better ask her. Bye. *[exits]*

Bear

Well, that's a **berry** good idea! Now where's that bee? *[Bee appears, buzzing]* Oh, hi there, Hannah. I have a question for you.

Hannah Honeybee

OK, but make it a quick quiz **bee**cause I'm bizz, bizz, bizzy!

Bear

Isn't it true that you bees don't need sunlight?

Honeybee

What?! Benjy, that's a silly question! If there weren't any sunlight, we wouldn't be able to find our way around, or find the flowers we need.

Bear

What do you need flowers for?

Honeybee

We need flowers because that's where we get the nectar to make honey! And flowers can't grow without sunlight.

Bear

I didn't know you needed flowers to make honey!

Honeybee

Not only that, we bees use the sun as a compass to find our way around! No sunlight?! I don't have time to listen to crazy questions like that! I'm **bizzzzy!** *[Bee buzzes off]*

Bear

Hmm...I guess if it were always dark it would make it pretty hard to see things and find your way around. But it would be a lot cooler! *[Turtle enters]*

Terry Turtle

Cooler?! It'd be downright freezing!

Bear

Oh, hi there, Terry Turtle. Don't you just hate this hot weather?

Turtle

I do not! Listen, Benjy. You're hot because you wear that heavy coat all the time. But we reptiles need the sunshine to keep warm.

Bear

You do?

Turtle

Sure. We can't make heat from the food we eat, the way you do. And you know what, Benjy? If there weren't any sun, it'd be so cold around here that even your fur coat wouldn't help. We'd all freeze, that's for sure.

Bear

Hmm. Well, maybe I was wrong. Maybe the sun really **is** important.

Turtle

Sure it is. We need it to stay warm, and we need the light so we can see. And we need it so the plants can grow and make food for us.

Bear

I guess a fella like me with a warm fur coat ought to just rest in the shade on a day like today.

Turtle

Good idea, Benjy. But first, come over here a minute. I've got something that will make you really cool.

Bear

Oh yeah? What is it? *[both exit back stage, then reappear with Benjy wearing sunglasses]* Hey, these are pretty cool!

Turtle

You do look cool! And now when it's really hot out, Benjy, you'll be able to just grin and **bear** it!

Bear

I guess so! Thanks, Terry.

Terry

Don't mention it. And now, I'm going to go catch some rays. See you later, Benjy.

Bear

Bye, Terry. See you later.

APPENDIX A

Environmental Learning for the Future (ELF) – *in your community?*

The Vermont Institute of Natural Science (VINS) initiated the ELF program in 1972 with a simple mission: to connect children to the natural world. ELF's strategy is also simple: to train parents and other community volunteers to teach nature education to children in their local elementary schools. ***Hands-On Nature*** is a collection of the workshops developed for the ELF program. ELF continues to grow, with more than 75 schools in Vermont participating as of 2000. The enthusiasm with which children, parents, and teachers embrace the program is contagious.

To participate in the ELF program, communities apply to VINS each year, demonstrating in their application a strong commitment by teachers, administrators, and community volunteers to carry out the program in their school. Throughout the school year, a VINS staff member meets with volunteers in each school (or sometimes in a nearby church, fire station, or library) for monthly training workshops, bringing all necessary materials and equipment. ELF volunteers are people from the community who are eager to learn more about natural history and to share their learning and their love of nature with children. During a training workshop, volunteers are led through an ELF unit, activity by activity, indoors and outdoors, as described in the chapters of this book. Volunteers learn firsthand what it feels like to do the activities that they will lead children through later in the month. After the workshop, volunteers meet with their teaching partners, often in consultation with the classroom teacher, to plan how they will present the unit to their students. Typically, classroom workshops last an hour to an hour and a half, though some programs, such as a visit to a nearby pond, may last much longer.

In each school, one or two volunteers take on the all-important task of being the coordinators for their community's ELF program. Coordinators find volunteers, set up the schedule of workshops, and act as liaison between the professional ELF teaching staff and the school. Materials such as slides or collections are provided by VINS or borrowed from state Fish and Wildlife departments or other local sources. Everyday objects like winter weeds or shoe boxes are brought from home. Puppets and murals are generally prepared by a couple of volunteers (often on a rotating basis) for use by everyone teaching ELF in a school.

The positive aspects of the ELF program are many. ELF emphasizes outdoor learning experiences – on school grounds or in nearby parks. ELF brings additional resources into schools and provides students with hands-on science learning, indoors and out, with relatively little financial investment. ELF activities reflect many of the science and collaborative skills called for by national and state standards. In Vermont, many principals report that ELF has helped raise students' science scores in their schools. Nationally, ELF has earned recognition by winning a Chevron Conservation Award and a National Arbor Day Award.

In addition, ELF invites a broad range of parents and community members into children's classrooms as partners in education. Through ELF, parents and other adult volunteers become integral parts of the learning community. Indeed, many volunteers find that their own excitement about what they learn is a prime motivation for being an ELF volunteer. Children receive the important message that many adults care about them and about their education. Through ELF, children learn that nature and science are not confined to a textbook or a classroom, nor dependent on special equipment, experts, and faraway field trips.

VINS has worked with ELF projects in Michigan, Tennessee, Colorado, Massachusetts, and New Hampshire. We have found that the program is easily adapted to meet the needs of children and schools in different areas of the country. Though the structure of the program may vary, the true secret of ELF lies in the transfer of enthusiasm, knowledge, and a sense of wonder from ELF trainers to ELF volunteers to children. If you are interested in finding out more about bringing ELF to your community, call the ELF Program Director at VINS, (802) 457-2779, or find us on the Internet at www.vinsweb.org.

APPENDIX B
Equipment – Where to Get It or How to Make It

The emphasis throughout this book is on simple, inexpensive materials and equipment that are readily available or easy to locate. A shelf full of arts and crafts or stationery supplies will meet the requirements of most activities. However, there are some shortcuts to making or finding the more complicated and specific items. Hopefully, the following hints will save time and frustration.

Animal Parts (skull, antlers, etc.) – Fish and Wildlife Department, local game warden, school and university biology departments, local collectors, hunters.

Bags, Mystery – sewn 8" (length) x 6" (width) dark cloth bags with drawstrings.

Bicycle Wheels (for spider web activity) – bicycle repair shop.

Bingo Cards – simple drawings or pictures of animals (2" x 2 1/2") glued onto 8 1/2" x 11" paper, four across and four down. Make as many copies as needed. Cut the copies into individual pictures and glue onto 8 1/2" x 11" boxboard, mixing up the order so each card is somewhat different.

Binoculars – Eagle Optics offers a catalog that includes a Binocular Buying Guide and many different brands of binoculars and accessories at discount prices. Eagle Optics, 716 S. Whitney Way, Madison, WI 53711. Toll Free: 1-800-289-1132.

Boxes, Bug – Frey Scientific (1-888-222-1332) carries these. Or use small glass jars or small plastic containers.

Boxes, Shoe (for dioramas, etc.) – shoe store or sports store. Tissue boxes also work well.

Cards, Paint Chip – hardware or paint store.

Cards, Task – photocopy the task card directly from the book, leaving room on the copy for the children to take notes. Or, copy the written instructions onto index cards, then cover with clear contact paper. Add a strip of masking tape on one edge if listed items need to be checked as accomplished. Replace tape after each activity.

Compasses – we use Silva Starter Type 1-2-3 compasses from Campmor, P. O. Box 700, Saddle River, NJ 07458. 1-888-CAMPMOR.

Discs (for clocks, spin the dial games, etc.) – pizza cardboard or heavy-duty paper plates.

Feathers – poultry farm, someone who raises chickens, or game bird hunters. Note: Special permits are needed for possessing many types of feathers. Contact your state wildlife agency.

Felt Board – cardboard or other solid, lightweight board covered with felt; just cover front of board, fold felt over edges, and tape to back.

Felt Board Cutouts – felt cut-outs don't stick well over an extended period; use masking tape. Also, oak-tag or heavy paper cut-outs with scotch tape on back side work well.

Hand Lenses – available from Delta Education, P. O. Box 3000, Nashua, NH 03061-3000, 1-800-442-5444. Add shoelace or nylon cord neck strings before using.

Jars (for insects and spiders, etc.) – save small plastic jars with lids or baby food jars. Punch holes in lids if live creatures are to be housed.

Mural – for easy transporting, use a white window shade (can be rolled or folded). For permanent display, use a large piece of cardboard and laminate or cover it with clear contact paper.

Mural Cutouts – simple illustrations or photographs from magazines can be covered with contact paper and used over and over again. Use a loop of masking tape stuck onto back of picture for attaching to mural.

Pictures (of animals, plants) – old nature or outdoor magazines. Senior center visitors or nursing home residents are often willing to cut out pictures. National Wildlife Federation stamps are also useful.

Pond/Stream Nets – 6" or 8" plastic kitchen strainers (do not use metal, it can damage amphibian skin).

Pond/Stream Pans – white plastic dishpans (for temporary holding of creatures). Cottage cheese container, for individual children to use.

Puppets – we use simple drawings glued onto boxboard or oak-tag, painted with markers or watercolors, and mounted on paint stirrer sticks or dowels (two drawings glued back-to-back so either side can face the audience). Many of the animals appearing as characters in the puppet shows are illustrated throughout the book.

Puppet Stage – table tipped on its side, bookcase, desk with sheet thrown over it, anything that hides the puppeteer and can hold a taped-up script. Science display boards or the boxes that flipchart paper comes in make nice portable puppet stages.

Puzzles – mount the full picture on boxboard, cover with contact paper, then cut out pieces.

Scents (for smelling games) – use natural extracts such as peppermint, lemon, almond; many people are allergic to perfumes and colognes.

Slides – (to borrow) Fish and Wildlife Department; libraries, schools, and universities; local nature lovers and travelers. (to buy) Laboratory of Ornithology, Cornell University, Ithaca, NY sells slides of birds. Many other professional organizations also sell slides (insects, mammals, plants, and more); find various slide collections on the Internet.

Specimens (mounted animals or insects, nests, nature collectibles) – Fish and Wildlife Department; schools and universities; local collectors (advertise to find them).

Thermometers – check with hardware stores for simple aluminum thermometers with °F and °C; also available from Carolina Biological Supply, 1-800-334-5551.

Track Prints (for track stories) – we use spongy (Dr. Scholl-type) shoe inserts cut to shape with the correct print outline and number of pads and toes, glued onto plastic or wooden blocks.

Tuning Forks – available from Frey Scientific 1-888-222-1332.

GLOSSARY

abscission layer
a zone of thin-walled cells that forms across the base of leaf stems and causes the leaf to fall off

alternate
bud, twig, or leaf arrangement on a plant in which there is only one bud, twig, or leaf at a node so that the twigs or leaves grow on alternating sides of the stem

amphibian
any member of the class of cold-blooded vertebrate animals that is characterized by having eggs laid in water that hatch into gill-breathing larvae and metamorphose into lung-breathing adults; includes frogs, toads, and salamanders

angiosperm
a plant with seeds enclosed in a mature ovary (fruit)

antenna
a jointed appendage used as a sensory receptor, usually occur in pairs on arthropods

anther
the upper pollen-bearing portion of the stamen

arthropod
an invertebrate with a segmented external skeleton and jointed legs; includes insects, spiders, ticks, millipedes, centipedes, and crustaceans

axis
an imaginary line through the earth's poles around which the earth rotates; also used to demonstrate its angle of tilt with respect to the sun

bird of prey
any member of a group of birds that kill and eat other animals for their food; characterized by a sharp, hooked beak and strong, sharp talons, as in hawks and owls (also called raptors)

bract
a specialized leaf or leaf-like structure on a plant, usually found at the base of a flower

browse
tender shoots, twigs, and bark of shrubs and trees eaten by animals such as deer and rabbits; the act of consuming this kind of food

brumation
a state of torpor found in amphibians in which they pass the coldest winter months

cambium
a layer of tissue in woody plants from which new bark and new wood originate

canines
the four pointed teeth occurring at the front corners of the upper and lower jaws of flesh-eating mammals, used for gripping and tearing flesh

capillary action
a process in which liquid is pulled into thin spaces due to the relative attraction of the molecules of the liquid for each other and between the molecules of the liquid and the material that surrounds the space

carnivore
an animal that consumes meat

castings
worm droppings; the nutrient-rich pellets of decomposed organic material that has passed through the worm's digestive tract

cephalothorax
a body part that is both head and thorax combined, as in spiders

chitin
the primary component of arthropod skeletons, which is tough, flexible, and resistant to most chemicals, including water

chlorophyll
the green pigment found in plants that is necessary for the process of photosynthesis

chrysalis
the pupa of a butterfly

cirrus
thin, wispy clouds made of ice crystals, found at very high altitudes

clitellum
a thickened section of a worm's body that secretes mucus and forms the egg sacs

composite
one of a family of flowering plants that produce many small flowers closely grouped into compact heads, such as dandelions, goldenrods, and sunflowers

compound eye

an eye composed of many individual facets, each capable of sight; common to insects

conifer

a cone-bearing plant such as pine, fir, spruce, hemlock, cedar, and redwood

consumer, primary

an animal that eats plants

consumer, secondary

an animal that eats other animals

countershading

a form of protective coloration in which the animal is two-toned, often light-colored underneath (making it camouflaged when seen from below against the light sky) and darker on its back (making it camouflaged when seen from above against the darker earth)

crepuscular

active in the twilight hours at dusk and dawn

cumulonimbus

thunderheads; dark, billowing, cumulus-type clouds towering to great heights, that often produce lightning and thunderstorms

cumulus

thick, billowy clouds with horizontal bases and upper parts piled up into heaps

deciduous

trees and shrubs that lose their annual growth of leaves

decomposer

an organism that causes the mechanical and chemical breakdown of dead plants and animals

decomposition

the process of rotting and decaying that causes the complex organic materials in plants and animals to break down into simple elements that can be returned to the atmosphere and soil

disruptive coloration

a form of camouflage in which dark and light bands, patterns, or patches of color serve to disrupt the outline of an animal, as in the stripes of zebras or the spots on a fawn

dormancy

a temporary state of inactivity for a plant or animal that enables it to endure a period of environmental stress such as extreme heat, cold, or scarcity of food

drone

the male of the honeybee and other bees

echolocation

a means of navigating or locating prey or obstacles by emitting high pulses of sound that are reflected back from objects in the environment

emergent

an aquatic plant that is rooted in a pond or stream bottom and has stems and leaves above the surface, such as grasses, sedges, rushes, and cattails

erosion

the wearing away and carrying off of rock, soil, sand, or other earth materials

exoskeleton

a hard external body-covering found in arthropods

filament

the stalk of a stamen bearing the anther

food chain

the transfer of food energy in sequence from plants to herbivores to carnivores

food web

a network of interconnected food chains within a community

frass

material resembling sawdust, made up of wood fragments that have been chewed by insects (and, in some cases, passed through the gut); often found in rotting logs

freeze

to become absolutely motionless, either through fear or caution, in order to escape detection by another animal

frost line

the maximum depth to which water freezes during the winter

fungi

a group of organisms including mushrooms, molds, smuts, rusts, yeasts, and mildews, all of which lack chlorophyll and subsist upon dead or living organic matter; important as decomposers

ground water

water found in porous rock and beds of sand and gravel beneath the earth's surface

gymnosperm

a plant with seeds that are not enclosed in an ovary; includes conifers such as pines and spruces

herbaceous
referring to any nonwoody plant

herbivore
an animal that eats plant material

hibernation
a prolonged state of torpor by which some animals cope with the stresses of winter; characterized by greatly reduced body temperature, heart rate, and breathing

humus
the organic portion of the soil, formed from decomposed plants and animals

hyphae
white, thread-like strands of fungi found in the soil or in rotting wood

igneous
rock formed from the cooling and crystallization of magma

incisors
teeth in the front part of the lower and upper jaws of mammals, adapted for cutting

infrared
heat rays; thermal radiation with wavelengths longer than that of red light

insectivore
any of a small group of mammals, including moles and shrews, that characteristically feed on insects

invertebrate
an animal that has no backbone, as in worms, insects, mollusks, and crustaceans

keratin
the primary component of horns, feathers, hairs, and nails; a protein that is very resistant to physical wear and chemical disintegration

larva
an immature and usually active feeding stage of an animal, unlike the adult in form

lateral bud
a bud that grows out from the side of a stem or twig

lenticels
small structures on the bark of a shrub or tree, usually in the form of horizontal slits or pores, that permit the exchange of gases

magma
molten rock found at great depth within the earth

mandible
in birds, the lower part of the beak; in insects, one of a pair of mouth appendages

masking
a form of camouflage in which animals disguise themselves by attaching material from their surroundings – such as twigs, petals, or leaf pieces – to their bodies or cases

matching color
a form of camouflage in which an animal's color resembles the color of the background against which it is most often seen

metamorphic
rock formed from igneous or sedimentary rock that has been transformed by intense heat, pressure, and/or chemical action

metamorphosis
a process by which an immature animal transforms to an adult through a series of developmental changes

metamorphosis, complete
the four-stage development of insects that includes egg, larva, pupa, and adult

metamorphosis, simple
the three-stage development of insects that includes egg, nymph, and adult

mineral
an inorganic crystalline material with a homogeneous chemical composition, such as quartz, copper, sulfur; the building blocks of rocks

molars
in mammals, the back, permanent teeth that have surfaces adapted for grinding

nictitating membrane
a transparent eyelid found in some animals including many frogs, beavers, cats, and birds

nimbostratus
low, dark, layered clouds that produce rain

nocturnal
active during the night

node
the region of a plant stem where one or more leaves or twigs arise

nymph
the immature stage of an insect that undergoes simple, or incomplete, metamorphosis

omnivore
an animal that eats both plants and animals; for example, raccoons, skunks, and humans

opposite
leaf, twig, or bud arrangement on a plant in which two buds, twigs, or leaves arise directly across from each other at a node

ovary
  the enlarged hollow part of the flower pistil that contains the ovules

ovipositor
  an egg-laying structure on the rear abdominal body segment of a female insect

ovule
  a plant structure that develops into a seed after the contained egg is fertilized

palmate
  a pattern of veins in leaves in which several main veins radiate outward from a common point at the base of the leaf so that it resembles a hand, such as that found in maple, sweet gum, and sycamore

parallel
  a pattern of veins in leaves in which the main veins run side by side from the base of the blade to the tip

pedipalps
  in spiders, a pair of jointed appendages on the cephalothorax that serve as sensory organs, and in males, for transferring sperm to the female. Male spiders' pedipalps have clubbed tips, while females' have pointed tips

petiole
  the slender stalk by which a leaf is attached to the stem of a plant

pheromone
  a chemical substance emitted by an animal and used for communication with members of its own species

phloem
  the conducting tissue in a plant that transports food produced in the leaf to all other parts of the plant

photosynthesis
  the production of sugars, by plants, from carbon dioxide and water in the presence of chlorophyll using sunlight as the source of energy

pinnate
  a pattern of veins in leaves with a large central vein running from base to tip and other veins branching off on either side, like the pinnae of a feather

pistil
  the central organ of a flower that contains the female parts: stigma, style, and ovary

pith
  soft, spongy tissue in the center of twigs and plant stems

pollen
  fine, yellowish powdery grains that contain the male germ cells of a plant

pollen basket
  on the hind legs of the honeybee, a concave area with hairs along both edges, used for holding pollen and carrying it back to the hive

pollination
  the transfer of pollen from the anther (male) to the stigma (female) of a plant for fertilization

predator
  an animal that hunts, kills, and eats other animals

prey
  an animal hunted for food

producer
  an organism that produces its own food (plants)

protective mimicry
  a form of camouflage in which a prey animal resembles a dangerous, unpalatable, or poisonous animal

pupa
  the third stage of complete metamorphosis in insects during which a larva transforms into an adult

raptor
  bird of prey; any member of a group of birds that kill and eat other animals for food; characterized by a sharp, hooked beak and strong, sharp talons, as in hawks and owls

reptile
  a member of the class of cold-blooded vertebrate animals, including lizards, snakes, turtles, and alligators, characterized by dry, scaly skin and eggs suited for development on land with membranes and shells to protect the embryo

revolution
  movement of the earth around the sun; orbit

rhizome
  an elongated, underground, horizontal stem that usually produces roots and sends up shoots

rotation
  the spinning of the earth around an axis

ruminant
  a herbaceous land animal with a compartmentalized stomach that allows progressive digestion as partially digested food (cud) is

regurgitated, chewed, and reswallowed; includes cattle, sheep, goats, deer, moose, and caribou

scavenger

an animal that feeds on dead organic matter, either plant or animal

sedimentary

rock formed by the compaction of layers of sediment composed of the weathered fragments of pre-existing rocks or organic remains such as shells and skeletons

sepal

leaf-like plant parts that enclose the flower bud

setae

tiny bristles, as those located on the underside of an earthworm's body

spinnerets

in spiders, finger-like appendages on the underside of the abdomen involved in silk production

spiracles

external openings in insects through which air enters the respiratory system

spore

a tiny reproductive cell as in the mosses, ferns, and fungi

stamen

the pollen-producing male organ of the flower, consisting of anther and filament

stigma

the most elevated part of a flower's pistil

stomata

minute openings in the outer surface of leaves and stems of plants through which water vapor and gases pass

straddle

the width of an animal's trail; the distance between the outside edges of opposing feet

stratus

low, layered, gray clouds that cover most of the sky in a thick blanket

stride

the distance between the prints of a walking animal or between sets of tracks in a running animal

style

a slender column of tissue that connects the ovary and the stigma of a flower pistil

submergent

an aquatic plant rooted in a pond or stream bottom that has completely submerged stems and leaves

subnivean

under the snow

surface tension

a property of liquids in which the surface layer has a stretched, elastic character that offers some resistance to penetration

syrinx

the vocal organ of birds in which sound is produced by the vibration of a bony ridge and certain membranes

talons

the claws on the toes of a bird of prey

terminal bud

a bud at the end of a plant stem or twig

thorax

the body part between the head and the abdomen of an insect to which the wings (when present) and legs are attached

transpire

to give off water vapor, as from the surface of leaves and other plant parts

tympanum

a thin membrane that vibrates when struck by sound waves

vertebrate

an animal having a segmental backbone or vertebral column; includes bony fishes, amphibians, reptiles, birds, and mammals

warning coloration

colors, such as the black and yellow of bees or the orange of efts, that warn other animals that this prey is dangerous or unpalatable

weathering

the physical and chemical breakdown of rocks into smaller fragments at the earth's surface

whorled

bud or leaf arrangement on a plant in which three or more buds or leaves arise at a single node and radiate from the stem

xylem

the vascular tissues that transport nutrient-rich water throughout the plant

yard

a sheltered area where animals, especially deer, congregate during severe winter weather

GENERAL BIBLIOGRAPHY

In addition to the Suggested References (works consulted) listed after the background section in each chapter, the following are wonderful resources for exploring a variety of natural history topics. And certainly many of the books listed as Suggested Reading for Children are also helpful resources for adults.

Field Guides

Audubon Society Pocket Guides (New York: Alfred A. Knopf) - color photographs plus descriptive text in a pocket-size guide. (e.g., *Familiar Flowers of North America, Familiar Mammals*)

Golden Field Guides (New York: Golden Press) - color illustrations with descriptive text. (e.g., *Birds of North America, Rocks and Minerals, Weather*)

Golden Guides (New York: Golden Press) - small, inexpensive, colorful, and handy for basic identification. (e.g., *Insects, Spiders and their Kin, Pond Life*)

Nature Study Guild Guides (Rochester, NY: Nature Study Guild) - pocket-size keys to identification. (e.g., *Track Finder, Winter Weed Finder*)

Peterson Field Guides (Boston: Houghton Mifflin) - comprehensive yet easy to use, with arrows that point out distinguishing features; lots of information and clear illustrations. (e.g., *A Field Guide to Insects, A Field Guide to Birds' Nests, A Field Guide to Eastern Butterflies*)

Stokes Nature Guides (Boston: Little, Brown) - a very accessible and informative series full of interesting background information, highly recommended. (e.g., *A Guide to Nature in Winter, A Guide to Observing Insects Lives, A Guide to Animal Tracking and Behavior, A Guide to Amphibians and Reptiles*)

Beyond Identification

Eastman, John. *The Book of Forest and Thicket: Trees, Shrubs, and Wildflowers of Eastern North America.* Mechanicsburg, PA: Stackpole Books, 1992.

Ehrlich, Paul R., David S. Dobkin, and Darryl Wheye. *The Birder's Handbook: A Field Guide to the Natural History of North American Birds.* New York: Simon & Schuster, 1988.

Imes, Rick. *The Practical Entomologist.* New York: Simon & Schuster, 1992.

Johnson, Charles W. *The Nature of Vermont.* Hanover, NH: University Press of New England, 1998.

Lawrence, Gale. *A Field Guide to the Familiar: Learning to Observe the Natural World.* Hanover, NH: University Press of New England, 1998.

Marchand, Peter J. *Life in the Cold: An Introduction to Winter Ecology.* 3rd edition. Hanover, NH: University Press of New England, 1996.

Rezendes, Paul. *Tracking and the Art of Seeing: How to Read Animal Tracks and Sign.* 2nd edition. New York: HarperCollins, 1999.

Wessels, Tom. *Reading the Forested Landscape: A Natural History of New England.* Woodstock, VT: The Countryman Press, 1997.

Additional Teaching and Planning Resources

Appelhof, Mary. *Worms Eat Our Garbage: Classroom Activities for a Better Environment.* Kalamazoo, MI: Flower Press, 1993.

Braus, Judy A., and David Wood. *Environmental Education in the Schools: Creating a Program that Works!* Washington, DC: North American Association For Environmental Education, 1993.

Caduto, Michael J., and Joseph Bruchac. *Keepers of the Earth: Native American Stories and Environmental Activities for Children.* Golden, CO: Fulcrum, 1988.

Cornell, Joseph Bharat. *Sharing Nature with Children.* Nevada City, CA: Ananda Publications, 1979.

Cvancara, Alan M. *Exploring Nature in Winter: A Guide to Activities, Adventures, and Projects for the Winter Naturalist.* New York: Walker, 1992.

Harlen, Wynne, ed. *Primary Science . . . Taking the Plunge.* Oxford: Heinemann, 1985.

Hogan, Kathleen. *Eco-Inquiry: A Guide to Ecological Learning Experiences for the Upper Elementary/ Middle Grades.* Dubuque, IA: Kendall/Hunt, 1994.

Hunken, Jorie, and The New England Wildflower Society. *Botany for All Ages: Discovering Nature Through Activities Using Plants.* Chester, CT: The Globe Pequot Press, 1989.

Johns, Frank A., Kurt Allen Liske, and Amy L. Evans. *Education Goes Outdoors.* New York: Addison-Wesley, 1986.

Kneidel, Sally Stenhouse. *Creepy Crawlies and the Scientific Method: Over 100 Hands-on Science Experiments for Children.* Golden, CO: Fulcrum, 1993.

Miller, Lenore Hendler. *The Nature Specialist: A Complete Guide to Programs and Activities.* Martinsville, IN: American Camping Association, 1996.

National Science Resources Center. *Resources For Teaching Elementary School Science.* Washington, DC: National Academy Press, 1996.

Parrella, Deb. *Project Seasons.* Shelburne, VT: Shelburne Farms, 1995.

Project Learning Tree. Washington, DC: American Forest Foundation, 1996. (available to participants of PLT workshops)

Ranger Rick's Nature Scope series. Washington, DC: National Wildlife Federation.

Russo, Monica. The Tree Almanac: A Year-Round Activity Guide. New York: Sterling, 1993.

Sheehan, Kathryn, and Mary Waidner, Ph.D. Earth Child: Games, Stories, Activities, Experiments & Ideas about Living Lightly on Planet Earth. Tulsa, Oklahoma: Council Oak Books, 1991.

Sisson, Edith A. Nature with Children of All Ages: Activities and Adventures for Exploring, Learning, and Enjoying the World around Us. Englewood Cliffs, NJ: Prentice-Hall, 1982.

Sobel, David, M.Ed. Beyond Ecophobia: Reclaiming the Heart in Nature Education. Great Barrington, MA: The Orion Society, 1996.

University of California-Berkeley: Lawrence Hall of Science. "Outdoor Biology Instructional Strategies." Hudson, NH: Delta Education.

Nature Craft and Project Books

Butterfield, Moira. How to Draw and Paint the Outdoors. Edison, NJ: Chartwell Books, 1994.

Diehn, Gwen, and Terry Krautwurst. Kid Style Nature Crafts: 50 Terrific Things to Make with Nature's Materials. New York: Sterling, 1995.

Drake, Jane, and Ann Love. The Kids' Summer Handbook. New York: Ticknor & Fields, 1994.

Rainis, Kenneth G. Nature Projects for Young Scientists. New York: Franklin Watts, 1989.

Tucker, Priscilla M. Basic Nature Projects: 101 Fun Explorations. Mechanicsburg, PA: Stackpole Books, 1995.

Read Alouds

Arnosky, Jim. Nearer Nature. New York: Lothrop, Lee, and Shepard, 1996.

Baylor, Byrd. I'm In Charge of Celebrations. New York: Simon & Schuster, 1986.

Curtis, Will. The Second Nature of Things: How and Why Things Work in the Natural World. Hopewell, NJ: The Ecco Press, 1992.

Fleischman, Paul. Joyful Noise: Poems for Two Voices. New York: Harper and Row, 1988.

Montgomery, Sy. Nature's Everyday Mysteries: A Field Guide to the World in Your Backyard. Shelburne, VT: Chapters, 1993.

Rood, Ronald. Ron Rood's Vermont: A Nature Guide. Shelburne, VT: The New England Press, 1988.

Teale, Edwin Way. Circle of the Seasons: The Journal of a Naturalist's Year. New York: Dodd, Mead, 1953.

Magazines/Journals

For Children

"Kids Discover" – 170 Fifth Ave., New York, NY 10010

"National Geographic World: The official magazine for junior members of the National Geographic Society" – Dept. EV, PO Box 370, Vandalia, OH 45377

"Odyssey: Adventures in science" – Cobblestone Publishing, 7 School St., Peterborough, NH 03458

"Owl: The discovery magazine for kids" – The Young Naturalist Foundation, Bayard Press, 25 Boxwood Lane, Buffalo, NY 14227

"Ranger Rick" – National Wildlife Federation, Mt. Morris, IL 61054

"Your Big Backyard" – National Wildlife Federation, Mt. Morris, IL 61054

"Zoobooks" – Wildlife Education, Ltd. 930 W. Washington St., San Diego, CA 92103

For Adults

"Audubon: The magazine of the National Audubon Society" – 700 Broadway, New York, NY 10003

"National Wildlife" – National Wildlife Federation, Mt. Morris, IL 61054

"Natural History" – American Museum of Natural History, Central Park West, New York, NY 10024

"Science & Children" – National Science Teachers Assn., 1840 Wilson Blvd., Arlington, VA 22201

"T.H.E. Journal: Technological Horizons in Education" – 150 El Camino Real, Tustin, CA 92680

"Weatherwise" – Heldref, 1319 18th St. NW, Washington, DC 20036

Suggested Reading for Children

These books complement the different topics addressed in each chapter of **Hands-On Nature**. We've noted age appropriateness (primary: 5-8 year olds; intermediate: 9-11; middle: 12-14; all: all ages) plus brief descriptions in parentheses after each citation. Topics are listed here according to the table of contents.

ADAPTATIONS

Amazing Insects

Cole, Joanna. *Magic School Bus Gets Ants in Its Pants*. New York: Scholastic, 1996. (intermediate – story, with good information about ants)

Cole, Joanna. *An Insect Body*. New York: William Morrow, 1984. (intermediate – focus on cricket; b/w photos; factual)

Fischer-Nagel, Heiderose, and Andreas Fischer-Nagel. *The Housefly*. Minneapolis: Carolrhoda, 1990. (intermediate – lots of color photos; vivid close-ups; full description of fly's life)

Mound, Laurence. *Eyewitness Insect*. New York: Alfred A. Knopf, 1990. (all – clear pictures; lots of information)

Snedden, Robert. *What Is an Insect?* San Francisco: Sierra Club Books, 1993. (intermediate – large color close-ups; information about life, movement, food)

Souza, D. M. *Insects around the House*. Minneapolis: Carolrhoda, 1991. (intermediate – chapter book; color photos; various insects)

Grasses

Chambers, Catherine, and Caroline Chambers. *Grasses* (Would You Believe It). Chatham, NJ: Raintree/Steck-Vaugn, 1996. (intermediate – introduces grasses and their many uses)

Lerner, Carol. *Seasons of the Tallgrass Prairie*. New York: William Morrow, 1980. (intermediate – line drawings; explores each season)

Rinkoff, Barbara. *Guess What Grasses Do*. New York: Lee & Shepard, 1972. (all – uses of grasses)

Sayre, April Pulley. *Grassland*. New York: Henry Holt, 1994. (intermediate – examines adaptations and uses of grasses)

Trost, Lucille Wood. *The Lives and Deaths of a Meadow*. New York: G.P. Putnam's Sons, 1973. (intermediate – story of a meadow over hundreds of years)

Hunter – Hunted

Bell, Simon M. *Birds of Prey*. New York: Somerville House, 1998. (intermediate – 3-D w/viewer; color photos)

Brooks, Bruce. *Predator!* New York: Farrar, Straus & Giroux, 1991. (intermediate – color photos; lots of predators, hunting strategies)

Patent, Dorothy Hinshaw. *Hunters and the Hunted: Surviving in the Animal World*. New York: Holiday House, 1981. (intermediate – b/w photos; good descriptions of predator-prey strategies)

Powell, Consie. *A Bold Carnivore: An Alphabet of Predators*. Emeryville, CA: Roberts Rinehart, 1995. (primary – great full-page illustrations of 26 predators and their prey)

Ryden, Hope. *Bobcat*. Toronto: General, 1983. (intermediate – b/w photos; factual account of bobcat)

Saunders, Susan. *Seasons of a Red Fox*. Norwalk, CT: Trudy Management, 1991. (primary – story of a red fox's first year of life; read-aloud)

Simon, Seymour. *Wolves*. New York: HarperCollins, 1993. (intermediate – clear, informative text and beautiful color photographs)

Teeth and Skulls

Blocksma, Mary. *Amazing Mouths and Menus*. New York: Prentice-Hall, 1986. (intermediate – description and illustrations of various mouths – mammal, bird, reptile)

Lauber, Patricia. *What Big Teeth You Have!* New York: Thomas Y. Crowell, 1986. (intermediate – drawings of different animals' skulls and teeth)

Parker, Steve. *Eyewitness: Skeleton*. New York: Alfred A. Knopf, 1988. (all – clear photos; lots of facts)

Parker, Steve. *Look at Your Body: Skeleton*. Brookfield, CT: Copper Beech Books, 1996. (intermediate – lots of drawings, photos; read-aloud for younger)

Simon, Seymour. *Bones: Our Skeletal System*. New York: Morrow Jr. Books, 1998. (intermediate – great, large, close-up photos; fairly simple text)

Beaks, Feet, and Feathers

Arnosky, Jim. *Crinkleroot's Guide to Knowing the Birds*. New York: Bradbury Press, 1992. (primary – colorful drawings; birding techniques; identification and natural history)

Bishop, Nic. *The Secrets of Animal Flight*. Boston: Houghton Mifflin, 1997. (intermediate – photos and drawings showing flight of birds, bats, insects)

Doris, Ellen. *Ornithology*. New York: Thomas & Hudson, 1994. (intermediate – basic birding; identification, habitats, migration; color photos)

Gallimard, Jeunesse, and Pascale de Bourgoing. *The Egg*. New York: Scholastic, 1992. (primary – photographs with overlays; simple text)

Garelick, May. *What Makes a Bird a Bird?* New York: Mondo, 1988. (primary – nice paintings; very colorful)

Schultz, Ellen. *I Can Read about Birds*. Mahwah, NJ: Troll, 1996. (primary – introduces many types of birds)

Snedden, Robert. *What Is a Bird?* San Francisco: Sierra Club Books, 1992. (intermediate – good color photos; feathers, flying, egg-laying)

Owls

Arnosky, Jim. *All about Owls*. New York: Scholastic, 1995. (primary – drawings by author; good introduction to owls)

Heinrich, Bernd. *An Owl in the House: A Naturalist's Diary*. Boston: Little, Brown, 1990. (intermediate – line drawings; story of an adopted owl; what to do with a baby bird)

Niemuth, Neal. *Owls for Kids*. Minnetonka, MN: NorthWord Press, 1995. (intermediate – photos and drawings; lots of facts)

Sattler, Helen Roney. *The Book of North American Owls*. New York: Clarion Books, 1995. (intermediate – many paintings; lots of natural history)

Selsam, Millicent E., and Joyce Hunt. *A First Look at Owls, Eagles, and other Hunters of the Sky*. New York: Walker, 1986. (primary – nice, clear drawings; factual)

Togholm, Sally. *Animal Lives: The Barn Owl*. New York: Kingfisher Books, 1999. (all – traces the life of a barn owl from winter to fall; read-aloud)

Yolen, Jane. *Owl Moon*. New York: Philomel Books, 1987. (all – story of a girl owling with her father on a cold winter night)

Thorns and Threats

Bennett, Paul. *Escaping from Enemies*. New York: Thomson Learning, 1995. (all – large color photos of various animal defenses)

Kaner, Etta. *Animal Defenses: How Animals Protect Themselves*. Buffalo: Kids Can Press, 1999. (intermediate – beautiful paintings; displays, camouflage, mimicry)

Parsons, Alexandra. *Eyewitness Jr.: Amazing Poisonous Animals*. New York: Alfred A. Knopf, 1990. (all – lots of information and facts)

Sowler, Sandie. *Eyewitness Jr.: Amazing Animal Disguises*. New York: Dorling Kindersley, 1992. (intermediate – easy to read text describes many animal disguises; color photos)

Frogs and Polliwogs

Grossman, Patricia. *Very First Things to Know about Frogs*. New York: Workman, 1999. (primary – easy reading; photos, stickers)

Himmelman, John. *A Wood Frog's Life*. Danbury, CT: Grolier, 1998. (primary – nice paintings; simple text - year of wood frog's life)

Lavies, Bianca. *Lily Pad Pond*. New York: E. P. Dutton, 1989. (primary – nice color photos)

Mara, William P. *The Fragile Frog*. Morton Grove, IL: Albert Whitman, 1996. (intermediate – factual; natural history of endangered Pine Barrens tree frog)

McClung, Robert M. *Peeper, First Voice of Spring*. New York: William Morrow, 1977. (all – line drawings; nice story of spring through fall)

Parker, Steve. *Frogs and Toads*. San Francisco: Sierra Club Books, 1994. (all – lots of photos, drawings, some lift the flaps)

HABITATS

Life in a Field

Carle, Eric. *The Very Quiet Cricket*. New York: Philomel Books, 1990. (primary – story; colorful cut-paper illustrations; how insects make sounds)

Carter, David A. *Over in the Meadow: An Old Counting Rhyme*. New York: Scholastic, 1992. (all – new pictures to accompany a traditional song/rhyme; all ages)

Fleming, Denise. *In the Tall, Tall Grass*. New York: Scholastic, 1991. (primary — simple rhyming words introduce a great variety of field inhabitants)

Koch, Maryjo. *Dragonfly Beetle Butterfly Bee*. San Francisco: Collins, 1996. (all – excellent illustrations as reference for kids; text more geared for adults)

Mound, Laurence. *Eyewitness Books: Insect*. New York: Alfred A. Knopf, 1990. (all – photos and drawings; types of insects, life cycles, senses, feeding)

Taylor, Kim, Jane Burton, and Barbara Taylor. *Look Closer: Meadow*. New York: Dorling Kindersley, 1992. (all – close-up photos; plants and animals found in meadows)

Forest Floor

Glaser, Linda. *Wonderful Worms*. Brookfield, CT: Millbrook Press, 1992. (primary – appealing pictures underground and above; simple text)

Himmelman, John. *A Slug's Life*. Chicago: Children's Press, 1998. (primary – clear, colorful illustrations and simple text; life cycle of a slug)

Johnson, Sylvia A. *Mosses*. Minneapolis: Lerner, 1983. (intermediate – examines mosses through color photos, diagrams, and text)

Mazer, Anne. *The Salamander Room*. New York: Alfred A. Knopf, 1991. (all – story; boy imagines converting his room into a forest for his pet salamander)

Pasco, Elaine. *Earthworms*. Woodbridge, CT: Blackbirch Press, 1997. (intermediate – photographs of worm anatomy, life cycle; simple experiments)

Ryder, Joanne. *Chipmunk Song*. New York: Lodestar Books, 1987. (primary – fanciful illustrations; poetic story of chipmunk preparing for winter)

Ryder, Joanne. *Snail in the Woods*. New York: Harper, 1979. (primary)

Soutter-Perrot, Adrienne. *Earthworm*. USA: American Education, 1994. (all – reference)

Rotting Logs

Cole, Joanna. *The Magic School Bus Meets the Rot Squad*. New York: Scholastic, 1995. (all – fanciful story that includes lots of facts about decomposers in a log)

Johnson, Jinny. *Children's Guide to Insects*. New York: Simon & Schuster, 1996. (all – photos and illustrations; good information about insects and spiders)

Julivert, Maria Angels. *The Fascinating World of Beetles*. New York: Barrons, 1995. (primary – illustrations of varieties of beetles; life cycle, body parts)

Pfeffer, Wendy. *A Log's Life*. New York: Simon & Schuster, 1997. (primary – paper sculpture illustrations; story of log's changing inhabitants as it decays)

Romanova, Natalia. *Once There Was a Tree*. New York: Dial Books, 1985. (primary – beautiful illustrations)

Ross, Wilda. *Who Lives in This Log?* New York: Coward, McCann, 1971. (intermediate)

Schreiber, Anne. *Log Hotel*. New York: Scholastic, 1994. (primary – simple text and pictures)

Animals in Winter

Bancroft, Henrietta. *Animals in Winter*. New York: HarperCollins, 1997. (primary – nice drawings; information on hibernation, migration, and other wintering strategies)

Berger, Melvin. *What Do Animals Do in Winter?: How Animals Survive the Cold*. Nashville, TN: Ideals Children's Books, 1998. (primary – easy reader; explains hibernation, migration, and other wintering strategies)

George, Jean Craighead. *Dear Rebecca, Winter Is Here*. New York: HarperCollins, 1993. (primary – wonderful illustrations; how bears, bees, wolves, mice overwinter)

George, Lindsay Barrett. *In the Snow: Who's Been Here?* New York: Greenwillow Books, 1995. (primary – beautiful color illustrations; readers guess animals from clues in the snow)

Hader, Berta, and Elmer Hader. *The Big Snow*. New York: Macmillan, 1948. (all – animals talk about their preparations for winter)

San Souci, Daniel. *North Country Night*. New York: Doubleday, 1990. (all – beautiful scenes of snow falling in woods, animals looking for food)

Selsam, Millicent E. *Where Do They Go? Insects in Winter*. New York: Scholastic, 1981. (intermediate – reference)

Tejima. *Fox's Dream*. New York: Scholastic, 1992. (all – beautiful woodblock prints; read-aloud story of a fox on a winter night)

Snug in the Snow

Branley, Franklin M. *Snow Is Falling*. New York: Thomas Y. Crowell, 1963. (primary – simple text; how snow is good for plants and animals)

Brett, Jan. *The Mitten*. New York: G.P. Putnam's Sons, 1989. (all – engaging illustrations; classic folktale about a lost mitten and all the animals that try to squeeze into it)

Lionni, Leo. *Frederick*. New York: Pantheon, 1969. (primary – story of mice living in a stone wall through the winter)

Miller, Edna. *Mousekin's Woodland Sleepers*. New York: Prentice Hall, 1970. (all – story introduces hibernators, migrators, dormant, and active animals in winter)

Ryder, Joanne. *Chipmunk Song*. New York: Lodestar Books, 1987. (primary – fanciful drawings; story of a chipmunk preparing for the winter)

White-tailed Deer

Arnosky, Jim. *All about Deer*. New York: Scholastic, 1996. (primary – detailed information about deer biology)

Arnosky, Jim. *Deer at the Brook*. New York: Mulberry Books, 1986. (primary – easy reader; nice close-up drawings of deer and fawns by a brook)

Kalbacken, Joan. *White Tailed Deer*. Chicago: Children's Press, 1992. (intermediate – examines the characteristics, habits, and habitats of deer)

Wolpert, Tom. *Whitetails for Kids*. Minnetonka, MN: NorthWord Press, 1990. (intermediate – color photos; describes birth of fawn and growth through first year)

Streams

Amos, William H. *Life in Ponds and Streams*. USA: National Geographic, 1981. (all – nice photographs of pond life)

Dunphy, Madeleine. *Here Is the Wetland*. New York: Hyperion, 1996. (primary – detailed paintings; informative look at the murky world of the wetland)

Parker, Steve. *Eyewitness Books: Pond and River*. New York: Alfred A. Knopf, 1988. (all – photos and drawings; freshwater plants and animals and their adaptations)

Paul, Tessa. *By Lakes and Rivers: Be an Animal Detective*. New York: Crabtree, 1997. (intermediate – vivid illustrations; simple text; wetland animal's tracks and habits)

Sayre, April Pulley. *River and Stream*. Brookfield, CT: 21st Century Books, 1996. (intermediate – information about different streams and what lives in them)

Stidworthy, John. *Ponds and Streams*. Mahwah, NJ: Troll, 1990. (intermediate – animals that live around ponds and streams)

Ponds

George, Lindsay Barrett. *Around the Pond: Who's Been Here?* New York: Greenwillow Books, 1996. (intermediate – realistic illustrations; readers guess the animal from clues given)

Lasky, Kathryn. *Pond Year*. Cambridge, MA: Candlewick Press, 1995. (intermediate – two friends explore a shallow, mucky pond through the year)

Porte, Barbara Ann. *Tale of a Tadpole*. New York: Orchard Books, 1997. (intermediate – nice illustrations of tadpole's development; story of family realizing tadpole has become a toad)

Rood, Ron. *Wetlands*. New York: HarperCollins, 1994. (intermediate – looks at the varied habitats within a pond using illustrations and text)

Rosen, Michael J. *All Eyes on the Pond*. New York: Hyperion, 1994. (primary – simple rhyming text; color illustrations of pond seen through the eyes of inhabitants)

Wyler, Rose. *Puddles and Ponds*. New Jersey: Julian Messner, 1990. (activity book; children explore puddles and ponds)

CYCLES

Insect Lives

Facklam, Margery, and Paul Facklam. *Creepy, Crawly Caterpillars*. Boston: Little, Brown, 1996. (primary – large paintings; two pages each about several common caterpillars)

Fischer-Nagel, Heiderose, and Andreas Fischer-Nagel. *Life Cycle of the Butterfly*. Minneapolis: Carolrhoda, 1998. (intermediate – many close-up photos; follows entire life cycle of a peacock butterfly)

Goor, Nancy, and Ron Goor. *Insect Metamorphosis: From Egg to Adult*. New York: Atheneum, 1990. (intermediate – lots of photos; factual)

Hariton, Anca. *Butterfly Story*. New York: Dutton Children's, 1995. (primary – nice paintings; simple text; good read-aloud)

Oxford Scientific Films. *The Butterfly Cycle*. New York: G.P. Putnam's Sons, 1977. (primary – stunning photos; very simple, read-aloud text)

Pringle, Laurence. *An Extraordinary Life: The Story of a Monarch Butterfly*. New York: Orchard Books, 1997. (intermediate – glorious color paintings; story of one monarch's migration)

Rosenblatt, Lynn M. *Monarch Magic: Butterfly Activities and Nature Discoveries*. New York: Workman, 1998. (intermediate – lots of photos, information, activities, maps)

Meet a Tree

Arnosky, Jim. *Crinkleroot's Guide to Knowing the Trees*. New York: Simon & Schuster, 1992. (primary – illustrated by author, very child-friendly guide)

Burns, Diane L. *Trees, Leaves and Bark*. Minnetonka, MN: NorthWord Press, 1995. (primary – identification of common trees; activities)

Dowden, Anne Ophelia. *The Blossom on the Bough: A Book of Trees*. New York: Thomas Y. Crowell, 1975. (intermediate – beautiful illustrations by author; good information)

Lavies, Bianca. *Tree Trunk Traffic*. New York: E.P. Dutton, 1989. (primary – photos and simple, story-like text describe various occupants of a tree)

Locker, Thomas. *Sky Tree*. New York: HarperCollins, 1995. (all – gorgeous paintings of a tree through the seasons, against different skies; great read-aloud)

Selsam, Millicent E. *Maple Tree*. New York: Morrow Jr. Books, 1968. (all – simple text, close-ups; life of a maple tree)

USDA Forest Service. *A Tree Hurts Too*. New York: Charles Scribner, 1975. (intermediate – illustrations of effects of tree wounds)

Seed Dispersal

Carle, Eric. *The Tiny Seed*. New York: Crowell, 1970. (primary – colorful cut-paper illustrations; story of a tiny seed that grows up and produces a big flower)

Cole, Joanna. *Magic School Bus Plants Seeds*. New York: Scholastic, 1995. (intermediate – story with good information about how plants grow)

Dowden, Anne O. *From Flower to Fruit*. Boston: Houghton Mifflin, 1984. (intermediate – explains reproductive cycle of flowering plant)

Gibbons, Gail. *From Seed to Plant*. New York: Holiday House, 1991. (primary – picture book; explains life cycle of plant, parts of flowers, how seeds travel)

Hickman, Pamela. *A Seed Grows: My First Look at a Plant's Life Cycle*. Buffalo: Kids Can Press, 1997. (primary – "House that Jack Built" style; also, find things on each page)

Lauber, Patricia. *Seeds Pop, Stick, Glide*. New York: Crown, 1981. (intermediate – great b/w photos - close-ups of seeds)

Fly Away or Stay?

Lerner, Carol. *Backyard Birds of Winter*. New York: Morrow Jr. Books, 1994. (intermediate – descriptions and illustrations of birds that stay; feeder projects)

Ross, Judy. *Nature's Children: Canada Goose*. Danbury, CT: Grolier, 1986. (primary – large color photos; all phases of goose's life)

Swinburne, Stephen R. *Swallows in the Birdhouse*. Brookfield, CT: Millbrook Press, 1996. (primary – nice watercolors & photos (3-D effect); lots of facts; birdhouse project)

Galls

Mound, Laurence. *Eyewitness: Insect*. New York: Alfred A. Knopf, 1990. (all – includes two pages with photos of galls)

Winter Twigs

Davis, Bette J. *Winter Buds*. New York: Lothrop, Lee & Shepard, 1973. (intermediate – drawings; detailed descriptions of twigs and buds)

Bird Songs

Mason, George F. *Animal Sounds*. New York: Morrow Jr. Books, 1948. (intermediate – mostly text; conversational tone; good chapters about bird song)

Procter, Noble. *Songbirds: How to Attract Them & Identify Their Songs*. Emmaus, PA: Rodale Press, 1988. (audio cassette included)

Inside a Flower

Burnie, David. *Eyewitness Explorer: Flowers*. New York: Alfred A. Knopf, 1992. (all – good color photos)

Burns, Diane L. *Wildflowers, Blooms and Blossoms*. Minnetonka, MN: NorthWord Press, 1998. (primary – identifies some familiar flowers; nice illustrations and activities)

Heller, Ruth. *The Reason for a Flower*. New York: Scholastic, 1983. (primary – colorful drawings; amusing poem about pollination, seeds, fruits, flowers)

Jordan, Helene J. *How a Seed Grows*. New York: Harper Collins, 1992. (primary – story-like; what seeds are and how to grow some)

Ladyman, Phyllis. *Learning about Flowering Plants*. New York: Young Scott Books, 1970. (intermediate – line drawings; parts of plants explained; some activities)

Dandelions

Busch, Phyllis. *Wildflowers and the Stories behind their Names*. New York: Charles Scribner, 1977. (middle – nice drawings and explanations of how some flowers got their names)

Hewit, Sally. *Plants and Flowers* (It's Science). Chicago: Children's Press, 1999. (primary)

Himmelman, John. *A Dandelion's Life*. Chicago: Children's Press, 1998. (primary – picture book)

Oxlade, Chris. *Flowering Plants* (Step by Step Science). Chicago: Children's Press, 1999. (intermediate – flower parts, pollination, and reproduction)

Selsam, Millicent E. *The Amazing Dandelion*. New York: William Morrow, 1972. (intermediate – detailed photos of every phase of dandelion life; lots of information)

Watts, May Theilgaard. *Flower Finder* (Nature Study Guild). New York: Warner Books, 1955. (intermediate – easy pocket-size dichotomous key – northeast area)

DESIGNS OF NATURE

Spiders and Webs

Carle, Eric. *The Very Busy Spider*. New York: Philomel Books, 1995. (primary – picture book by *Very Hungry Caterpillar* author)

Climo, Shirley. *The Cobweb Christmas*. New York: Harper Trophy, 1986. (primary – sweet story of spiders' decoration for an old woman's Christmas)

Cole, Joanna, and Ron Broda. *Spider's Lunch: All about Garden Spiders*. New York: Grosset and Dunlap, 1995. (primary – profiles life of a garden spider)

Fowler, Allan. *Spiders Are Not Insects*. Chicago: Children's Press, 1996. (primary – emphasizes differences between spiders and insects)

Gibbons, Gail. *Spiders*. New York: Holiday House, 1994. (primary – characteristics, behavior, habitats of different spiders)

Kalman, Bobby. *Web Weavers and Other Spiders*. New York: Crabtree, 1996. (primary – color pictures; good overview of spider characteristics and habitats)

Winer, Yvonne, and Karen Lloyd Jones. *Spiders Spin Webs*. Watertown, MA: Charlesbridge, 1998. (primary – illustrations and verse, visual field guide)

Variations on a Leaf

Beame, Rona. *Backyard Explorer Kit*. New York: Workman, 1989. (primary – life cycles of trees, how leaves function, leaf and tree projects)

Johnson, Sylvia A., and Yuko Sato. *How Leaves Change*. Minneapolis: Lerner, 1986. (intermediate – describes structure and purpose of leaves and ways they change)

Lauber, Patricia, and Holly Keller. *Be a Friend to Trees* (Let's Read-And-Find-Out). New York: HarperCollins, 1994. (primary – photosynthesis explained in simple language; beauty and usefulness of trees)

Marzollo, Jean, and Judith Moffat. *I Am a Leaf* (Hello Reader! Science Level 1). New York: Cartwheel Books, 1999. (primary – easy text describing life cycle and function of a leaf)

Cones

Arnosky, Jim. *Crinkleroot's Guide to Knowing the Trees*. New York: Simon & Schuster, 1992. (primary – woodsy character [Crinkleroot] tells differences between deciduous trees and conifers)

Burnie, David, and Peter Chadwick. *Eyewitness: Tree*. New York: Alfred A. Knopf, 1988. (all – good color photos; life of tree examined)

Snow and More

Blanchard, Duncan C. *The Snowflake Man: A Biography of Wilson A. Bentley*. Blacksburg, VA: McDonald & Woodward, 1998. (intermediate and up – story of remarkable man who studied and photographed flakes)

Hader, Berta. *The Big Snow*. New York: Aladdin Paperbacks, 1993. (primary – how animals prepare for the winter; simple text and pictures)

Keats, Ezra Jack. *The Snowy Day*. New York: Viking Press, 1981. (primary – small boy's experience of snowy day)

Krensky, Stephen. *Snow and Ice* (Do It Yourself Science). New York: Scholastic, 1994. (primary – easy experiments and facts)

Martin, Jacqueline Briggs. *Snowflake Bentley*. Boston: Houghton Mifflin, 1998. (primary – beautifully illustrated picture book; story)

Tresselt, Alvin. *White Snow, Bright Snow*. New York: William Morrow, 1989. (primary – fun, magic, mystery of a snowfall)

Tracks and Traces

Bowen, Betsy. *Tracks in the Wild*. Boston: Houghton Mifflin, 1988. (primary – illustrations and quotes, nice block prints)

Miller, Dorcas. *Track Finder*. Rochester, NY: Nature Study Guild, 1981. (intermediate – pocket-size field guide)

Paul, Tessa. *In Fields and Meadows*. New York: Crabtree, 1997. (intermediate – Animal Trackers Series; field animal dwelling places and tracks)

Paul, Tessa. *In Woods and Forests*. New York: Crabtree, 1997. (intermediate – Animal Trackers Series; woodland animal dwelling places and tracks)

Pearce, W. J. *In the Forest* (Nature's Footprints). Morristown, NJ: Silver Burdett, 1990. (primary – follow animal tracks to observe habits and habitats of forest animals)

Winter Weeds

Brown, Lauren. *Wildflowers and Winter Weeds*. New York: W. W. Norton, 1976. (intermediate and up – wonderful line drawings; clear, useful key and descriptions of many plants)

Embertson, Jane. *Pods: Wildflowers and Weeds in Their Final Beauty*. New York: Charles Scribner, 1979. (intermediate and up – beautiful photos assist in identification)

Miller, Dorcas. *Winter Weed Finder*. Rochester, NY: Nature Study Guild, 1989. (all – sketches; simple key for common, nonwoody plants in winter)

Camouflage

Arnosky, Jim. *I See Animals Hiding*. New York: Scholastic, 1995. (primary – nice watercolors; simple story of camouflage)

Bailey, Jill, and Tony Seddon. *Mimicry and Camouflage*. New York: Facts on File, 1988. (intermediate – Nature Watch Series)

Duprez, Martin. *Animals in Disguise*. Watertown, MA: Charlesbridge, 1994. (intermediate – good illustrations and informative text)

Selsam, Millicent E. *Backyard Insects*. New York: Scholastic, 1991. (primary – great photos of insect disguises and warning coloration)

Sowler, Sandie. *Eyewitness Jr.: Amazing Animal Disguises*. New York: Dorling Kindersley, 1992. (intermediate – color photos; easy to read text describes a variety of animal disguises)

Honeybees

Cole, Joanna, and Bruce Degen. *Magic School Bus: Inside a Beehive*. New York: Scholastic, 1998. (primary – imaginative and informative story in which the school bus turns into a beehive and the students into honeybees)

Crewe, Sabrina. *The Bee*. Chatham, NJ: Raintree/Steck-Vaugn, 1997. (intermediate – good information on bee life cycle with illustrations and photos)

Fischer-Nagel, Heiderose, and Andreas Fischer-Nagel. *Life of the Honeybee*. Minneapolis: Carolrhoda, 1986. (intermediate – wonderful close-up photos inside a hive; detailed information about bees)

Fowler, Allan, Mary Nalbandian, and Robert Hillerich. *Busy, Buzzy Bees*. Chicago: Children's Press, 1996. (primary – Rookie Read-About Science Series)

Gibbons, Gail. *The Honey Makers*. New York: William Morrow, 1997. (primary – introduction to honey bees; informative picture book)

Polacco, Patricia. *The Bee Tree*. New York: Philomel Books, 1993. (primary – good story & illustrations; Grandpa captures, releases, and follows bees)

EARTH AND SKY

Finding Your Way

Gibbons, Gail. *Planet Earth: Inside Out*. New York: William Morrow, 1995. (intermediate – clear illustrations of earth's anatomy from core to mountaintop)

Kjellstrom, Bjorn. *Be An Expert with Map and Compass: Complete Orienteering Handbook*. Foster City, CA: IDG Books Worldwide, 1994. (intermediate and up – introduction to map and compass, with activities)

Sweeney, Joan. *Me on the Map*. Dayton, MD: Dragonfly, 1998. (primary – introduction to maps and geography)

Taylor, Barbara. *Maps and Mapping*. New York: Kingfisher Books, 1993. (primary – what maps are and how they are made; some related activities)

Erosion

Bourgeois, Paulette et al. *The Amazing Dirt Book*. New York: Perseus Press, 1990. (primary – information and activities examining the many aspects of dirt)

Bramwell, Martyn. *Eyewitness: The Visual Dictionary of the Earth*. New York: Dorling Kindersley, 1993. (all – photos, drawings, and diagrams; good information on earth's physical features and processes)

Peters, Lisa Westberg. *The Sun, the Wind and the Rain*. New York: Henry Holt, 1990. (primary – the making of a mountain, and shaping by sun, wind, and rain)

Schmid, Eleonore. *The Living Earth*. San Francisco: North South Books, 1994. (primary – soils, processes, soil ecology)

Snedden, Robert. *Rocks and Soil*. Chatham, NJ: Raintree/Steck-Vaugn, 1998. (intermediate – experiments on erosion included)

Pebbles and Rocks

Baylor, Byrd. *Everybody Needs a Rock*. New York: Aladdin Paperbacks, 1987. (primary – thoughtful and funny story describing 10 rules for finding a special rock)

Gans, Roma, and Holly Keller. *Let's Go Rock Collecting*. New York: HarperCollins, 1997. (intermediate – the oldest things you can collect are rocks)

Gibbons, Gail. *Planet Earth: Inside Out*. New York: Mulberry Books, 1998. (primary – Earth's formation; three types of rock)

Hiscock, Bruce. *The Big Rock*. New York: Atheneum, 1988. (intermediate – story and illustrations trace origins of big rock in Adirondack Mountains)

Oldersaw, Cally. *3-D Eyewitness: Rocks and Minerals*. New York: Dorling Kindersley, 1999. (intermediate – 3-D mirror device makes rocks easy to see)

Symes, R. F., and Colin Kezts. *Eyewitness: Rocks and Minerals*. New York: Alfred A. Knopf, 1988. (intermediate – rock formation, erosion, mining)

Breath of Life

Ardley, Neil. *The Science Book of Air*. New York: Harcourt Brace, 1991. (intermediate – nice, clearly-illustrated experiments)

Branley, Franklyn. *Air Is All Around You*. New York: Thomas Y. Crowell, 1986. (primary – experiments; air is in everything)

Miller, Christina G. *Air Alert: Rescuing the Earth's Atmosphere*. New York: Atheneum, 1996. (intermediate – examines sources and consequences of pollution)

Oxlade, Chris. *Science Magic with Air*. Hauppage, NY: Barron Juveniles, 1994. (intermediate – magic tricks that teach properties of air)

Rius, Maria. *Air* (The Four Elements). Hauppage, NY: Barron Juveniles, 1985. (primary)

Sound Symphony

Branley, Franklin M., and Giulio Maestro. *Flash, Crash, Rumble and Roll*. New York: HarperCollins, 1999. (primary – how thunderstorms form; how sound travels)

Cole, Joanna, and Bruce Degen. *Magic School Bus in the Haunted Museum: A Book about Sound*. New York: Scholastic, 1995. (primary – sound and how it is made)

Oxlade, Chris. *Science Magic with Sound*. Hauppauge, NY: Barron Juveniles, 1994. (intermediate – magic tricks to illustrate properties of sound and how it is made)

Sabbeth, Alex. *Rubber Band Banjoes and a Java-Jive Bass: Projects and Activities on the Science of Music and Sound*. New York: John Wiley & Sons, 1997. (intermediate – projects; basics of sound and music)

Showers, Paul. *Listening Walk*. New York: HarperCollins, 1991. (primary – young girl and father take a listening walk)

Water, Water Everywhere

Ardley, Neil. *The Science Book of Water*. New York: Harcourt Brace, 1991. (intermediate – simple experiments demonstrate properties of water)

Berger, Melvin, and Gilda Berger, eds. *Water, Water Everywhere: A Book about the Water Cycle*. Broomall, PA: Chelsea House, 1998. (primary – water cycle story)

Cole, Joanna. *Magic School Bus Wet All Over: A Book about the Water Cycle*. New York: Scholastic, 1996. (primary – Ms. Frizzles's class travels through the water cycle)

McKinney, Barbara Shaw. *A Drop around the World*. Nevada City, CA: Dawn, 1998. (primary – great illustrations; travel story)

Wind and Clouds

Ardley, Neil. *The Science Book of Weather*. New York: Gulliver Books, 1992. (all – clear illustrations, information, weather instruments to make)

De Paola, Tomie. *The Cloud Book*. New York: Holiday House, 1985. (primary – introduction to 10 types of clouds; weather changes)

Dewitt, Lynda. *What Will the Weather Be?* New York: Harper Trophy, 1993. (primary – basic characteristics of weather)

Dorros, Arthur. *Feel the Wind*. New York: Harper Trophy, 1990. (all – what makes wind; weathervane instructions)

English, Karen. *Big Wind Coming!* Morton Grove, IL: Albert Whitman, 1996. (primary – story of a wind storm)

Farndon, John. *Eyewitness Explorers: Weather*. New York: Dorling Kindersley, 1998. (all –different aspects of weather)

Ludlum, Dr. David. *Clouds and Storms*. New York: Alfred A. Knopf, 1995. (all – beautiful pictures; National Audubon Society Pocket Guide)

Shearer, Alex. *Professor Sniff and the Lost Spring Breezes*. Danbury, CT: Orchard Books, 1996. (intermediate – fiction; fun story about wind)

Simon, Seymour. *Weather*. New York: HarperCollins, 1993. (intermediate – clear explanation of winds, clouds, and how winds create different cloud types)

Wyatt, Valerie. *Weatherwatch*. New York: Perseus Press, 1990. (all – weather wisdom, facts, and projects)

Sun Power

Branley, Franklin M. *What Makes Day and Night*. New York: HarperCollins, 1986. (primary – NASA photos and drawings explain Earth's rotation)

Branley, Franklin M., and Giulio Maestro. *Sunshine Makes the Seasons*. New York: Harper Trophy, 1986. (primary – simple explanation of how sun and earth's tilt make seasons)

Gibbons, Gail. *The Reasons for the Seasons*. New York: Holiday House, 1996. (primary – pictures and simple explanations; reasons for seasons; changing activities of animals)

Hillerman, Anne. *Done in the Sun*. Santa Fe: Sunstone Press, 1983. (intermediate – introduction to sun as renewable energy source; simple experiments and craft projects)

Simon, Seymour. *The Sun*. New York: Mulberry Books, 1989. (intermediate – dramatic photographs; simple, informative text)

INDEX

Italicized page numbers indicate illustrations.

A

abscission layer, 157, 192
acorns, *136*
adaptations, 11
air, 266-272
algae, 115
alternate bud pattern, 157
amphibians, 58, 66
 see also frogs; salamanders; toads
angiosperms, 135, 198
annuals (plants), 172, 222
antennae, 12, 123
 of honeybee, 235
anther, 171, 176
 of grasses, 19
anthocyanin, 192
antlers, 51, 100
ants
 carpenter, 80
 scent trails of, 244
 as seed dispersal agents, 135
 strength of, 12
arctic tern, 142, 244
arthropods, 12, 52, 73, 122, 184
assessment of learning, 5
attitude toward the environment, modeling, 6
axis (earth's), 295

B

ballooning (spiders), 185
bark beetle, 80, *84*
bark (tree), 129
barn owl, sense of hearing of, 24
 see also owls
barn swallow, *141*, 146
bat
 diet of, 31
 and echolocation, 24, 244, 274
 hibernation of, 88, 91
 migration of, 87
 as pollinator, 172
 teeth of, 31
beaks, *39*
 of owls, 45
bear, 88, 89
beaver
 feet of, 115
 incisors of, 31
 overwintering of, 89
bee bread, 234
beehives, 234-236
bees *see* honeybee
beggar ticks, 135, *136*
behavioral expectations of students, 6-7
biennials (plants), 172, 222-223

Bingo, card sample, *92*
binocular vision, 24, 32, 46
bird feeders, 42, 202
birds
 adaptations of, 38-44
 diagram of leg and foot bones, *44*
 endangered, 42
 feet, *40*
 migration of, 141-149, 165, 244
 as seed dispersal agents, 135
 songs, 164-170
 feather-dropping strategy of, 52
 in winter, 88, 145
 see also individual birds
black bear, 88, 89
black-eyed Susan, *66*
black fly larva, *108*
bluebird, 147
blue jay, 145
bobcat
 hunting strategy of, 25
 teeth of, 31
body posture, of animals, 52
boundaries, establishing for lessons, 7
bounding track pattern, *216*
box elder bug, *13*
bracts, 176, 222
breathing, 266
brook trout, 91
brown-headed cowbird, 147
brown thrasher, 147, 164
browse, *103*
browsing, 19, 101, 158
brumation, 58
buds, 157
bullfrog, 58-59
bundle scars (leaf), 158
burdock, 135
butterfly, 12, 123
 see also monarch butterfly

C

caddisfly, 12
 larva, 107, *108*, 229
cambium, 129
camouflage, 228-233
canines (teeth), 31
capillary action, 129, 279
caribou, 87, 244
carnivores, 24
 teeth of, 31
 see also predation; predators
carotene, 192
castings (worm), 74
catbird, 147
caterpillar
 diagram of, *125*
 see also monarch butterfly, caterpillar
cats, 31, 52
 track pattern of, 215
 see also bobcat

caverns, 251
cedar waxwing, 147
cellulose, 80
centipede, *80*
cephalothorax, 184
chestnut, *55*
chickadee, *88*
    beak of, 39
    and galls, 151
    in winter, 141, 145
chipmunk, 81, *87*, 88
    in winter, 91
chitin, 13
chlorophyll, 67, 130, 191-192, 295
chrysalis, 122
cirrocumulus clouds, 287
cirrus clouds, 287
clams, 114
clay, 261
climate, 296
clitellum, 74
clouds, 278, 286-294
    types of, 287-*288*
clover, 66, *70*
coal, 261
cobweb, 184, *185*
cockroach, hearing and, 274
cocoon, *87*, 122
collecting when outdoors, 7-8
color blindness, 45
compass, 245, 247
    holding a, *244*
    making a, 248
    parts of, *247*
    rose, *245*
composite plants, 151, 176
cones, 198-204
    blue spruce, *198*
    eastern hemlock, *198*
    eastern white cedar, *201*
    loblolly pine, *201*
    white pine, *199*
    see also conifers
conifers, 109, 129, 135, 157
    see also cones
consumers (in food chain), 67
countershading, 229
coyote
    diet of, 32
    hunting strategy of, 25, 66
cranberries, 135-136
crane fly larva, 108
crayfish, 52, 59, 114
crepuscular animals *see* nocturnal animals
cricket, *12*, 123
crossbill, beak of, 39
cross-pollination, 172
crystals, 205, 278
cumulus clouds, 287
cycles, 121

D

daddy longlegs, 80
damselfly, 12, *114*
    gills, 115
    nymph, 114
dandelion, 176-181
    floret, *178*
    flower, *176*
    plant, *177*
    pollen, *177*
    seed, *176*
deciduous trees, 129-130, 192
deciduous weeds, 223
decomposers, 74, 80
decomposition, 73-74, 80-81, 114, 130
deer, 100-106
    buck, *104*
    doe, *101*
    sense of hearing of, 24
    teeth of, 31, 101
    track pattern of, 215-*217*
    in winter, 88, 91, 158, 228
    yards, 88, 101
defense strategies, 51-57
designs of nature, 183
direction, sense of, 244-250
discipline of students, 6-7
disruptive coloration, 228
dog family
    eating habits of, 31
    track pattern of, 215, *221*
dolphin, 244, 274
dormancy, 87-88
dragline (spider), 185
dragonfly, 12, 115, 123
    nymph, 59, 107, 114, 115, 123
drawing (student), 5
drones, *234*
duck, webbed feet of, 11, 39

E

eagle
    beak and feet of, 39
    wings of, 40
ears
    animal, 274
    human, *273*-274
    owl, 45, 46
earthworms, *74*, 87
eastern phoebe, 147
eastern towhee, 147
echoes, 273-274
echolocation, 24, 244, 274
ecosystem, balanced/healthy, 25
education *see* environmental education
eft, *74*
elephants, communication of, 274
emergent plants, 115
engraver beetle, 84
entomologists, 16

environmental education
    importance/purpose of, 1-2
    variety in, 2
Environmental Learning for the Future (ELF) program, 1, 302
equipment for teaching, 4, 303-304
ermine, 89, *97*
    see also weasel
erosion, 19, 107, 251-259
evaporation, 278, 295
evening primrose, *223*
evergreens see conifers
exoskeleton, 13, 73, 122
eyelids
    of birds, 40
    of owls, 45
eyes
    of birds, 40, 45
    compound, 12, 235
    position of, 24, 32, 40, 46

F

fantasies, *see under* stories
fears (students' and teachers'), 6
feathers, *38*
    diagram of, *41*
    owl, 45
field (life in), 66-72
filament (flower), 171
firefly, 123
fish, 115, 267
    see also salmon; trout
fisher
    hunting strategy of, 25, 52, 80
    track pattern of, *217*
flicker, 80
flight
    of insects, 13
    see also birds
floret, 176, *178*
flowers, 171-175
    diagram of, *171*
    of grasses, 19
    see also dandelion
fly, mouth of, 12
flycatcher, beak of, 39
flyways, 142
foliage, 192
food chain, 31, 67, 295
food web, 67, 108
forces of nature, 243
forest floor, 73-79
fossils, 258-*259*
fox, *25*, 66, 91
frass, 80
freezing rain, 206, 209, *211*
frogs, *58, 62*
    adaptations of, 58-63
    eggs, *58*
    life cycle of, *59*
    and oxygen uptake, 267

    taste of, 51
    in winter, 87, 89, 91
frost line, 87
fruit, 135, 172
    seedless, 136
fruit fly, 274
fuel, 295
fungi, 73, 80, 136
funnel web, 184, *185*

G

galloping track pattern, 215, *216*
galls, 150-156
    goldenrod ball, *150*
    pine cone willow, *151*
    spindle, *153*
geese, 89
germination, 136
gills, 59, 115, 266
glaciers, 251
golden-crowned kinglet, *142*, 145, 147
goldenrod, 66, *266*
goldenrod ball gall fly, 150-151
grains, 21
granite, 258
grasses, 19-23, 52
    diagram of, *20*
    poem to distinguish, 21
grasshoppers, 12
    hearing and, 274
grasslands, 66
graupel, 206
gravity, as erosion agent, 252
grazing, 19
great horned owl, 25, *46*, 52, 91
    see also owls
green frog, 58
ground water, 251, 278-279
group discussions (class), 5
grouse
    tail of, 40
    in winter, 95, 145, 158
gymnosperms, 135, 198
    see also conifers
gypsy moth, life cycle of, *122*

H

hail, 206, 209, *211*
hand lenses, 4, 303
hare see snowshoe hare
hawk
    beak and feet of, 39
    in fields, 66
    wings of, 40
hawthorn, *52*
hearing, sense of, 24
    see also sound
heartwood, 129
herbaceous plants, 222

herbivores
    adaptations to grasses, 19
    defined, 24, 67
    teeth of, 31
heron, legs of, 39
hibernation, 58, 87-88
honeybee, 12, *234-241*
    dances, *235-236*, 244
    development of, *239*
    diagram of, *235*
    in winter, 88, 89, 91
honeycomb, 236
horns, 51
house wren, *164*
humans
    ears of, *273-274*
    teeth of, 32
hummingbird, 147
    beak of, 39
    as pollinator, 172
humor, importance of, for teachers, 6
humus, 73, 81
hunter/hunted *see* predation
husks, prickly, 51
hydroelectric power, 295
hyphae, 73, 80, 252

I

ice, 279
identifying natural objects, 7
igneous rocks, 258
incisors, 31
infrared radiation, 295
insectivores, teeth of, 31
insects
    adaptations of, 12-18
    compared to spiders, 184
    diagram of, *13*
    life cycles of, 122-128
    overwintering of, 87, 122
insulation, 267
Io moth, *229*

J

jewelweed, 135
journals, 5

K

keratin, 38, 45
kestrel, *67*

L

lady's slipper, 136
larva, 122, 151
    see also under individual insects
lateral buds, 157
learning, assessing, 5

leaves, 191-197
    compound, *195*
    diagram of, *192*
    scars of, 157-158
lenticels, 158
life cycles *see* cycles
lightning fires, 19
lignite, 261
limestone, 258, 261
lizards, regeneration of, 52
logs, 80-86

M

magma, 258
magnetic field, 245
mandible (honeybee), 235, 236
map, mental, 244
marble, 258, 261
masking (camouflage device), 229
materials for teaching, 4, 303-304
mayfly, 12, 123
    nymph, *107*-108
meadowsweet, *224*
metamorphic rocks, 258
metamorphosis
    of frogs, 58-63
    of insects, 122-123
migration, 87, 244-245
    of birds, 141-149, 165, 244
milkweed, 66, 172, *183*
    seed, *267*
millipede, 74, *81*
minerals, 258-259
mink, track pattern of, *221*
mite, 74, 153
molars, 31
mole, 24, 31, 74
molting
    birds, 38
    insects, 122
monarch butterfly, *11*
    caterpillar, 52, 122
    coloration of, 229
    migration of, 87, 89, 244
moose, 88
mosquito
    hearing of, 274
    larva, 114, 115
    mouth of, 12
mosses, 81
moth, 123
    adaptation of, 13
    cocoon, *87*
mountains, 252
mourning dove, 147
mouse, *66, 97*
    track pattern of, 215, *221*
    woodland jumping, 32, 88
mural
    animals in winter, *90*
    pond, *117*

tracks and traces, *218*
    water, water everywhere, *280*
mushrooms, 73, 77, 80, 81

N

natural forces, 243
natural selection, 11
navigation, 244-250
nettle (stinging), 51
nictitating membrane (eyes), 40
nimbostratus clouds, 287
nocturnal animals, 24, 45, 46
nodes (grass), 19
nuthatch
    beak of, 39
    in winter, 145
nymph, 59, 122
    see also under individual insects

O

oak galls, 151
obsidian, 258
odor
    as defense, 52
    of flowers, 172
    of twigs, 158
omnivores, 24
    teeth of, 31-32
opossum
    behavior of, 24, 52
    teeth of, 32
opposite bud pattern, 157
orbit *see* revolution (earth's)
orb web, 184-*185*
osprey, feet of, 39
ostrich, legs of, 39
otter, hunting strategies of, 25
outdoor lessons, planning, 9
ovary, 135, 171-172
overgrazing, 19
ovipositor, 12, 151
ovule, 135, 171-172, 176
    of conifers, 199
owls, 45-50, 66
    beaks and feet of, 39
    defense strategy of, 52
    hunting in snow, 95
    skull, *45*
    see also barn owl; great horned owl
oxygen, 266-267
    from trees, 130
    in ponds, 114-115
    in streams, 107

P

palmate pattern (leaves), *191*, 192
panda, teeth of, 32
pappus, 177
parallel pattern (leaves), *191*, 192

parasites, 151
parrot, feet of, 39
partridge, 40
passerines, 164
patterns in nature *see* designs of nature
peacock, tail of, 40
peanuts, 136
    diagram of, 137
peat, 261
pebbles, 258-265
pedipalp, 184
pellets (owl), 45, *48*
perennials (plants), 172, 223
petiole, 157, 192
pheromones, 123
phloem, 129
photosynthesis, 67, 114, 130, 191, 266, 279, 295
    diagram of, *193*
physical world, 243
pillbug, 81
pinnate pattern (leaves), *191*, 192
pistil, 171, 176
    of grasses, 19
pitch (sound), 274
pith, 158
plantain, *70*
poem
    metamorphosis, 63
    to distinguish grasses, 21
poison ivy, 51
poisonous animals/plants, 51, 74
pollen *see* pollination
pollen basket (bees), 235
pollination
    of conifers, 198-199
    of dandelions, 176
    of flowers, 172
    of grasses, 19
polliwog, 59, 115
pollution
    air, 267
    water, 279
Polyphemus moth, 229
ponds, 114-119
porcupine, 24, *51*, 52
    track pattern of, *215, 221*
portfolios (students'), 5
posture (body), 52
power, fuel for, 295
precipitation, 278, 295
    see also rain; snowflakes
predaceous diving beetle, 59, 115
predation, 24-30
predators, 13, 24-25
    birds as, 39, 45
    see also carnivores; predation
prey, 12, 24
    see also predation; predators
producers (in food chain), 67
protective mimicry, 229
pumice, 258
pupa, 122, 151

puppet shows, 4
    amazing insects (adaptations), 17-18
    animals in winter, 93-94
    beaks, feet, and feathers, 43-44
    bird songs, 169-170
    breath of life, 271-272
    camouflage, 232-233
    conducting, 4
    cones, 203-204
    dandelions, 181
    erosion, 256-257
    finding your way, 249-250
    fly away or stay (bird migration), 148-149
    forest floor, 78-79
    galls, 155-156
    honeybees, 240-241
    insect lives, 127-128
    inside a flower, 175
    life in a field, 71-72
    owls, 49-50
    pebbles and rocks, 264-265
    ponds, 118-119
    rotting logs, 85-86
    seed dispersal, 139-140
    snow and more, 213-214
    snug in the snow, 98-99
    spiders and webs, 189-190
    streams, 112-113
    sun power, 300-301
    teeth and skulls, 36-37
    thorns and threats, 56-57
    tracks and traces, 220-221
    variations on a leaf, 196-197
    water, water everywhere, 284-285
    white-tailed deer, 105-106
    wind and clouds, 293-294
    winter twigs, 162-163
    winter weeds, 226-227
puzzles
    thorns and threats, *53*
    insect life cycles, *124*

Q

quartzite, 258, 261
Queen Anne's lace, 66, *222*
queen bee, *234*
questions (students'), 6
quills, 24, 51

R

rabbit, *28*
    incisors of, 31
    sense of hearing of, 24
    track pattern of, 215-*217*
raccoon
    diet of, 31
    dormancy of, 88
    home of, 80
    track pattern of, *215*

radiation (solar), 295
rain, 251, 287
    freezing, 206, 209
raptors, 39, 45
    rehabilitation of, 48
rattlesnake, 51, 274
red eft, *74*
red-spotted newt, 74, 115
red-winged blackbird, 147
regeneration powers of animals, 52
release ceremonies, 8
respect, encouraging in children, 6-7
revolution (earth's), 295, *296*
rhizomes, *19*
riffle beetle, 107
robin, 147
rocks, 258-265
    cycle diagram, *263*
rodents, incisors of, 31, 52, 101
roots
    as erosion agents, 252
    grass, 19
    tree, 129-130
    weed, 223
rose bush, thorns of, 11
rotation (of earth), 295
rotting logs, 80-86
royal jelly, 234
ruminants, 100

S

salamanders, 51, 87, 115
salmon, 244, *266*
sandstone, 258, 261
sapwood, 129
scavengers, 80
seasons, cause of, 296-297
sedimentary rocks, 258
seed, defined, 135
seed dispersal, 135-140, 267
    of conifers, 199
    of grasses, 20
    of weeds, 222-223
self-pollination, 172, 176
sensitive plant, 52
sepal, 172
setae, 74
shadows/shade, 296
shale, 258, 261
sheet web, 184, *185*
shrew, 74, 95
    poisonous bite of, 51
    teeth of, 31
sight, sense of, 295
    in predators, 24, 45
    see also binocular vision; vision
silence, importance of, in lessons, 8
silkworm, 244
skulls, 31, *32*-37
skunk, 52, 229
    in winter, 88, 91

slate, 258, 261
sleet, 206, 209, *211*
smell, sense of, in prey species, 24
snag, 80
snails, 87, 115
snakes, 51, 66, *67*, 81, 115
 in winter, 87, 91
 see also rattlesnake
snow, 95-99, 209, *211*
 see also snowflakes
snowflakes, 205-214
 crystal types, *205-206*
 hidden images in, *212*
 making paper, *208*
 see also snow
snowshoe hare, 11, 24
 in winter, 89, 158, *228*
solar power, 295
songbirds *see* birds, songs
sound, 273-277
sowbug, *80*, 81
sparrow, 40
spider crabs, 229
spiders, 12, 184-190
 diagram of, *186*
 fishing, *115*
 garden, *184*
 regeneration power of, 52
 on rotting logs, 81
 venom of, 51
spindle galls, *153*
spinnerets, 184
spiracles, 12, 125, 266-267
spores, 80
spring peeper, 58-59, *274*
springtail, 73-74, *76*
squirrel
 flying, 80
 teeth of, 31
 track pattern of, *215*
 in winter, 88, 91, 95
stalactites/stalagmites, 251
stamen, 171, 176
stigma, 171, 176
stinkbug, life cycle of, *123*
stings (insect), 12, 235
stomata, 191-192
stonefly, 123
 nymph, 107-*108*
stories
 be a tree fantasy, 163
 engraver beetle, 84
 gall fantasy, 154
 rabbits and foxes, 29-30
 snowflake fantasy, 212
 students', 6-7
straddle, 216
stratus clouds, 287
strawberry plant, *121*

streams, 107-113
stride, 216
style (flower), 171
submergent plants, 115
subnivean layer, 95
sugar maple, seeds of, *267*
sun, 295-301
 rays of, *295*
surface tension, 114
swim bladders, 267
syrinx, 164

T

tadpole, 59, 115
talons, 39, 45
taproot
 dandelions,177
 trees, 129-130
teeth, *31-37*
terminal buds, 157
thorax, 12
thorns, 11, 51
thunderheads, 287
timothy grass, *70*
titmouse, beak of, 39
toads, 51, 58-*59*, 87
tracks/traces (animal), 215-221
 diagram of, *216*
transpiration, 130, 192, 279
trees, 129-134, 192
 decomposition of, 80-81
 diagram of parts, *129*
 water and, 279
 see also twigs
tree swallow, 147
trout, 91
turkey, tail of, 40
turtles, 87, 91, 115, *296*
tusks, 51
twigs, 157-163
 diagram of, *158*
 in winter, *161*
tympanum, 123, 274

U

ultrasonic frequencies, 274

V

Vermont Institute of Natural Science (VINS), 1, 302
viburnum, berries, *136*
viceroy butterfly, 229
vision, 295
 of birds, 40
 see also binocular vision; sight, sense of
vocal cords, 274
vole, *95*
vulture, wings of, 40

W

waddling track pattern, 215, *216*
walking/trotting track pattern, 215, *216*
warning coloration, 229
wasp
> diagram of, *13*
> digger, 244
> sting of, 12

water, 251, 278-285
> cycle, *278-279*, 295
> *see also* precipitation

water boatman, *115*
water flea, 115
water penny, *107*
water scorpion, 12, 115
water strider, 107, *111*, 114, 115
water tiger, *115*
wax glands (honeybee), 235
weasel, *97*
> eating habits of, 31
> track pattern of, 215-216
> in winter, 89, 95, 228

weather, 286-287
> preparing class for, 9
> vanes *see* wind, vanes

weathering, 252, 258
webs (spider), 184-190
> types of, *185*

weeds, 136
> in winter, 222-227

whales, 87, 274
whirligig beetle, 114, 115
whorled bud pattern, 157
wind
> and clouds, 286-294
> erosion, 251-252
> force scale, *292*
> power, 295
> prevailing, 286
> as seed dispersal agent, 20, 135, 177, 267
> vanes, *286*, 289

winter
> animals in, 87-94, 100-101
> > *see also* snow; twigs

wireworm, *81*
wolf, hunting strategy of, 25
woodchuck, 88, 89
woodpecker, 80, 89, 147, 151
> tail of, 40

woolly bear caterpillar, *65*, 91

X

xanthophyll, 192
xylem, 129

Y

yarrow, 66, *70*

Notes

Notes

Notes

Hands-On Nature Revision Team

(front row) Jenepher Lingelbach, Amy Powers,
Charmaine Kinton, Lisa Purcell, Elizabeth Cooper,
Deb Parrella

(back row) Susan Sawyer, Nicole Conte, Bonnie Ross,
Karen Murphy, Chris Runcie, Melissa Kendall

(not pictured: Jenna Guarino, Carol Curran)

Vermont Institute of Natural Science

Hands-On Nature is published by the Vermont
Institute of Natural Science in Woodstock, Vermont.
The book's activities have been used with great success
in classrooms, homeschool and afterschool programs,
children's clubs, nature camps, and teacher training
programs.

 The Vermont Institute of Natural Science is
a non-profit membership organization whose
mission is *to protect Vermont's natural
heritage through education and research designed
to engage individuals and communities in the active
care of their environment.*

For more information about the Vermont Institute
of Natural Science and its programs call 1-802-457-2779
or visit us online at www.vinsweb.org.